WITHDRAWN

SPECIAL CONCRETES
CONCRETE PRO...

PORTLAND CEM...
ASSOCIATION

SPECIAL CONCRETES AND CONCRETE PRODUCTS

PORTLAND CEMENT ASSOCIATION

JOHN WILEY & SONS, INC.
NEW YORK   LONDON   SYDNEY   TORONTO

Copyright is claimed until 1980.
Thereafter all portions of this work covered
by this copyright will be in public domain.

Copyright © 1975 by Portland Cement Association

All rights reserved. Published simultaneously in Canada.

No part of this book may be reproduced by any means,
nor transmitted, nor translated into a machine language
without the written permission of the publisher.

Library of Congress Cataloging in Publication Data

Portland Cement Association.
   Special concretes, mortars, and products.

   "Third in a series of five textbooks that comprise the
National concrete technology curriculum."
   Includes index.
   1. Concrete. I. Title.
TA439.P852   1975   620.1'36   74-28255
ISBN 0-471-67432-X

Printed in the United States of America

10 9 8 7 6 5 4 3 2 1

ACKNOWLEDGMENTS

This series is a national program to educate persons for employment in the concrete industries, sponsored by the Portland Cement Association, the National Ready Mixed Concrete Association and the American Concrete Institute in cooperation with the Office of Education, U.S. Department of Health, Education, and Welfare.

PROJECT STAFF

George R. White, *Project Director*
J. David Jackson, *Senior Curriculum Editor*
David Anderson, *Associate Editor*
William Perenchio, *Technical Advisor*
Elizabeth A. Craig, *Administrative Assistant*
Cynthia A. Holck, *Editorial Assistant*

TECHNICAL REVIEWERS

The staff acknowledges the many excellent contributions by outstanding people in the concrete field and the advisory group that made up the administrative team. Of special significance were contributions by Howard Wiechman, John Seeger, John Metcalf, Dr. J. R. D. Brown and others who helped launch this undertaking. Special mention must go to the instructors and students who tested the trial materials at the pilot schools and whose comments and suggestions have been incorporated in these volumes.

DISCLAIMER STATEMENT

This work was developed under a grant from the U.S. Office of Education, Department of Health, Education and Welfare. However, the opinions and other content do not necessarily reflect the position of the agency, and no official endorsement should be inferred.

This publication is based on the facts, tests, and authorities stated herein. It is intended for the use of professional personnel competent to evaluate the significance and limitations of the reported findings and who will accept responsibility for the application of the material it contains. The Portland Cement Association and publication co-sponsors disclaim any and all responsibility for the application of the stated principles or for the accuracy of any of the sources other than work performed or information developed by the Association.

Caution: Avoid prolonged contact between unhardened (wet) cement or concrete mixtures and skin surfaces. To prevent such contact, it is advisable to wear protective clothing. Skin areas that have been exposed to wet cement or concrete, either directly or through saturated clothing, should be thoroughly washed with water.

PREFACE

Like most construction products in today's business world, concrete and concrete products can be made to meet exacting standards and new designs. New materials are available in many sizes, shapes, colors, and textures for special uses ranging from superlightweight and insulative concrete to dense structures capable of radiation shielding.

This book describes many ways that concrete can be used to make our lives more rewarding by creating an environment unknown to previous civilizations.

Special Concretes

Lightweight concrete is used today in practically every application of normal-weight concrete where a reduction in dead weight leads to significant savings in the cost of the structure. Lightweight concrete is not just one product—it is a spectrum of concretes that yield a variety of results and fill a number of needs. It can be a gaseous or foam concrete that uses specially prepared chemicals; it can be a no-fines concrete that uses ordinary gravel or crushed stone, a normal-weight aggregate concrete with an excessive amount of entrained air, or a concrete that is made from lightweight aggregates. The absorptive lightweight aggregates used affect the strength, durability, workability, finishability, and economy of the product and require special consideration in mix design and handling.

Insulating concretes are specially designed to provide insulation against heat, cold, fire, and sound. Structural properties are secondary to their insulation value.

Heavyweight concrete has been used for years as counterweights for lift bridges. Today it is being used for shielding against radiation from sources such as industrial radiography, X-ray and gamma-ray therapy, and nuclear reactors.

Nonplastic mixes are very dry concretes that require tamping or special compaction. Their major use is in the manufacture of precast concrete products.

Decorative Concretes

Advancing technology has given designers in the field of decorative concrete the ability to combine rugged strength, attractive appearance, and easy upkeep. The variety of textures, forms, and patterns possible in this medium afford esthetic freedom that is not possible with other building materials. This section of the book shows the use of decorative concrete and deals primarily with surface treatment methods. The use of various form liners is also discussed. The latter part of the section deals with pigments used to color the concrete and various paints that can be used to decorate concrete.

Reinforced Concrete

Most of the concrete used today is a composite material of steel and concrete called reinforced concrete. Only the right kind of steel, in the right place, properly bonded to the concrete can fulfill the demands of modern construction. The methods of reinforcement are described in Section Three. This section also discusses precast and prestressed concrete. Concrete is not always cast in the position where it will be used (cast-in-place). It can be cast in a factory or in special forms at the jobsite and moved into place after it has hardened.

The advantages and disadvantages of precast concrete are discussed. Precasting is an important part of the "systems building" concept—a construction technique in which design, production, and erection are pre-engineered. The potentials of prestressing are explained with examples and detailed equations. Pretensioning and posttensioning are introduced as the two basic methods of stressing concrete.

Concrete Masonry

Building shelters by placing one unit on top of another is one of the oldest construction methods. Eventually it was discovered that if a cementing material is placed between the small units they stay in place much better. Advances both in the cementing material and in the units themselves have evolved into concrete masonry construction. The basic properties of block strength, bond, and flexural strength must be properly developed when masonry walls are constructed.

When modular units are put together to form a larger unit, they must interact properly to produce the desired results. The final chapter in this section deals with masonry construction from footings to flashings.

Concrete Pipe

Concrete pipe is used throughout the world for irrigation, drainage, sewerage projects, culverts, and pressure water-supply mains. Its manufacture is traced from the early days of the United States to the highly automated quality-control-conscious plants of today. Various methods of pipe manufacture are currently in use. The products required for a given project are developed by the pipe industry from a suitable mix using appropriate methods of manufacture to produce the most economical and satisfactory pipe for the intended use.

There are various means available for burying and bedding concrete pipe. A pipeline is no more effective than the tightness of its joints. Special problems such as jacking, unstable materials, and high ground water are sometimes encountered. Before a concrete pipeline can be turned over for use, it must be thoroughly tested for infiltration and exfiltration. All of these details of construction are covered in this section.

Soil-Cement

Soil-cement is a mixture of soil, cement, and water that is compacted and allowed to cure. It is an excellent low-cost material for roads, streets, parking areas, dams, and ditches, for example. The required cement, water, and compacting are necessary for each soil sample. After the proper proportions are established, they must be carefully maintained in production. The manner in which the correct proportions are obtained and maintained during production, construction, and control testing are discussed in this section.

Fire Resistance

The fire resistance of concrete became apparent when sections of cities burned to the ground leaving only concrete structures standing. Procedures were developed to give a structural member a fire rating based on the number of hours the member would withstand a "standard fire." Some factors that increase the fire rating are greater mass, calcareous aggregates, increased moisture content, and some restraint during testing. This section discusses all these matters.

Special Concretes and Concrete Products is the third in a series of five books that comprise the National Concrete Technology Curriculum. The U.S. Office of Education funded the preparation of these instructional materials to assist in meeting the construction industry's continuing need for the development of personnel with a fundamental knowledge of concrete technology. The first book in the series is *Principles of Quality Concrete*; the second is *Basic Concrete Construction Practices*. The two which follow this text are: *Administrative Practices in Concrete Construction* and *Concrete Inspection Procedures*. The Series is supplemented by a *Laboratory and Exercise Manual on Concrete Construction* and *Instructor's Guides*.

PORTLAND CEMENT ASSOCIATION

CONTENTS

SECTION I / SPECIAL CONCRETES 1

Chapter 1   Lightweight and Lightweight Structural Concrete 7
Chapter 2   Insulating Concrete 23
Chapter 3   Heavyweight Concrete 39
Chapter 4   Nonplastic Mixes 49

SECTION II / DECORATIVE CONCRETES 67

Chapter 5   Architectural Concrete 69
Chapter 6   Stucco, Plaster, Marblecrete, and Sgraffito 89
Chapter 7   Pigments, Paints, and Stains 115

SECTION III / REINFORCED CONCRETE, PRECASTING 127

Chapter 8   Reinforced Concrete 129
Chapter 9   Placement of Steel 173
Chapter 10   Precast Concrete 217
Chapter 11   Prestressed Concrete 235

SECTION IV / CONCRETE MASONRY 269

Chapter 12 Manufacturing, Materials, and Materials Handling 271
Chapter 13 Shapes, Sizes, and Construction Design 293
Chapter 14 Mortars for Concrete Masonry 314
Chapter 15 Using Concrete Masonry 325

SECTION V / CONCRETE PIPE 343

Chapter 16 The Manufacturing of Concrete Pipe 345
Chapter 17 Concrete Pipe Construction 365

SECTION VI / SOIL-CEMENT 393

Chapter 18 Soil-Cement 395
Chapter 19 Using Soil-Cement 417

SECTION VII / FIRE RESISTANCE OF CONCRETE 453

Chapter 20 Fire Resistance of Concrete 455

Glossary 469

Index 475

SECTION ONE
SPECIAL CONCRETES

CHAPTER 1
LIGHTWEIGHT AND LIGHTWEIGHT STRUCTURAL CONCRETE

USES OF LIGHTWEIGHT CONCRETE

Structural lightweight concrete is used in practically every application of normal-weight concrete where a reduction in deadweight leads to significant savings in cost of structure. Lightweight concrete may cost more per cubic yard, but the structure may cost less as a result of reduced deadweight. Less dead load may lead to reduced sizes in many sections and, therefore, may require less concrete and reinforcing steel. Overall reduction in building weight generally reduces foundation costs.

Some outstanding projects built with high-strength, structural, lightweight concrete are:

The thin-shell roof of the TWA terminal building at Kennedy International Airport, New York (Fig. 1-1).

Floor slabs and beams of Marina Towers and Lake Point Tower, Chicago (Fig. 1-2).

The Broadmoor Hotel's International Center, Colorado Springs (Fig. 1-3).

The Assembly Hall at the University of Illinois, Urbana.

One Shell Plaza, Houston.

Figure 1-1

Lightweight concrete is used in prestressed members, hyperbolic paraboloid roofs, multistory frame and floor jobs designed by either elastic or ultimate-strength methods, long-span folded plates or barrel shells, floating docks, barges (Fig. 1-4), bridge decks, and many other projects.

HISTORY

One of the earliest uses of reinforced lightweight concrete was in the construction of ships and barges by the Emergency Fleet Building Corporation of World War I. Investigation of various aggregates for this program led to the selection of the type of aggregate developed by Stephen J. Hayde in 1917. Hayde had observed that certain raw shales and clays would expand into a tough, hard, lightweight aggregate when treated with heat under controlled conditions in a rotary kiln. Eight plants were subsequently licensed under the Hayde patent to produce an expanded shale aggregate called "Haydite."

About the same time, F. J. Straub developed the use of cinders as an aggregate for concrete masonry units. Following World War I, the lead-

Figure 1-2

Figure 1-3

Figure 1–4

ing lightweight aggregate was coal cinders, and the major use was in Mr. Straub's "cinder block." In 1923, expanded slag was produced commercially, and it has since been used extensively in the manufacture of concrete masonry units, in precast structural concrete products, and in cast-in-place concrete.

Haydite, together with cinders, pumice, scoria, and expanded slag, was used extensively by the concrete masonry industry and in occasional structural applications. The Park Plaza Hotel in St. Louis, built during the 1920s, is a fine example of the early use of reinforced lightweight concrete.

Early in the 1930s, the San Francisco-Oakland Bay Bridge became a reality, and lightweight concrete for the upper roadway of the double-deck structure was one of the keys to its economic feasibility. Structural lightweight-aggregate concrete was selected for the deck, which is still in service today with only a minimum of maintenance to the roadway.

In World War II, history repeated itself with the construction of concrete ships to conserve steel and, once again, expanded shale and clay aggregates were used.

After World War II, in 1948, the first commercial expanded shale aggregate was developed in eastern Pennsylvania. A coal-bearing shale was used, which produced a highly satisfactory lightweight aggregate in conjunction with the operation of a heating plant. Subsequently, the sintering method was used for the production of expanded clays and shales in different parts of the country.

Shortly after World War II, a National Housing Agency survey of potential lightweight aggregate for home building use was conducted. Attention was directed to the fact that structural concrete could be made from certain types of lightweight aggregate: the rotary kiln expanded shales, clays and slates, the sintered clays and shales, and the expanded slags.

In the early 1950s, several structural applications of lightweight concrete attracted the interest of the construction industry to the economy of lightweight concrete. It was used extensively in steel frame buildings for floors and interior walls and was used later in suspension bridges.

These and other structural applications stimulated research by several organizations in order to develop more information on the properties and potential economics of structural lightweight concrete. As a result of these developments, aggregate plants were built in various parts of the United States and Canada. High-grade, structural lightweight aggregate is now available in most parts of North America.

WHAT IS LIGHTWEIGHT CONCRETE?

Lightweight concrete is not just one item; it is a spectrum of different concretes with a variety of characteristics and it fills a number of needs. It can be a gaseous or foam concrete using specially prepared chemicals; a no-fines concrete using ordinary gravel or crushed stone on a gap-graded basis; a normal-weight aggregate concrete with an excessive amount of entrained air; or a concrete that is made using lightweight aggregate.

Specifications for lightweight aggregate demand that it have a unit weight of 70 lb per cu ft or less; normal-weight aggregate weighs in the neighborhood of 100 lb per cu ft. However, all the materials that meet this very general requirement are capable of producing a "spectrum" of concrete weights, varying from a low of 15 lb per cu ft to a high of 120 lb per cu ft.

LIGHTWEIGHT CONCRETE SPECTRUM

In considering the spectrum of concrete weights made from various lightweight aggregates, we will start at the low end of the scale with the "super" lightweight aggregates: vermiculite and perlite. They are capable of producing a highly insulative concrete with compressive strengths ranging from 200 to a maximum of 1000 psi. This concrete is used as an insulative roof fill over a structural system or as fireproofing. Vermiculite and perlite also find widespread use in making lightweight plaster.

Next, with strengths from 1000 to 2000 psi and unit weights between 50 and 85 lb per cu ft, are fill concretes that have some insulative value. Depending on the materials and the techniques of using them, these concretes may have properties of finishability or wearability and, at the high end of their range, they may be used in making small precast products.

At the high end of the scale is structural concrete ranging in unit weight from 85 to 120 lb per cu ft and capable of developing compressive strengths from 2500 to more than 6000 psi.

LIGHTWEIGHT AGGREGATES

Lightweight aggregate is any solid material used in a concrete mix that weighs less than the usual aggregates; examples are sand, gravel, and crushed stone. According to specification, it has a bulk density of 70 lb per cu ft or less.

Lightweight aggregates can be considered in several categories:

Natural aggregate.
By-product aggregate.
Processed aggregate.
Chemical or foamed concrete.

Natural Aggregate

Natural lightweight aggregates are pumice, scoria, and diatomite.

Pumice is mined in California, Oregon, Washington, Nevada, and New Mexico. Mining is done by excavation with a tractor-scraper. The pumice is crushed and thoroughly washed to remove anything adhering to it and anything that might pass a $1/8$-in. sieve. The final product will pass a $3/8$-in. sieve and will be retained in a $1/8$-in. sieve.

It is white or gray to yellow in color and has a frothlike appearance. It is porous but yet has firmness. Pumice is composed of acidic volcanic glass with fragments of rhyolite, perlite, quartz, feldspar, and hornblende.

Scoria is a volcanic lava that is basically uncontaminated. It resembles industrial cinders in texture in that it is angular and hard. It ranges in color from red to black. Like pumice and tuff, it has generally poor concrete-making properties and seldom produces high strengths.

Diatomite is a soft, porous aggregate. It is very absorptive and can be easily broken with the fingers. In appearance, it is white and angular.

By-Product Aggregate

By-product lightweight aggregates include cinders, expanded blast-furnace slag, other industrial slags, and sintered fly ash.

Cinders are a result of high-temperature combustion of coal or coke. The cost of trucking cinders is so high that they are not used as aggregate much, except very close to the production site. They are not as readily available today as they were when coal and coke were used more extensively.

Over 25 million tons of iron blast-furnace slag are processed annually for construction aggregates, but only 3 million tons are expanded. Slag consists of silicates, aluminosilicates of lime, and other bases developed simultaneously with the production of iron in a blast furnace.

Expansion or "foaming" is caused by bringing the molten slag into contact with controlled quantities of water. This process utilizes steam or compressed air. After expansion the expanded slag is crushed, screened, and stockpiled. Expansion produces angular particles that vary in color from dark to light gray or cream with an occasional gray particle. Expanded slag is limited in that it can only be produced at the site of a blast furnace.

Fly ash aggregate is produced by spraying fly ash with water in a drum-type mixer. This pelletizes it. The size of the pellets can be regulated by controlling the speed of the mixer.

The pellets are spread over the bed of a grate. Fuel enough to start combustion is added. Fly ash has enough carbon to support combustion once it has started. The combustion gets rid of carbon and drives off sulfur and other volatile compounds, giving a relatively inert product.

Sintered fly ash has a brick-hard inner core covered with a hard outer skin. It is smooth and less water absorbent than other lightweight aggregates. It is crushed to a fine or medium size, depending on the use to which it will be put.

Processed Aggregate

One method of manufacturing aggregates is the expansion or "bloating" method. Stephen J. Hayde patented the rotary kiln method under the name "Haydite" in 1917. In this method raw clay, shale, or slate is fed into the upper end of a kiln after which it travels slowly to the lower or burning end. Gases formed in the raw material expand in the burning zone under temperatures of 1800 to 2200°F. Myriads of tiny cells are formed in the mass. At the lower end of the kiln the mass is generally discharged into a rotary cooler. After cooling it is crushed, screened, and stockpiled. If the raw materials are presized, the crushing operation may be bypassed. The process causes a vitreous membrane to surround the particles.

Rotary kiln expanded shale, clay, or slate may vary in shape from slightly angular to well rounded. Internally, they are composed of uniform-sized cells. They have good to excellent concrete-making properties and can achieve high strength with reasonable cement factors. These are the principal materials used as aggregate today in making structural lightweight concrete, and about 75 to 85% of all structural lightweight concrete uses these aggregates. They are also used in making quality lightweight concrete block. Their color varies, depending on the raw material, from tan to gray or from dull orange to light pink.

In the sintering process for shale and clay, the raw materials are crushed, screened, and mixed with fuel before being spread to a depth of 8 to 12 in. on a moving grate. The grate passes through an ignition hood where the fuel is fired. To insure combustion, air is blown or sucked through the bed.

This process causes the mass to expand and the particle surfaces to fuse. This forms a sintered cake, which is then cooled by a water spray or allowed to cool naturally. Subsequently, the mass is broken up and finally crushed, screened, and graded. Because of the crushing process, most aggregate particles are rough and angular. Fly ash, which was discussed under by-product aggregates, is a sintered aggregate.

Vermiculite is a form of mica. It looks like thousands of thin sheets of paper in a chunk. Between the thin sheets are trapped water molecules. When this is exposed to heat at 2000°F, the water turns to steam and the sheets move apart. The granules expand to 12 to 15 times their original size. Tiny cells of dead air are formed that provide most of the insulating properties of this type of aggregate. The shiny surfaces of the particles reflect radiant heat, which contributes to their insulating properties.

Raw vermiculite ore is worked at the mine because shipment is costly due to its high bulk. At the mine it is crushed, cleaned, dried, and sized. Then it is transported to the processor.

The finished aggregate is brown to buff in color and has a pearly luster.

Perlite is volcanic lava or glass. It pops like popcorn when it is heated because it also has water trapped in its cells.

The processing of perlite starts with crushing the ore to sand size. It is then dried, screened, and blended before being put into horizontal- or vertical-type furnaces and subjected to temperatures of 1500 to 2000°F. Perlite usually expands to 10 times its original size but can expand anywhere from 4 to 20 times. The color changes from light gray or glossy black to almost pure white under processing.

Fines are removed in a cyclone separator, leaving frothlike particles that are fragile and irregular in shape.

Chemical or Foamed Concrete

One new area of concrete development is that of chemicals introduced into the mix that cause the concrete to expand or foam. These products are manufactured under various trade names and give concrete 28-day compressive strengths ranging from 50 to 3600 psi. Their densities range from 15 to 150 lb per cu ft.

The foaming agents are water-soluble plastics that must be mixed in a special mixer. The high volume of air incorporated into the concrete consists of a homogeneous bubble or cellular structure that is light in weight.

These special concretes offer qualities of moisture resistance, good insulation, fire resistance, and noncorrosion of steel.

DIFFERENCES BETWEEN LIGHTWEIGHT AND OTHER AGGREGATES

Size

The maximum sizes of lightweight aggregates are smaller than normal weight aggregates. Perlites and vermiculites pass a #4 screen. The expanded shales and slags have a top size of $3/8$ to $3/4$ in. A few expanded shales are available in 1-in. size.

Shape

Most lightweight aggregates are more angular in shape and rougher than normal weight aggregates. Some tend to be irregular with harsh, pitted surfaces. However, changes in production processes have led to improved particle shape and texture.

Specific Gravity

Although the specific gravity of lightweight aggregates is very low, it cannot be accurately determined. It varies with the size of the aggregate —larger pieces have a lower specific gravity than smaller pieces.

Absorption

Lightweight aggregates, because of their porosity, generally absorb more water. Accurate determination of absorption is impossible, but values range from 5 to 20% by weight for structural lightweight aggregates and higher for insulative lightweight aggregates.

Strength

Lightweight aggregates are weaker than normal-weight aggregates. However, their strength varies from type to type.

HANDLING LIGHTWEIGHT AGGREGATES

Certain characteristics of lightweight aggregates make handling and use more difficult than with other aggregates. These characteristics are:

1. Difficulties in determining specific gravity and absorption values. It it impossible to use conventional mix designs based on water-cement ratios. The actual net water-cement ratio cannot be definitely established.
2. Low specific gravity and variation in specific gravity make segregation more common. It is desirable to handle coarse and fine sizes of lightweight aggregate separately and to have them damp or moist when transporting to minimize wind loss.
3. Lightweight aggregates are more susceptible to breakage than others, therefore, handling, dropping, and running over them with machinery should be kept to a minimum.

PROPERTIES A READY-MIX PRODUCER WANTS IN A MIX

An ideal lightweight concrete will have the following qualities.

1. Uniformity of composition and properties.
2. Suitable strength and other design properties.
3. Low weight, in order to save weight in the structure.
4. Thermal insulation properties.

5. Small, well-dispersed voids inside, with a minimum of external voids that would have to be filled in.
6. Firm, hard individual particles that will withstand handling and mixing.
7. Particles that will bond with cement paste but will not react with cement or reinforcing steel.
8. Good resistance to weathering, moisture, insects, and fungi.

Within the lightweight concrete spectrum can be found a concrete to fill almost any construction need. There are some 250 plants producing lightweight aggregate in the United States and Canada with an annual production of 15 to 20 million tons. There are close to 150 different trade names for various aggregates, which can cause some confusion to engineers and potential users. However, most of the producers are thoroughly familiar with their product, and the different types of aggregate are represented by associations and institutes ready to assist and give information. Research and development are important functions of these associations, promising new and interesting developments in the field of lightweight aggregate concrete.

PROPERTIES OF LIGHTWEIGHT AGGREGATES

In general, those aggregate characteristics that influence properties of normal-weight concrete also influence properties of structural lightweight concrete. More consideration, however, is generally given to such factors as bulk unit weight, absorption, and particle shape, size, and surface texture for lightweight aggregates. These factors affect the strength, durability, workability, finishability, control, and economy of structural lightweight concrete.

Table 1–A
PROPERTIES OF FOAMED CONCRETES

Celluar Product	Density (lb/cu ft)	Compressive Strength (28-Day)	Thermal Conductivity ("K" Factor)
Aerofill	25–100	50–1500 psi	0.65–3.65
Betocel	20–75	100–1300	0.50–1.60
Calsi-Crete	35 (approx.)	650	0.81
Durox	30–45	285–1000	0.57–1.05
Elastizell	25–150	100–3600	1.1–6.0
Mearlcrete	15–50	50–950	0.45–1.4
Thermo-Con	45–50	500	0.8–1.3

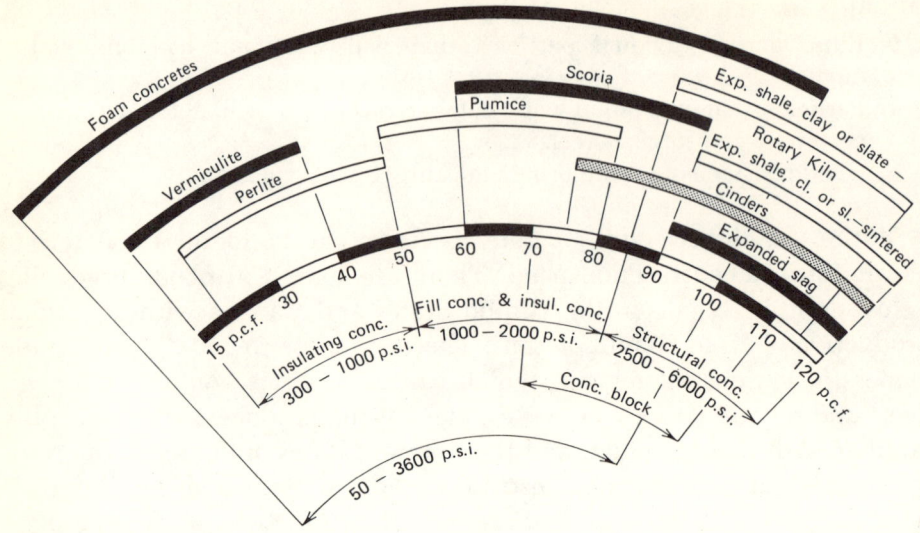

Figure 1–5 Spectrum of lightweight concretes.

Requirements for lightweight aggregates for use in structural concrete are outlined in ASTM C330, *Standard Specification for Lightweight Aggregates for Structural Concrete*.

Bulk Unit Weight
Expanded shale, clay, slate, fly ash aggregate, and slag usually have loose bulk weights varying from 30 to 70 lb per cu ft. The dry-rodded weight of air-cooled slag ranges from 75 to 90 lb per cu ft, and normal-weight natural aggregates vary in dry-rodded weight between 90 and 110 lb per cu ft. The low unit weights of lightweight aggregates account for the low unit weights of structural lightweight concretes. ASTM C330 places the following upper limits on the unit weights of lightweight aggregates: fine, 70 lb per cu ft; coarse, 55 lb per cu ft; and combined fine and coarse, 65 lb per cu ft.

Absorption
Normal weight aggregates usually absorb 1 to 2% water by weight of dry aggregate. These aggregates usually contain some interior moisture at the time of batching and, during the mixing operation, they absorb very little additional water. The amount of mixing water required can be adjusted readily to compensate for absorption in normal-weight concrete.

In contrast, most lightweight aggregates can absorb 5 to 15% water by weight of dry material. In rare cases this can amount to as much as 250 lb of water per cu yd of concrete if total absorption is reached. During mixing, allowance should be made for the water demand of the aggregates so that the mixture does not stiffen and become unworkable during the interval between mixing and placement. Often, lightweight aggregate is wetted beforehand to diminish absorption of water before mixing into concrete.

Gradation
Gradation of lightweight aggregates should conform to the requirements of ASTM C330. Well-graded aggregates have a minimum void content and require a minimum of cement paste to fill the voids.

Bulk Specific Gravity
The specific gravity of an aggregate is its weight compared to the weight of an equal volume of water. Bulk specific gravity for lightweight aggregates is generally between 1.0 and 2.4 as compared to 2.4 to 2.9 for most normal-weight aggregates. The bulk specific gravity for any particular aggregate will usually increase as particle size decreases.

PLASTIC STATE OF LIGHTWEIGHT CONCRETE

Unit Weight
Unit weight of lightweight concrete in the plastic state may range from 90 to 120 lb per cu ft. Differences in unit weight are due mainly to variations in composition, absorption, air content, and bulk specific gravity of lightweight aggregates.

Workability
A good workable mix of lightweight concrete should have the same general characteristics as a workable mix of normal-weight concrete. When troweled, it should show a sufficient amount of mortar to coat and carry the coarse aggregate, but the fine mortar should not be excessive. Coarse aggregate particles should be evident in the concrete but should not stand apart or segregate from the mortar. Mortar should be present in just a sufficient amount to hold the coarse aggregate in suspension and to provide mobility to the mixture.

A mix using sand and gravel or stone will usually comply with these requirements with from 35 to 45% sand and 55 to 65% coarse aggregate.

Lightweight aggregate (except when coated) generally has a slightly angular particle shape throughout its range of sizes, and the maximum size of coarse aggregate is usually not over ¾ in. These characteristics require a greater percentage of fine aggregate sizes in the mix in order to provide proper workability. The mix designer will find that 48 to 60% of lightweight fines passing the #4 sieve combined with 40 to 52% of coarse aggregate will usually provide desired results. These proportions are by volume.

In most cases, the aggregate shape and texture necessitate the use of a lubricating admixture such as an air-entraining agent or a water reducer-retarder to improve workability. Most lightweight concrete contains an air-entraining agent for this purpose. Entrained air in lightweight concrete acts somewhat like tiny ball bearings might; that is, it allows the aggregates to slide over one another easily, thereby making the concrete more plastic.

The amount of entrained air in structural lightweight concrete should be sufficient to provide good workability of the plastic concrete and adequate freeze-thaw resistance of the hardened concrete. *This amount is generally between 4 and 8%.* Water reducer-retarders are also used by many producers as a plasticizer.

In determining the proper consistency for lightweight concrete, one should remember that the much lighter weight of the wet concrete is not affected by gravity to the same extent as normal-weight concrete. Thus, lightweight concrete does not slump the same amount as normal-weight concrete at the same consistency. Lightweight aggregate concrete having equal workability and placement properties will have a 2- to 3-in. less slump than normal aggregate concrete. As a matter of fact, properly designed and mixed lightweight concrete having a slump of 3 or 4 in. is extremely workable and mobile, and such slumps are rarely necessary or desirable. A slump of 3 in. or less produces best results in finishing lightweight concrete. Greater slumps may cause segregation and floating of the coarse aggregate, delay finishing operations, and result in rough, uneven surfaces.

Vibration can be used effectively to consolidate either lightweight or normal-weight concrete. For lightweight concrete, vibration frequencies in excess of 7000 rpm—about the same frequencies commonly used for normal-weight concrete—are recommended. The length of time for proper vibration varies, depending on the mix proportions. *Excessive vibration causes segregation by forcing large aggregate particles to float to the surface.*

Effect of Mix Design on Yield

Yield is sometimes described as the volume of concrete produced by a given amount of materials. This definition of yield, also called the "return," can be derived from the following equation.

$$\frac{\text{Total weight of all materials in batch}}{\text{Weight of concrete, in lb per cu ft.}} = \frac{\text{volume of concrete produced}}{\text{per batch}}$$

Even under well-controlled conditions, a loss of up to 2% in actual volume of air-entrained concrete and from 1 to 1½% of non-air-entrained concrete can be expected for concrete in place. This is the result of a partial loss of entrained air, evaporation and absorption of water, bleeding, settlement, and other factors on the job site.

Unless changes of aggregate properties are anticipated, for example, as a result of heavy rain or apparent stockpile segregation, it is normally sufficient to check the yield of the concrete about once every 50 cu yd.

If the unit weight of the concrete mix varies more than 2 lb per cu ft from the established fresh weight, the change is usually due to changes in air content or in moisture, and will become apparent with slump and air content determinations. If not, there may be a change in aggregate gradation or unit weight, possibly in the specific gravity of the aggregate, or in the handling and batching procedures. In such cases, adjustment of the mix may be necessary if the minimum specifications for the mix are being approached.

The moisture demands of lightweight aggregates will vary with the type and brand of aggregate. The total amount and rate of absorption are important in determining a mix design that will prove satisfactory in terms of yield, strength, and workability. This information, as well as trial mix proportions, can generally be obtained from the aggregate producer. In order to control yield, it is necessary to know and control the extent of saturation of the aggregates. Lightweight aggregates are often wetted before shipment to minimize segregation and loss of fines. When the shipment arrives at its destination, it is often sprinkled and allowed to stand overnight before unloading. In other cases, a spray bar over the conveyor belt is used or additional water is added at the mixer.

The air content of lightweight concrete is as high as, and sometimes higher than, that of conventional concrete in order to promote workability and accommodate the generally smaller maximum size aggregate. This has a considerable bearing on yield if the air content varies significantly from the design values. Air content can be checked by the volumetric method (ASTM C173). If frequent checks of the unit weight indicate variations of more than 2% from the established fresh unit weight value, the air content or aggregate batch weights should be adjusted to obtain the proper unit weight.

Initial air content should be increased to compensate for losses that will occur while in transit and, also, as much as possible for air that will be lost due to job site practices.

Sometimes long-haul, in-transit mixing can cause as much as a 2% loss in air content accompanied by a slump loss. This results in an increase in plastic weight, a harsh mix, and a loss in yield. With a mixing procedure following the practice recommended in ASTM C94, and with frequent unit weight, air content, and slump determinations, proper yield can be maintained.

Under-yielding mixes are costly, since more cement is used per delivered cubic yard than is necessary. This also results in shortage claims by the contractor who will require financial adjustment from the ready-mix producer. Over-yielding mixes result in lower compressive strengths due to the reduced cement content. Proper control is an adjunct to good construction.

To maintain yield, the ready-mix producer manufacturing structural lightweight concrete should obtain a workable mix design; keep close watch over the unit weight of aggregates, the air content and uniformity of materials; and keep his stockpiles of aggregates uniformly moist and free of contamination. The order of introduction of materials into the mixer is important due to the high absorption of lightweight aggregate particles. Although several procedures have proven successful, the one most commonly employed introduces first ½ to ⅔ of the mixing water and all of the aggregates; then, after 5 to 10 revolutions, the cement and the remainder of the water. Any admixtures such as an air-entraining agent are then added and the required number of revolutions is carried out.

The contractor should also be aware of some facts involving yield of lightweight concrete. If the contractor does not schedule concrete deliveries accurately, yield will vary due to loss of air and water as the trucks wait at the job site.

HARDENED STATE OF LIGHTWEIGHT CONCRETE

Compressive Strength
Lightweight concrete with 28-day compressive strengths of 3000 to 4500 psi can generally be produced with cement contents of 425 to 800 lb per

cu yd, depending on the particular lightweight aggregate being used. Certain lightweight aggregates can be used to make concretes with strengths of 7000 to 9000 psi, with cement contents of 565 to 940 lb per cu yd. Table 1-B shows strengths for different ranges of cement content.

Flexural and Tensile Strength

Moist-cured specimens of lightweight and normal-weight concretes of equal compressive strength have approximately equal flexural and tensile strengths.

In one test series using commercial aggregates, the split-cylinder tensile strength of air-dried lightweight concrete varied from about 70 to 100% of that of normal weight concrete of equal compressive strength.

Modulus of Elasticity

Normal weight concrete has a modulus of elasticity of 3 to 6 million psi, depending on the compressive strength. The modulus of elasticity of lightweight concrete is generally 20 to 50% lower than that for normal-weight concrete of equal strength, with the greater difference occurring in the high strength range. An approximate relationship can be used to estimate the modulus of elasticity (E_c) for concrete:

$$E_c = 33 w^{3/2} \sqrt{f'_c}$$

in which w is the unit weight of the concrete in lb per cu ft and f'_c is the compressive strength as determined for 6×12-in. cylinders. This empirical formula is reasonably reliable for concretes with compressive strengths of 3000 to 5000 psi. Figure 1-6 shows this relationship graphically. Note that the modulus of elasticity for most lightweight concretes is between 1.5 and 3.5 million psi. For important work, the modulus of elasticity should be determined by tests of the concrete in question.

Bond to Reinforcing Steel

The strength of the bond between concrete and deformed steel reinforcement is principally a function of the compressive strength of the concrete.

Table 1-B

Compressive Strength (28-day, psi)	Cement Content (lb per cu yd)
2500	425 to 700
3000	475 to 750
4000	550 to 850
5000	650 to 950

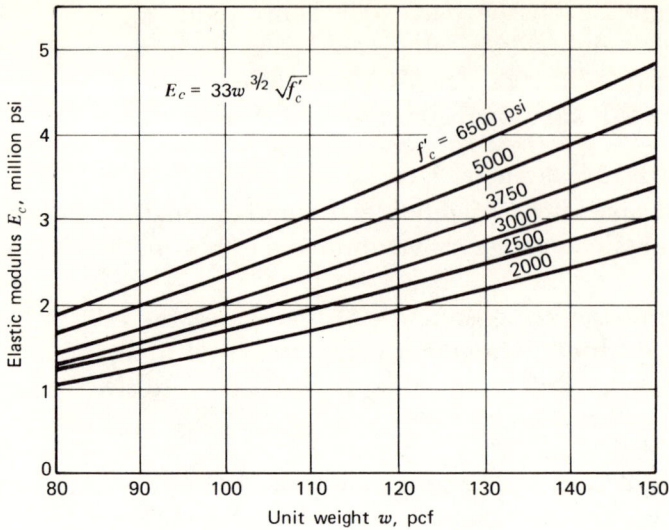

Figure 1-6 Elastic modulus as a function of strength and weight of concrete.

Normal-weight and lightweight concretes of equal compressive strength have comparable bond strengths.

Drying Shrinkage
Drying shrinkage of lightweight concrete made and cured at normal temperatures is generally slightly greater than that of normal-weight concrete because of the generally higher water and air contents. The difference in shrinkage is usually less than about 30% and, in some cases, there is little or no difference. High-strength lightweight concrete (7000 to 9000 psi) has about the same shrinkage as comparable normal-weight concrete.

Creep
The range of creep of lightweight concrete is about the same as the range of creep of normal-weight concrete. The average ultimate creep of lightweight concrete, however, is generally slightly greater than that of normal-weight concrete. Creep is dependent upon magnitude of stress, strength of concrete, age at loading, time after loading, method of curing, and moisture condition of the concrete. When precise knowledge of creep is required, tests should be performed on the concrete in question.

Freeze-Thaw Resistance

The resistance of lightweight concrete to the action of freezing and thawing is dependent upon the same factors that affect freeze-thaw resistance of normal-weight concrete—entrained air, water-cement ratio, and moisture condition of the concrete.

Use of intentionally entrained air increases the freeze-thaw resistance of concrete made with lightweight aggregates, especially if the aggregates are in a soaked condition at the time of mixing. Resistance to freezing and thawing of some air-entrained lightweight concretes is equal to that of air-entrained normal-weight concrete. The amount of intentionally entrained air required for adequate durability of lightweight concrete is about the same as that required for normal-weight concrete.

The effect of water-cement ratio on the durability of lightweight concrete (even though it cannot be accurately determined) is approximately the same as for normal-weight concrete. Reducing the water-cement ratio results in an improvement in durability.

Moisture condition of lightweight aggregates at the time of mixing has a significant effect on the freeze-thaw resistance of concrete. The influence of moisture condition of aggregate is not as pronounced for air-entrained concretes.

To evaluate freeze-thaw resistance of lightweight aggregates, laboratory freeze-thaw tests of concrete should be used, supplemented by field performance records. This is the same procedure generally used for evaluating normal-weight aggregates.

Resistance to Deicer Scaling

Concrete made with lightweight aggregates can be made resistant to the effects of deicer chemicals by the use of entrained air, low water-cement ratio, and adequate curing followed by several weeks of air drying prior to the application of deicers.

Thermal Insulation

Because thermal insulation varies inversely with unit weight, lightweight concrete has better thermal insulation properties than normal-weight concrete.

Use of Normal-Weight Fine Aggregates

Normal-weight fine aggregates are often used to partially or completely replace the lightweight fine aggregates in lightweight concrete mixtures. The principal reason for doing this is economy. Use of normal-weight fine aggregates increases the modulus of elasticity; improves the strength, workability, and finishability; and generally decreases the water required for a particular slump. However, the use of sand fines also increases the unit weight of the concrete from 10 to 20 lb per cu ft.

MIX DESIGN

In the preceding discussions on workability and yield, some of the special considerations necessary in mixing and handling lightweight concrete were discussed. These same factors cause methods of mix design of lightweight structural concrete to differ from those for ordinary concrete. Water-cement ratio, for instance, cannot be used in designing the mix because the variable absorption of the various aggregates makes the ratio impossible to calculate.

There are three methods for designing lightweight concrete: the absolute volume, the specific gravity factor, and the volumetric methods.

The absolute volume method depends on the determination of the absorption and specific gravity of lightweight aggregates. With presently known testing methods, these determinations are hard to make and are not entirely accurate. For this reason, the absolute volume method is seldom used in the field.

The specific gravity factor method is recommended when prior experience with the materials involved is not available. It uses a cement content to unit weight relationship to select proportions for concrete with a given strength. This method, which is set forth in ACI 211.2-69, *Recommended Practice for Selecting Proportions for Structural Lightweight Concrete*, is also of value for all lightweight aggregate mixes and for lightweight mixes where natural sand is used as, or as part of, the fine aggregate.

The volumetric method is easy to use and gives dependable results. Because of this, it is the most commonly used in the field. It is a trial batch method based on using known volumes of aggregates and cement content that experience has shown will yield the required strength. Mixing water is then added until desired slump and workability are obtained.

CHAPTER 2
INSULATING CONCRETE

Insulating concretes are concretes specially designed to provide insulation against heat, cold, fire, and sound. Structural properties are considered secondary to insulation value.

Concrete with good thermal insulation properties will result in lower air conditioning and heating costs and greater comfort for those occupying the building. Although an insulating concrete is not damaged by water or high humidity, its insulating value (as with other insulations) is greatly reduced while it is wet (Fig. 2-1). Obviously, good insulation depends on keeping moisture out of the dried insulating concrete by appropriate construction techniques, including vapor barriers and venting channels for drying.

Insulation from sound is controlled by weight, airtightness, stiffness, and edge-fixing conditions. The aim is to reduce stiffness without reducing strength. Generally, increasing the weight of concrete will decrease the sound transmitted through it, but lightweight concretes are porous and, therefore, have the capability of absorbing sound. Normally, however, sound insulation is accomplished by higher-density concrete (75 lb per cu ft and up), whereas thermal insulating concretes are considered to be those below 50 lb per cu ft.

One characteristic of lightweight concrete is its potential for rupture

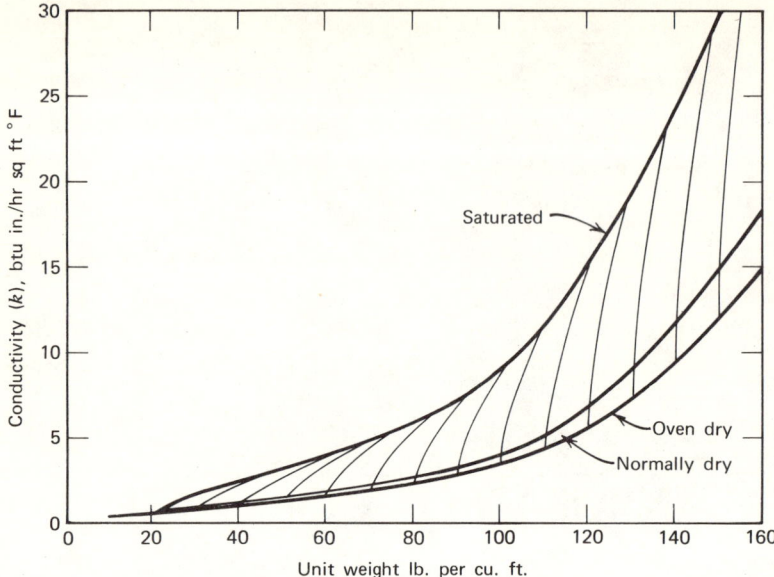

Figure 2-1 Typical increases in conductivity and unit weight as the moisture content increases.

and indentation. This makes it easy to damage, but it also means that it can be nailed and treated much like wood.

Some of the uses of insulating concretes are:

1. Light, structural roof decks.
2. Insulating fill over metal decking.
3. Floor fills.
4. Insulating fill over structural concrete.
5. Non-load-bearing fill in bridges and tunnels, for example.
6. Underground thermal conduit linings.
7. Firewalls.

One of the problems encountered in mixing lightweight insulating concrete is the measurement of mixing water because lightweight aggregates are extremely absorbent. The amount of water absorbed varies with the aggregate, the mix, and the amount of time the ingredients are left in the mixer. Water is assimilated into the lightweight aggregate particles without changing their volume or the volume of the concrete. The active water (that which is over and above the absorbed water) per cubic yard controls the slump and establishes the strength when mixed with a particular weight of cement.

TYPES OF INSULATING CONCRETES

Group I—aggregate type. Concrete made predominantly with low-density mineral aggregates, such as expanded perlite, vermiculite, slag, shale, clay, or slate.

Group II—cellular type. Concrete made by forming a cement matrix around air voids that are generated by preformed foams or special foaming agents, with or without the addition of mineral aggregates.

Some of the properties of perlite, vermiculite, and foam concretes that make them excellent insulating materials are:

1. They screed smooth without steel troweling.
2. They are completely inorganic and cannot rot or decay.
3. They are not damaged structurally by water.
4. They resist attack by rodents and vermin.
5. They last the life of the building.
6. Their strengths increase with age.
7. They are easy to handle.
8. There is little shrinkage.
9. They are incombustible.

GROUP I—AGGREGATE TYPE

Aggregate-type insulating concretes are those using expanded materials such as perlite or vermiculite aggregates, calcined or sintered materials such as blast-furnace slag, clay, shale, etc., or processed natural minerals such as pumice, scoria, or tuff. *Standard Specification for Lightweight Aggregates for Insulating Concrete* (ASTM C332) covers materials.

Vermiculite and Perlite

Vermiculite and perlite are normally delivered in 4-cu ft bags. To get an idea of the difference in weight between these aggregates and sand, a bag of perlite weighs about 32 lb while the same volume of sand weighs 400 lb.

Vermiculite, perlite, expanded clay, shale, slate, and slag give concrete superior heat and sound insulating properties. Depending on the density of the concrete, thermal conductivity will range from 0.69 to 2.70 Btu/hour/sq ft/degree F for a wall 1 in. thick. The lighter the concrete, the lower the conductivity.

Vermiculite and perlite concretes make excellent buffers between com-

Table 2–A
CONDUCTIVITIES (k) AND INSULATIONS (1/k)
FOR CONCRETE[a] WALLS 1 INCH THICK
AT 45° F MEAN TEMPERATURE

Concrete Density (lb per cu ft)	Conductivity k (Btu/hr/sq ft/°F)	Resistivity 1/k
20	0.69	1.45
30	0.95	1.05
40	1.21	0.83
50	1.50	0.67
60	1.80	0.56
70	2.24	0.45
80	2.70	0.37
90	3.30	0.30
100	4.00	0.25
110	5.20	0.19
120	6.90	0.14
130	9.04	0.11
135	10.40	0.096
140	11.80	0.085
145	13.30	0.075
150	15.00	0.067

[a] Portland cement concretes made with normal-weight and lightweight aggregates and cellular types—moist cured and subsequently air dried.

bustible roofing materials and materials used in the interior of a building because they are noncombustible. They have fire resistance ratings as good as dense concrete of high structural value. The use of these aggregates in concrete lowers insurance rates because of their known heat resistance. Vermiculite and perlite are relatively high-priced aggregates initially. However, significant savings are possible in the cost of steel because of the reduction in the weight of the concrete.

A 4-in. thickness of 1:6 mix (1 bag of cement to 6 cu ft of aggregate) perlite concrete, for example, provides insulation values equal to 1 in. of good insulation board or a regular sand-and-gravel concrete wall approximately 40 in. thick. The perlite concrete weighs 20 to 50 lb per cu ft, while the sand-and-gravel concrete weighs 140 to 150 lb per cu ft. Slag, clay, and shale concretes weigh 60 to 120 lb per cu ft.

A recommended mix for roof decking using vermiculite aggregate is 1:4 (1 bag of cement to 4 cu ft of aggregate). This concrete will have a density of 35 to 40 lb per cu ft and give a compressive strength of 350 to 500 psi. It will have indentation resistance of 410 to 515 psi and a rupture modulus of 180 to 205 psi.

A recommended mix for roof insulation using vermiculite aggregate is 1:6 or 1:8. When the air temperature during placing and curing is 40°F and the roof live load is less than 30 lb per cu ft, the 1:8 mix is recommended. Approximately 75 to 100 gal of water per cu yd will be necessary with vermiculite aggregate. Table 2-B shows average densities and proportions for Group I concretes.

Water and cement should be placed in the mixer and blended into a smooth slurry first. After the vermiculite aggregate is added, the mixing should be limited to the time required to obtain thorough mixing and proper fluidity. The maximum mixing time recommended is 5 min.

For truck mixing, water and cement are first put in the mixer. Then the drum is rotated slowly while the aggregate is added. Rotation should be continued 1 min after all the aggregates are in the mixer, but it should be discontinued during the trip to the job. At the job site the drum should be rotated at its fastest speed until the concrete is uniform and flows freely from the mixer. In a rotating-paddle mixer, the increased speed of rotation will entrain enough air to reduce the quantity of air-entraining agent needed.

The most common perlite concrete mixes are (Tables 2-C and 2-D).
1. 1:4 for stronger concrete.
2. 1:6 for roof decks.
3. 1:8 for greater insulation value.

These perlite concrete mixes may appear too wet by normal standards, but a slump of 7 in. is normal. If it is too dry or too stiff, extra water should not be added to produce a proper mix. Instead, the mixing time should be increased.

An air-entraining agent will cause tiny air bubbles to form. This will improve workability and control the water content and insulating value of the concrete. It also reduces density and increases yield. Excessive air entrainment, however, will reduce the strength of the concrete.

Steps in transit mixing of perlite concrete are:

1. Determine the load of the mixer.
2. Add to the mixer the correct amount of water for the load.
3. Add the correct amount of air-entraining agent and cement to the mixer and mix until a slurry is formed.
4. Add the required number of bags of perlite aggregate.

Table 2-B
AVERAGE PROPORTIONS AND DENSITIES FOR AGGREGATE TYPE INSULATING CONCRETES

Aggregate Type	Proportions (bags cement per cu ft aggregates)	Cement Factor		Density (lb per cu ft)	
		Pounds per Cubic Yard	Bags per Cubic Yard	Wet	Oven-dry
Vermiculite	1:4	635	6.75	57–63	35–40
	1:6	425	4.5	48–54	25–30
	1:8	330	3.5	46–52	20–25
Perlite	1:4	635	6.75	50	36
	1:6	425	4.5	40	27
	1:8	320	3.4	36	22
Pumice	1:4–1:9	470–310	5.0–3.3	80–75	63–57
Expanded slag	1:5–1:9	470–280	5.0–3.0	85–70	75–60
Expanded shale, clay and slate	1:3–1:14	510–190	5.4–2.0	83–60	72–54

Table 2-C
MATERIALS FOR 1 YARD OF PERLITE CONCRETE

Mix Ratio (bags cement per cu ft aggregate)	Cement (lbs)	Perlite (cu ft)	Water (gal)	Air-Entraining Agent[a] (pints)
1:4	635 (6.75 bags)	27	61	6¾
1:5	510 (5.40 bags)	27	59½	6¾
1:6	425 (4.50 bags)	27	54	6¾
1:7	360 (3.85 bags)	27	54	6¾
1:8	320 (3.38 bags)	27	54	6¾

[a] Available from Perlite Institute members (courtesy, Perlite Inst.).

5. If the drum is to be rotated during transit, it should be done at slow speed for no longer than 5 min.
6. Upon arrival at the job, the mixer should be operated at top speed for at least 5 min and not more than 10 min.
7. Discharge in the normal manner after checking wet density for conformity with specifications.
8. Rotate the drum at idle speed when discharging the last of the load to ensure complete discharge.
9. Do not wash out the drum between loads.

The first load of a mixing truck can discharge as much as ⅓ yd short because this much can coat the inside of the truck. Subsequent loads will be complete because the water added at the beginning of each mix washes the retained concrete from the walls of the truck and combines it with the next batch. After discharging the last load, water may be added to the revolving drum to wash the coating out.

Water, air-entraining agent, and cement can be put in the mixer at the batch plant and the perlite added at the job site. In this case mixing should be at high speed until the desired density and consistency are reached.

To check the yield at the job site, several unit weights of fresh concrete should be obtained. The truck can be equipped with a 10-lb scale and a gallon container. After taking the weight per gallon, multiply it by 202.5 to obtain the weight per cubic yard. Divide this weight by the total weight per cubic yard of all the ingredients and multiply this by 100 to obtain the percent yield.

Placing and screeding should immediately follow mixing. No change in consistency should be allowed.

Table 2-D
MIX DATA FOR PERLITE CONCRETE FLOOR FILLS

Cement (lb)	Perlite (cu ft)	Heavy Aggregate (cu ft)	Air-Entraining Admixture	Water-Cement Ratio (by wt)	Cement Content (lb/cu yd)	Density		Compressive Strength (28-day, psi)
						Wet	Dry	
100	3	2[a]	6	0.75	5.82	82	72	1060
100	3	2[b]	4	0.75	6.12	66	54	1220

Available from Perlite Institute, Inc., members (courtesy, Perlite Inst.).
[a] Sand.
[b] Rotary Kiln expanded shale.

Curing vermiculite and perlite concretes is the same as for other concretes. Perlite concrete usually has enough water content to make sprinkling unnecessary during curing. When roofing is to be placed over the concrete, the surface should be substantially dry first. It should withstand foot traffic and normal roofing operations.

These concretes should not be placed when the temperature of the air is below 30°F, or when it is expected to be below 30°F within the first 12 hours after placement, or if the base for the concrete is covered with standing water, ice, or snow. It may be placed by standard concrete equipment or specially designed pumping equipment. When the grade on which it is to be placed exceeds 30° inclination, it may be sprayed.

No-Fines Concrete

Group I includes no-fines concrete, which is concrete consisting of coarse aggregate, cement, and water. Fines are omitted entirely giving it a porous, open-textured quality. It has a high proportion of interconnected voids but practically no fine capillary pores. Moisture will not be transferred by capillary attraction.

No-fines concrete weighs about two-thirds of dense concrete made with the same aggregate. Shrinkage is about half that of dense concrete of the same aggregate, although the rate of shrinkage is more rapid. About 50 to 80% of the total shrinkage occurs in the first 10 days in no-fines concrete, whereas only about 20 to 30% of the total shrinkage of dense concrete occurs in that time.

Aggregate for no-fines concrete can be any sound aggregate, free from clay or dust, which is fairly round or cubical in shape. If more than 10% by weight is soft, friable, thin, flaky, elongated or laminated, the aggregate should not be used. Washed gravel and crushed limestone are ideal aggregates. Generally, no-fines aggregates will measure from 3/8 to 3/4 in. Lightweight aggregates used in no-fines concrete at a 1:6 mix ratio provide exceptional thermal insulation and lightness, but do not have good strength.

One hundred pounds of cement to 10 cu ft of aggregate gives a volume of 1:8 no-fines concrete. Mixes of 1:9 or even 1:10 and 1:12 with heavy aggregates are used satisfactorily.

No-fines aggregate must be sufficiently coated with cement grout so that all the pieces will lock together. Too much water will wash the cement off the pieces of aggregate and weaken the mix. A water-cement ratio of 0.35 to 0.40 should be used, but this will vary with the type of aggregate used.

Walls are usually designed with no reinforcing steel when using no-fines concrete because bonding is difficult. If steel is used, it should be coated with cement grout before the concrete is placed. The bonding strength will still be only about half of what it would be with regular concrete.

Rapid placement is important. Within 20 min after mixing, the mix becomes difficult to work. It is not desirable to add extra water to improve workability. Transit-mixing trucks should be equipped to introduce water in transit so mixing can be started en route.

No-fines concrete does not flow as easily as dense concrete. Bridges and voids tend to form over such obstructions as protruding nails. Compaction, in the form of light rodding, is all that is needed to fill the voids.

No-fines concrete made with pumice, expanded slag, or expanded shale has been made with 30 to 35 gallons of water per cu yd (not including absorbed water), with air contents of 20 to 35% and 470 lb of cement per cu yd.

Other Aggregates

Pumice and expanded natural minerals such as slag, shale, clay, and slate are often used in insulating concretes. Average proportions and densities for insulating concretes using these aggregates are tabulated in Table 2-B.

GROUP II—CELLULAR TYPE

History

In 1914, the first patent on a method to foam concrete was applied for. It entailed the use of aluminum powder to generate hydrogen gas in a cement slurry. Previous to that, inventive minds had tried egg whites, yeast, and other unusual methods of adding air to concrete.

Hydrogen peroxide and sodium or potassium hypochlorite were patented soon after 1914. This method causes internal gas generation. It requires close control over the quantities of reactants added to the mix as well as the temperature and time of mixing. It is usually necessary to trim and level the dimensions after reaction is completed.

In 1923, a patent was granted using aluminum powder in moist-cured and autoclaved concretes containing lime and burned oil shale. In its early development, air entrainment was accomplished by adding neutralized resins or synthetic surface active agents to which foam stabilizers were added. Foaming was generated by high mechanical agitation.

The foam process now in widest use was patented about 1931. It involves cellular concretes prepared by foaming methods, followed by curing in high-pressure steam. Patents were issued in 1929 on foaming agent formulations, mixing procedures, and mixing machines.

Preformed foam was developed in the early 1950s following the development of simple, but reliable and easily controlled, foam-generating equipment and highly refined foam stabilizers. The method was low in cost and provided for accurate density control. Since the 1950s, research has been carried on in the use of cellular concrete in structural and architectural applications. It is now the method most widely used for low-density cast-in-place roof decks, filling voids, tunnel back-fill, and for insulating underground heat distribution lines.

Scandinavia first used cellular concrete in the 1930s. Now, four out of five buildings erected in Sweden use cellular concrete.

Uses

Cellular or foam concrete can be used in the following ways.

1. Insulating fill.
2. Roof deck installed over forms, metal roof decking, or reinforcing wire mesh.
3. Insulation under ice skating rinks or underground heating systems.
4. Void filler pumped into abandoned underground pipes and trenches.
5. Roof construction (ideal for this because of plasticity, incombustibility, permanence, and low cost).
6. Insulating boilers, tanks, piping and conduits.
7. Topping over precast elements.

Table 2–E
PROPERTIES OF CELLULAR CONCRETE

Cellular Product[a]	Density (lb/cu ft)	Compressive Strength (28-day, psi)	Thermal Conductivity k	Cast-in-place Cellular Concrete[b]	Precast Cellular Products[b]
Aerofill	25–100	50–1500	0.65–3.65	X	X
Calsi-Crete	34.5	650	0.81		X
Cantilite	20–70	100–580	0.40–2.30		X
Elastizell	20–80	100–800	0.40–2.80	X	X
Mearlcrete	20–50	250+	0.60–1.30	X	X
	75–115	800–3500	—	X	X
Poretherm	30	200	0.65	X	X
Siporex	25–44	—	0.69–1.06		X

[a] This is a partial list of cellular products.
[b] X indicates the manufacturer of cast-in-place concrete and/or precast cellular concrete products.

General Properties (Table 2–E)
Cellular or foam concrete has bubbles or cells that vary from almost microscopic to about the size of a grain of medium sand. They are bound together with portland cement alone, cement-lime, cement-sand, or cement-pozzolan. Conventional lightweight aggregate is not used. Foam must have sufficient stability to maintain its structure until the cement sets. This combination of ingredients produces the following characteristics.

1. Plasticity.
2. Incombustibility.
3. Permanence.
4. Low cost (because no costly aggregate is required).
5. Improvement with age.
6. Noncompressibility.
7. Does not absorb humidity.
8. Easily pumped.
9. Fluidity from air cells gives workability.
10. Freedom from "pop-outs" because there is no coarse aggregate.

Thermal Properties
Thermal conductivity in cellular concrete (as with all insulating concretes) is proportional to density. The lighter the material, the better it will insulate. K factors (Btu/hour/square foot/degree F/inch thickness, at 75°F) range from 0.0169 for a 20 lb per cu ft mix to 3.30 for a 90 lb per cu ft mix. A wall of 40 lb per cu ft concrete will have the same insulating value as a brick wall 3 times as thick or as a dense concrete wall 8 times as thick.

Thermal expansion is negligible (0.000004 in. per in. from a $-58°F$ to a $+212°F$).

Air Entrainment
Normal air entrainment is less than 7% by volume. In cellular concrete the voids may occupy anywhere from 25 to 80% of the total volume. This makes a lightweight concrete with incomparable workability. Lack of large-sized aggregate does away with the problem of segregation.

Density
Cellular concrete can be produced with a dry density of as little as 20 lb per cu ft when portland cement alone is used with no mineral aggregate. On the other end of the density scale, conventional aggregate will pro-

duce a density of 110 lb per cu ft or more in cellular concrete, with strengths and other properties varying accordingly.

Strength

The 28-day compressive strength varies, depending on the density of the concrete. It can be as low as 50 and as high as 1500 psi. A standard cellular concrete with a dry density of 20 lb per cu ft made with 470 lb of portland cement can have a 28-day compressive strength of only 95 psi. Concrete with the same cement content and natural sand in a 1:4 ratio with a 75 lb per cu ft density has a strength of 740 psi. A mix containing 550 lb of cement will have a strength of 1300 psi, and a mix with a density of 110 lb per cu ft can offer strengths up to 2500 psi.

The flexural strength of cellular concrete is generally between $\frac{1}{5}$ and $\frac{1}{3}$ of its compressive strength.

The compressive strength of insulating concrete is not as important as impact and dimple resistance (surface hardness). Cellular concrete uses 550 to 650 lb of cement per cu yd. The paste on the top is richer in cellular material because the larger air cells are squeezed out of the top $\frac{1}{16}$ to $\frac{3}{16}$ in. when it is steel troweled. This results in a good hard-wearing surface which is especially resistant to freeze-thaw cycles.

Resistance

High water resistance is due to the uniform nature of the cells and to the fact that they are not interconnected so that water cannot flow by capillary action. This advantage can be improved on in some mixes by adding 2% of a plastic emulsion to give each air cell an individual waterproof coating. In cellular concrete the water absorption is about one-fourth that of common brick.

Frost resistance is also high. Cellular concrete walls will not sweat in winter when the exterior surfaces are given a weather-protective treatment of stucco or masonry paint. The hard-wearing surface produced by steel troweling resists freeze-thaw cycles.

Cellular concrete is completely inorganic and wholly noncombustible. A wall 8 in. thick will endure fire for a period of about 8 hours. Insurance rates on buildings using cellular concrete can be considerably lower than for most other building materials. The material also resists fungus, vermin, and rot.

Shrinkage

Shrinkage is higher in cellular concrete without aggregates than in ordinary concrete. The normal range is 0.2 to 0.6% for moist air-cured cellular concretes. Autoclaved cellular concretes have substantially lower shrinkage.

Floors do not need expansion joints around the perimeter of the room where the concrete meets the plate because cellular concrete shrinks enough when drying to pull away from the plate. Since the shrinkage is permanent, no expansion or growth will occur in the future. Shrinkage increases with a reduction in the density. The resultant cracking can almost be eliminated by using resin-coated glass fibers in the mix.

Mix Methods

Chemical Method. The chemical reaction method of foaming concrete using admixtures to generate gas bubbles is suitable for plant precasting. The admixtures may react chemically with each other or with certain compounds already present in the cementitious slurry. Aluminum powder is the most widely used agent in the commercial production of this type of concrete. It reacts with the alkalies in the cementitious mixture to generate hydrogen. According to the patent literature, hydrogen peroxide and bleaching powder may be used to generate oxygen, but they are not used commercially.

Mechanical Method. Mechanical methods are easier to control than the chemical methods and are usually more economical. They incorporate completely neutral foams into cement paste or mortar using high-speed mixers or compressed-air generators. One method requires only a high-speed mixer. High shearing action is provided by operating a standard paddle-type mixer at high speed. Some people prefer a horizontal propeller-type mixer. Some use blades consisting of ¾ in. of heavy wire mesh rotating at 55 to 60 rpm.

One method for producing low-density concrete with little or no aggregate is for the mixing water and foaming agent to be added first to the mixer to produce a foam. Then the portland cement (and sand if needed) are added. The paddles are rotated slowly while blending. For denser concretes with more aggregate, the process is reversed.

Another mechanical method is the excess water method where a thin slurry of cementitious and aggregate materials is prepared with a quantity of water several times greater than the requirements for an equal volume of conventional concrete. The concrete is cured in high-pressure steam (autoclaved) and allowed to dry. Finely divided air voids will form due to the evaporation of the excess water. Foamed concretes made by this method are of low density and are often called "light-lime" concretes because lime is usually used as the cementitious ingredient.

Foamed Method. ASTM standards for foam quality are being written. However, the following types of foaming agents have been used in the process of making cellular concretes, but not all are satisfactory:

1. Detergents (sodium lauryl sulfate, alkyl aryl sulfonate, etc.).
2. Resin soap and vegetable or animal glue.
3. Saponin.
4. Alkylated naphthalene sulfonate and degenerated glue, butylated naphthalene sulfonate and waterglass, and isopropylated naphthalene sulfonate and calcium chloride or high alumina cement.
5. Proprietary neutralized resin.
6. Hydrolized protein such as keratin (hoof and horn meal), cattle hooves and fish scales, blood and saponin, and casein.
7. Protein derivatives modified to provide the high film strength and toughness necessary to permit extended mixing and pumping of the concrete.

Mix-Foam Method. In the mix-foam method a foaming agent may be added to the mixture after all other ingredients (including an accelerator) have been mixed to a fairly thin slurry. Molds are therefore built to be almost watertight to prevent subsidence due to leakage of water through the bottoms of the molds. The effects of mixing speed, character of mix ingredients, type of mixer used, and the length of mixing time, for example, will determine the amount of foaming agent required to obtain a given density.

Prefoamed Method. Preformed foam is usually based on hydrolized protein, free from starches and sugar. It is chemically processed from protein waste material from packing houses (hooves, hair, horn, and blood) or from fish scales. The liquid concentrate is delivered to the site in drums. After dilution with water it is combined with 90-psi compressed air to foam 20 to 30 times the solution volume. For low-density concretes, continuous mobile machines have been developed that combine the cement and water and then continuously blend this slurry with preformed foam under controlled conditions to produce cellular concrete. It is then pumped through a hose as far as 1000 ft horizontally or 250 ft vertically. The foam generator is calibrated to supply foam at a given rate. It discharges foam into a standard mixer where it increases the volume of the concrete in accordance with the amount added. Each air bubble is surrounded by a tough protein membrane that insures its stability while mixing and handling. The membrane will eventually break down, so mixing and placing should be completed within 1 hour. The use of high-early-strength cement insures rapid setting and stability. Good results can also be obtained from standard portland cement and 2% calcium chloride by weight of cement.

Placing

When using foam concrete, the temperature must not be 40°F or less and falling, although it may be placed if the temperature is 40°F and rising. It should not be placed in a heavy rainfall.

Curing

Insulating and fill concretes are usually given a period of moist curing (one to seven days) and then allowed to air dry, prior to the application of a moisture-proofing material. The time required for satisfactory air drying is shortest in materials of lowest density.

CHAPTER 3
HEAVYWEIGHT CONCRETE

Heavyweight concrete is not new; it has been used for many years as a counterweight for lift bridges. Today, however, its most important application is as a protective shield from X-rays and gamma rays in industrial radiographic and medical therapy installations, and from neutrons and gamma rays in nuclear reactors and nuclear accelerators.

X-rays and gamma rays differ from each other only in source and have identical properties, which are dependent on the energies of the individual photons making up the beam. They readily penetrate opaque matter, and the required thickness of a shield is usually determined by the amount and character of the gamma rays. At the gamma ray energy range (below about 1 million volts and above about 10 million volts) the attenuation (absorption) of a shield is strongly dependent on the atomic number of the shield material. Lead has the highest atomic number of any commercially practicable material and finds extensive use, particularly in low-energy (low-voltage) installations. However, in the middle-energy range, where shields are the least effective, the attenuation is roughly proportional to the weight per square foot of shield and is relatively insensitive to the shield material. A thick shield of light material is roughly equivalent to a thin shield of heavy material if the weight per unit area traversed is the same.

Concrete, because of its low cost, easy handling, and good structural properties, is an excellent shielding material if space is available for the extra thickness as compared to lead. The required concrete thickness can be reduced by a factor of 1.5 to 2.5 by the use of high-density (heavy) aggregate. In addition, heavy aggregates may also be desirable for neutron shielding.

Neutrons play a key role in atomic reactors. They are high-energy particles with no electrical charge and have great penetrating power. Unlike the case with gamma rays, the reaction of neutrons with matter does not always vary with atomic number from one element to another. A particular isotope may be highly absorptive while its neighbors on either side are relatively transparent. Also, the absorption is usually accompanied by highly penetrating secondary gamma rays, and the energy and abundance of these vary from one absorber to another. Generally, therefore, a reactor shield must be suited to the unique properties of the particular reactor. A brief general discussion of the neutron attenuating process, however, should be useful.

The fast neutrons from the source must first be slowed down. This is accomplished through inelastic and elastic scattering (as a result of collision with atomic nuclei in the shield). Only iron and elements with atomic numbers higher than iron are effective as inelastic scatterers. This, in addition to their high density, is the reason for using iron- or barium-bearing aggregates in neutron shields.

By contrast, very light elements are most effective in the billiard-ball-type elastic scattering. Hydrogen is best because its atomic mass is about the same as that of a neutron, although oxygen is also quite effective. Both occur together as water (H_2O) in hydrated portland cement gel, as relatively free water in concrete pores, and as an ingredient in some hydrous aggregates. Oxygen also is an ingredient of cement and many aggregates. A silica-aggregate concrete is almost one-half oxygen.

Since the absorption of neutrons may take place well inside the shield, the resulting gamma rays present an important problem. The light isotope of boron B-10 has a particularly high neutron absorbing capacity and, in addition, the capture results in 0.48-million-volt gamma rays of relatively low penetrating power.

An effective neutron shield may contain an ingredient such as iron to bring about inelastic scattering, water to promote elastic scattering and neutron capture, and perhaps boron to minimize the secondary gamma ray problem.

AGGREGATES

As for normal-weight concrete, aggregates must be clean and free from fine material, oil, and foreign coatings to insure proper bonding with the cement matrix. Aggregates should be roughly cubic in shape and free from flat or elongated particles.

If there is too great a difference in specific gravities of aggregates used in a mix, there may be segregation. For example, steel punchings sink in a silica sand mortar. Where steel shot or steel punchings are used as the heavy aggregate, the mix must be kept stiff.

Heavyweight aggregates can be naturally occurring ores and minerals, steel shot or steel punchings, or manufactured products such as ferrophosphorus. Steel aggregates must be used when concrete, weighing in excess of 300 lb per cu ft, is required (Table 3-A).

The materials discussed below include (1) natural and synthetic aggregates containing either iron or barium, (2) hydrated natural aggregates, and (3) aggregates containing boron. The first group includes materials that are effective absorbers of X-rays and gamma rays because they contain relatively large proportions of atoms of high or moderately high atomic weight. The second group includes aggregates containing bound water, which are effective in attenuating neutrons. The third group includes aggregates containing boron, which effectively absorb thermal neutrons without producing penetrating, high-energy gamma rays.

Heavy Aggregates

Hematite has a hardness of 5 to 6 on Mohs' scale (it can be scratched by hard steel) and a specific gravity of 5.26 in the pure mineral. It varies in color from bright red to dull red to steel gray. Its luster ranges from metallic to submetallic or dull. Hematite ores dust in handling. They vary from one deposit to another and even within a deposit.

Ilmenite has a hardness of 5 to 6 and a specific gravity of 4.72 ± 0.04 in the pure mineral. Its color is iron black with metallic to submetallic luster. It is slightly magnetic. Ilmenite ore is one of the most widely used heavy aggregates.

Magnetite has a hardness of $5\frac{1}{2}$ to $6\frac{1}{2}$ and a specific gravity of 5.17 with metallic luster. Magnetite ores are associated with metamorphic, igneous, or sedimentary rocks. Magnetite ore is one of the most widely used high-density aggregates.

Ferrophosphorous is a by-product of the production of phosphorous.

Table 3–A
COMPOSITION AND SPECIFIC GRAVITY OF HEAVYWEIGHT AGGREGATES

Predominant Constituent	Fixed water[a] Percentage by Weight	Class of Material	Chemical Composition of Principal Constituents	Aggregate Specific Gravity	Aggregate Bulk Unit Weight (lb/cu ft)	Concrete Unit Weight (lb/cu ft)
Limonite	8–9	Crushed stone	$(HFeO_2)_{1/x}(H_2O)_y$	3.4–4.0	130–150	180–210
Goethite	10–11	Crushed stone	$HFeO_2$	3.4–3.8	130–140	180–200
Barite	0	Gravel or crushed stone	$BaSO_4$	4.0–4.6	145–160	210–230
Ilmenite	[b]	Crushed stone	$FeTiO_3$	4.3–4.8	160–170	220–240
Hematite	[b]	Crushed stone	Fe_2O_3	4.9–5.3	180–200	240–260
Magnetite	[b]	Crushed stone	$FeFe_2O_4$	4.2–5.2	150–190	210–260
Ferrophosphorus	0	Synthetic	Fe_nP	5.8–6.8	200–260	255–330
Steel punchings or shot	0	Synthetic	Fe	6.2–7.8	230–290	290–380

[a] Water retained or chemically bound in aggregates at 100°C.
[b] Test data not available. Aggregates may be combined with limonite to produce fixed-water contents varying from about ½ to 5%.

42　SPECIAL CONCRETE

It has been used as coarse and fine aggregate in radiation-shielding concrete to produce concrete weighing as much as 300 lb per cu ft. It contains approximately 70% iron. The coarse aggregate is reported to degrade easily, however. Its specific gravity is approximately 5.72 to 6.8.

Ferrophosphorous may radically affect the setting time of concrete and will generate flammable gases that develop high and dangerous pressures if confined. In one construction job, ferrophosphorous produced enough gas in the concrete to cause a fire that burned the forms away. Ferrophosphorous sealed in steel containers has exploded because of the development of high hydrogen pressures. Ferrophosphorous aggregates should be tested carefully before they are used.

Barite

Barite (also known improperly as baryte) has a hardness of 3 to $3\frac{1}{2}$ and a specific gravity of 4.50 in the pure mineral. It is the most common barium mineral and the major barium ore. It occurs in veins running through many kinds of rocks, in sedimentary rocks and in clays formed by the solution of sedimentary rocks. It is often weathered and fragile. It is difficult to find well-grained barite coarse aggregate in the United States.

Hydrous Aggregates

Aggregates processed from natural ores that contain water of hydration are used in shielding concrete to attenuate neutrons. At high temperatures the water of hydration is lost from the aggregates. Some of these hydrous aggregates are described here.

Most bauxite comes from the West Indies or South America. It is the raw material used in the production of aluminum. It is relatively light, soft, and porous, but it may have a combined water content as high as 25%.

Some rocks classified as serpentine are not suitable for shielding concrete since they are structurally weak, light in density, soft, low in water of hydration, or high in asbestos fiber. Good serpentine has 10 to 13% of water and has a specific gravity of 2.4 to 2.6. It has the capability of retaining most of its water of hydration at temperatures up to about 1000°F.

Limonite and goethite are the hydrous iron oxides most used in shielding concrete. Limonite is a general name for hydrous iron oxides of variable composition. Their specific gravities run from 2.7 to 3.8 with variation in water of hydration, most of which is lost at temperatures above about 400° to 575°F. Limonite tends to have a heavy dust coating and a preponderance of fines. The water of hydration in some ores is 11 to 12% while others contain almost no water. It is not dehydrated until the temperature reaches 350° to 400°F. Limonite is often found in combination with hematite and magnetite. This combination has a higher specific gravity but has less water of hydration. The hardness of goethite is 5 to 5½; and the specific gravity is 4.28±0.01 in the pure mineral and 3.3 to 4.3 in massive goethite. It contains fixed water in a range from 8 to 12%.

Boron-Containing Aggregates (Table 3–B)
The unusually high capture capability of boron-10 permits its use in relatively small quantities. Boron is most frequently added as a partial replacement of aggregate by borate minerals or synthetic boron frits. Both methods cause some retardation of set of concrete, which can be counteracted by the use of calcium chloride. Experience recorded in the United States suggests that the higher cost of synthetic frits may be counterbalanced by the uniform composition of the frits, which permits effective control of the properties of the concrete.

Since boron is added to concrete in small quantities, it does not affect the shield density much. It may reduce the required thickness and hence the shield weight.

Colemanite is a calcium borate mineral mined in California and Turkey. It has a specific gravity of 2.4 and is one of the two least soluble of the natural borates. The boron content of pure colemanite is about 50%. However, the colemanite available may have only 15% while the better quality ore has around 30%. Water of hydration is present.

Portland cement concrete containing colemanite is somewhat unpre-

Table 3–B
BORON ADDITIVES

	Specific Gravity	Boron Content of Pure Mineral	Source
Colemanite	2.4	50.9%	California, Turkey
Boron frits	2.4–2.6	50%	Synthetic
Ferroboron	5.0	11%	Synthetic
Tourmaline	2.98–3.20	3.9%	Madagascar

dictable. The setting behavior depends on temperature and the chemical composition of the colemanite, which may vary from one batch to another, and on the possible presence of sodium impurities. Sometimes calcium chloride (2% by weight) will offset the retardation, but it cannot be used with prestressing steel.

Borocalcite is available in Europe, but not in the United States, since it is a hand-picked product of Turkey. It is probably ulexite or colemanite or both. It is more pure and less soluble than colemanite, therefore more consistent results can be obtained with it. Pure borocalcite contains 52% boron, but commercial grades have about 40 to 50%.

Boron frits are clean, colorless synthetic glasses obtained by fusing and quenching boric acid, calcium carbonate, silica, and possibly alumina. Boron frits can be purchased which have a specific gravity of 2.4 to 2.6. Cheaper ones contain about 50% boron and 25% sodium. The sodium cannot easily go into solution in the plastic portland cement because it is bound within the fused glass complex. More expensive boron frits are practically sodium free and contain over 55% boron. Boron frits can be purchased passing the No. 4 sieve, or in graded sizes. It costs several cents per pound more in graded sizes. The uniform composition and quality often make boron frits preferable to colemanite.

Boron carbide is extremely expensive, so it is seldom used for borating concrete. It can be obtained in various compositions with a technical grade having 65 to 76% boron content. It comes in graded sizes passing the No. 4 sieve and has a specific gravity of 2.5. It is hard, chemically stable, strong, and has a high melting point.

Ferroboron is the only boron additive that increases the unit weight of concrete. The others tend to decrease density. It is obtainable in various boron contents and sizes from 1 in. down to fine sand. Material passing the No. 50 screen, but not the No. 100 screen, is often used to facilitate mixing with the ingredients of the concrete mix. Ferroboron contains about 11% boron and 85% iron. Its specific gravity is about 5.0.

Handling Heavy Aggregates

Heavyweight aggregates must be stored on a concrete floor, out of contact with other aggregates. Contamination of the material can lead to a leakage of rays through the concrete when it is processed. Also, some heavyweight aggregates, including limonite, barite, and boron minerals, tend to break or to dust. Colemanite must be handled with extra care since as it breaks up easily.

MIXING

Not more than 1 cu yd of heavyweight aggregate should be batched at a time into a truck because of the weight of the material. Batching more than 1 cu yd at a time leads to segregation and can make the material difficult to mix.

Once the mixer is moving, it should not be stopped; hence, most job specifications require mixing at the job site. Heavyweight concrete tends to climb up the sides of the transit mixer and can so imbalance the truck that it could turn over on a sharp corner. With heavyweight aggregates, stopping and starting causes extra wear and tear on equipment.

To avoid segregation, the mixing water is added when mixing is started and must be kept to a minimum. A slump test is usually impractical, but a visual test that controls the slump at 3 in. or less is necessary to avoid segregation and prevent the sinking of heavier particles in the mortar. Usually, a slump of 2 to 3 in. is required; a lower slump will minimize segregation.

PLACING

Conventional methods of mixing and placing are often used. However, precautionary measures should be taken to avoid overloading the mixer, especially with very heavy aggregates such as steel punchings. Thorough consolidation by vibration is required.

The prepacked (prepakt) method (Table 3-C) is used to minimize the segregation of coarse aggregate, especially steel, and to produce concrete of a uniform density and composition. This method is also useful for placing concrete in confined areas and around embedded items. In this method the coarse aggregate or combinations of several types of material are preplaced in the forms. Grout is then forced under pressure into the spaces surrounding the aggregates.

Pumping heavyweight concrete through a steel pipeline may be advantageous in locations where space is limited. Because of the higher unit weights, heavyweight concretes cannot be pumped as far as normal-weight concretes using the same equipment.

ADMIXTURES

Heavyweight concrete should not be made with air-entrained cement. In fact, most literature suggests that an air-detraining agent should be

Table 3–C
GRADING REQUIREMENTS FOR COARSE AGGREGATE AND GROUT SAND FOR PREPLACED AGGREGATE CONCRETE

Sieve Size	Percentage Passing	
	Grading 1 For 1½ in. (38.1 mm) Maximum Size Aggregate	Grading 2 For ¾ in. (19.0 mm) Maximum Size Aggregate
Coarse Aggregate		
2 in. (50.8 mm)	100	—
1½ in. (38.1 mm)	95–100	100
1 in. (25.4 mm)	40–80	95–100
¾ in. (19.0 mm)	20–45	40–80
½ in. (12.7 mm)	0–10	0–15
⅜ in. (9.5 mm)	0–2	0–2
Fine Aggregate		
No. 8 (2.38 mm)	100	—
No. 16 (1.19 mm)	95–100	100
No. 30 (595γ)	55–80	75–95
No. 50 (297γ)	30–55	45–65
No. 100 (149γ)	10–30	20–40
No. 200 (74γ)	0–10	0–10
Fineness Modulus	1.30–2.10	1.00–1.60

on hand at all times in case a small amount of air is entrained in the mix.

Sometimes the use of a retarder that does not entrain air is wise. This reduces the required amount of mixing water, thereby giving denser concrete.

PROPERTIES OF HEAVYWEIGHT CONCRETE

The properties of heavyweight concrete in both the plastic and hardened states can be tailored to meet job conditions and shielding requirements by the proper selection of materials and mix proportions.

Except for unit weight, physical properties of heavyweight concrete are generally similar to those of normal-weight concrete. For any particular set of materials, strength is a function of water-cement ratio. Strengths equal to or greater than those of normal-weight concrete can be achieved with structurally strong aggregates. Typical unit weights of concrete that are made with common heavyweight aggregates are shown in Table 3-A. Strength is usually not a problem since sections of radiation shields are made thick and the mixes have a high cement content. Massive walls with strengths of 2000 psi are adequate, while structural concrete must have strengths from 3000 to 5000 psi. However, the concrete for radiation shielding must resist cracking during service life to avoid impairment of shielding effectiveness.

Nuclear shields are exposed to high temperatures and to thermal cycling. Concrete loses 20 to 50% of its compressive strength when heated to 600 or 800°F for long periods. Structural members should not be exposed to temperatures of more than 500 to 600°F. Concrete is almost completely dehydrated at 800°F. The exposure to fluctuating temperatures causes a greater loss of strength than exposure to constant high temperatures. Tensile, bond, and flexural strengths are affected more by high temperatures than is compressive strength.

CHAPTER 4
NONPLASTIC MIXES

INTRODUCTION

The terms dry tamp, no-slump concrete, nonplastic mixes, and dry mixes all refer to a type of concrete with a very stiff consistency—that is, with a slump less than 1 in. This type of concrete is generally used in the manufacture of precast concrete products. These include concrete block of a variety of materials and shapes, prestressed concrete products, concrete pipe, and precast units of many sizes and shapes. These products are made by a number of different manufacturers throughout the world. They are manufactured with a variety of machinery using different techniques and materials to produce a great variety of products.

FUNDAMENTALS OF NO-SLUMP CONCRETE

The same fundamental principles that apply to plastic concrete are true for no-slump concrete. Water-cement ratio still determines the strength, while durability is a function of water-cement ratio and entrained air content of the concrete. Workability is still determined by the grading of the aggregates, the amount of cement paste in the mix, and the percentage of fine aggregate in the concrete.

The great advantage of nonplastic mixes is their ability to provide high strength at low cost. This is possible because a low-slump mix requires less water; therefore, a low water-cement ratio can be produced with a minimum amount of cement.

The advantages of dry mixtures have been known for some time. In fact, when concrete was first used in the construction industry, the recommended consistency was that of moist earth. This was tamped into place in small layers using a heavy tamper operated by hand. The foundations, retaining walls, dams, roads, and other structures constructed by these methods about 50 years ago have excellent records of strength and durability. These first structures were fairly massive in size.

When reinforced concrete came into use, forms became narrower and concrete sections became thinner. The stiff mixes were difficult, if not impossible, to place under these conditions. As a result, concrete with more plastic properties became popular. Many users reasoned that if plastic concrete was good, sloppy concrete would be even better. As a result, many structures failed or deteriorated before their time. Experience has proven that the dryer the mix, the stronger the concrete becomes. However, low-slump, dry mixes were seldom used because of the difficulty in placing and compacting them.

Then it was discovered that these stiff, unworkable mixes completely changed character when they were shaken or vibrated, and the consistency changed from an unmanageable mass to a plastic, semifluid material. The rapid vibration causes this effect by reducing the friction between particles. When the vibration stopped, the mix immediately became a dense solid mass. In addition to benefits of low cost and high strength, it was discovered that vibration provided greater density and homogeneity in the concrete, improved the bond with reinforcement, provided greater bond at construction joints, and reduced shrinkage.

When it was realized that vibration was one of the greatest advancements in concrete technology, devices for imparting this motion to all types of concrete handling, placing, and products manufacturing equipment were developed.

There is a limit, of course, to how dry a concrete mix can be. As much damage can be done by trying to place concrete that is too dry as by having it too wet. There should be enough cement paste to fill the voids between the aggregates plus a small excess. In addition, the principles of good concreting must also be followed: aggregate gradation, water-cement ratio, the manner in which the concrete is handled and placed, and the way the vibrator is used are all important.

Special types of equipment are used in various applications and industries for batching, mixing, handling, placing, and finishing no-slump concrete.

There is no one combination of ingredients or one method of proportioning that is ideal for no-slump concrete for all applications. Manufacturing of block, pipe, prestressed and precast concrete, and on-site construction all require different mixes. Even within a particular industry, mix design varies from one type of manufacturing or vibration equipment to another. As a result, each manufacturer generally develops his own mix design using the raw materials that he has available to achieve the results he wants. For this reason, no-slump concrete is usually designed for a particular application, and little general information is available. However, in the following sections we will discuss some of the general properties of no-slump concrete and at the end of the chapter illustrate one basic method of designing a mix.

CONSISTENCY AND WORKABILITY

Although the material has a very dry consistency, workability is still necessary. Workability of concrete is the ease with which it can be mixed, placed, consolidated, and finished. If a no-slump concrete was used to pave a driveway or a sidewalk using the conventional mixer and the standard methods of placing, consolidation, and finishing, it would prove to be entirely unworkable. Using the special equipment available at a manufacturing plant, however, the same mix would prove very satisfactory to produce concrete products. Workability for no-slump concrete, therefore, is the ease with which it can be mixed, placed, consolidated, and finished with the equipment and methods to be used on the job.

The strength and durability of no-slump concrete depend on the same factors that determine the strength and durability of ordinary concrete—water-cement ratio and air entrainment. Proportions should be selected to provide adequate durability and strength with satisfactory economy. Maximum size of aggregate for the application should be chosen for economy, and the stiffest consistency that can be effectively compacted should be used to provide maximum strength. Water-cement ratios for no-slump concretes are determined by the application, just as is done with ordinary concrete. Table 4-A lists maximum permissible water-cement ratios for different applications and severe exposure conditions.

The consistency or workability of no-slump concrete cannot be measured in the same way as concrete with a normal consistency, because the slump cone is not designed for extremely stiff mixes. Concrete that measures zero slump would have poor workability if compacted by hand. If vibration were used, however, such concrete might be considered to have excellent workability characteristics. The range of workable mixtures can, therefore, be widened by adoption of compaction techniques that impart greater energy into the mass to be consolidated. This principle of greater energy is used in five types of devices for measuring the consistency of stiff mixes: the Vebe apparatus, the Powers' remolding test, the compacting ratio method, the compacting factor method, and the Thaulow drop table or flow test.

WATER-CEMENT RATIO

The first step in designing the mix is to select a water-cement ratio that will provide the strength and durability needed for application. This can be estimated in advance by using Tables 4-A and 4-B, which were developed on the basis of past experience. Table 4-B provides a rough estimate of the water-cement ratio required to produce a given strength in either air-entrained or non-air-entrained concrete. Experience or trials with the materials to be used will provide more accurate values. Using these tables, it is possible to estimate a water-cement ratio that would produce both the required durability and strength. If the water-cement ratios derived from the two tables are not the same, the lower of the two should be selected to insure that the concrete will meet specifications.

The amount of water needed to produce a mixture of the desired con-

Table 4–A
ACI RECOMMENDED MAXIMUM PERMISSIBLE
WATER-CEMENT RATIOS FOR CONCRETE IN SEVERE EXPOSURES[a]

Type of Structure	Structure Wet Continuously or Frequently and Exposed to Freezing and Thawing[b]	Structure Exposed to Seawater or Sulfates
Thin sections (railings, curbs, sills, ledges, and ornamental work) and sections with less than 1 in. cover over steel	0.45	0.40[c]
All other structures	0.50	0.45

[a] Adapted from "Recommended Practice for Selecting Proportions for Concrete" (ACI 211.1-70).
[b] Concrete should also be air-entrained.
[c] If sulfate-resisting cement (Type II or Type V of ASTM C 150) is used, permissible water-cement ratio may be increased by 0.05.

Table 4–B
COMPRESSIVE STRENGTH OF
CONCRETE FOR VARIOUS WATER-CEMENT RATIOS[a]

Water-Cement Ratios By Weight	Probable Compressive Strength (at 28 days, psi)	
	Non-Air-Entrained Concrete	Air-Entrained Concrete
0.31	6800	5500
0.35	6000	4800
0.44	5000	4000
0.53	4000	3200
0.62	3200	2600
0.71	2500	2000

[a] These average strengths are for concretes containing ASTM Type I portland cement. Strengths are based on 6×12 in. cylinders, moist-cured under standard conditions for 28 days. See "Method of Making and Curing Concrete Compression and Flexure Test Specimens in the Field (ASTM C31)."

sistency is influenced by the maximum size, particle shape and grading of the aggregate, and by the amount of entrained air. As long as there is less than about 650 lb of cement per cubic yard, the quantity of water per yard for a desired consistency is relatively unaffected by the amount of cement. In mixtures richer than this, however, mixing water requirements may increase significantly as cement content is increased. Table 4-C gives mixing water requirements for different consistencies and maximum sizes of aggregates. Like the other tables used in proportioning, this one is also based on experience. The quantities of water shown are accurate enough for preliminary estimates of proportions but will vary somewhat with the particular materials being used.

AGGREGATE PROPORTIONS

To produce concrete at the lowest cost with the greatest strength and the minimum amount of mixing water, the largest proportion of coarse aggregate should be used as long as workability is satisfactory. This quantity of coarse aggregate can best be determined in the laboratory with an adjustment in the field or plant if required. If such data is not available, Table 4-D provides a good estimate of the amount of coarse aggregate for various concretes in reinforced construction. The concrete can be made stiffer or more plastic by varying the amount of coarse aggregate per unit volume of concrete. As greater amounts of coarse ag-

Table 4-C
APPROXIMATE MIXING REQUIREMENTS FOR DIFFERENT CONSISTENCIES AND MAXIMUM SIZES OF AGGREGATES[a]

Description	Consistency				Relative Water Content (percent)	Water (gal per cu yd) for Indicated Maximum Sizes of Coarse Aggregate				
	Slump (in.)	Vebe (sec)	Drop Table (revolutions)	Compacting Factor		⅜ in.	½ in.	¾ in.	1 in.	1½ in.
					Non-air-entrained concrete					
Extremely dry	—	32–18	112–56	—	78	36	34	32	30	28
Very stiff	—	18–10	56–28	0.70	83	38	37	34	32	30
Stiff	0–1	10– 5	28–14	0.75	88	40	39	36	34	32
Stiff plastic	1–2	5– 3	14– 7	0.85	92	42	40	37	36	33
Plastic	3–4	3– 0	<7	0.91	100	46	44	41	39	36
Flowing	6–7	—	—	0.95	106	49	46	43	41	38
Approximate amount of entrapped air in non-air-entrained concrete (percent)						3	2.5	2	1.5	1
					Air-Entrained Concrete					
Extremely dry	—	32–18	112–56	—	78	32	30	28	27	25
Very stiff	—	18–10	56–28	0.70	83	34	32	30	28	27
Stiff	0–1	10– 5	28–14	0.75	88	36	34	32	30	28

Stiff plastic	1–2	5– 3	14– 7	0.85	92	37	36	33	31	29
Plastic	3–4	3– 0	<7	0.91	100	41	39	36	34	32
Flowing	6–7	—	—	0.95	106	43	41	38	36	34
Recommended average total air content, percent[b]						8	7	6	5	4.5

[a] These quantities of mixing water are for use in computing cement factors for trial batches. They are for reasonably well-shaped angular coarse aggregates graded within limits of accepted specifications.
If *more* water is required than shown, the cement factor, estimated from these quantities, should be increased to maintain desired water-cement ratio, except as otherwise indicated by laboratory tests for strength.
If *less* water is required than shown, the cement factor, estimated from these quantities, *should not* be decreased except as indicated by laboratory tests for strength.

[b] For consistencies below 1 in. slump, the volume of air entrained by either an air-entraining cement or the usual amount of air-entraining admixture used for more plastic mixtures may be significantly lower than those shown. For these mixtures, it is recommended that the air content resulting from the use of air-entraining cement or the usual amount of air-entraining admixture per unit of cement for more plastic mixtures be accepted as adequate for insuring durability. In the absence of such information for a particular air-entraining admixture, the amount to use per unit of cement can be determined on a trial mix having a slump in the 3 to 4 in. range, or by determining the amount needed to obtain 19 ± 3% air in mortar prepared in accordance with ASTM C185.

Table 4–D
VOLUME OF COARSE AGGREGATE PER UNIT
VOLUME OF CONCRETE[a] (FROM ACI 211.1-70)

Maximum Size of Aggregate (in.)	Volume of Dry-Rodded Coarse Aggregate Per Unit Volume of Concrete for Different Fineness Moduli of Sand			
	2.40	2.60	2.80	3.00
3/8	0.50	0.48	0.46	0.44
1/2	0.59	0.57	0.55	0.53
3/4	0.66	0.64	0.62	0.60
1	0.71	0.69	0.67	0.65
1 1/2	0.75	0.73	0.71	0.69

[a] Volumes are based on aggregates in dry-rodded condition as described in "Method of Test for Unit of Aggregate" (ASTM C29). These volumes are selected from empirical relationships to produce concrete with a degree of workability suitable for usual reinforced construction.

gregate per unit volume are used, the concrete will become stiffer. For the flowing and plastic consistencies, the volume of coarse aggregate would be about that shown in Table 4-D. For the stiffer consistencies—those requiring vibration—the amount of coarse aggregate that can be accommodated increases rather sharply in relation to the amount of fine aggregate required. Table 4-E shows this type of relationship. Note that for plastic mixes, the value shown in Table 4-E is 100% for all aggregate sizes. For stiffer mixes, however, the values run from 104 up to 190%. This means that a greater percentage of coarse aggregate should be used for the stiffer mixes. The information contained in these two tables provides a basis for selecting the right amount of coarse aggregate for the first trial mixture. Adjustments in this amount may be necessary in the field.

CONSISTENCY TESTS FOR STIFF CONCRETE

Those who work extensively with no-slump concrete know that some mixes respond readily to vibration, while other concretes with the same external appearance are virtually impossible to compact using the same vibrating units.

Some of the ways that have been used to measure the consistency of the stiffer, less plastic types of concrete are described below. Although these methods measure the reaction of concrete to compaction, none of them is an absolute measure of concrete workability. They are termed

Table 4-E
VOLUME OF COARSE AGGREGATE PER UNIT
VOLUME OF CONCRETE FOR DIFFERENT CONSISTENCIES[a]

Description	Consistency				Volume of Dry-Rodded Coarse Aggregate Per Unit Volume of Concrete for Maximum Size of Aggregate Shown (expressed as a percentage of the values shown in Table 4-D)				
	Slump (in.)	Vebe (sec)	Drop Table Revolutions	Compacting Factor	3/8 in.	1/2 in.	3/4 in.	1 in.	1 1/2 in.
Extremely dry	—	32–18	112–56	—	190	170	145	140	130
Very stiff	—	18–10	56–28	0.70	160	145	130	125	125
Stiff	0–1	10– 5	28–14	0.75	135	130	115	115	120
Stiff plastic	1–2	5– 3	14– 7	0.85	108	106	104	106	109
Plastic	3–4	3– 0	<7	0.91	100	100	100	100	100
Flowing	6–7	—	—	0.95	97	98	100	100	100

[a]Based on tests of non-air-entrained concrete made with a natural sand having a fineness modulus of 2.90 and a rounded gravel, containing some crushed over-size. Maximum sizes used where 3/8 in., 3/4 in., and 1 1/2 in. Values for 1/2 in. and 1 in. are interpolated.

Three slump ranges used in these tests: 0-1in., 1-2 in., and 4-5 in. Other values obtained by interpolation and extrapolation.

It is assumed, for the purpose of this method, that the multiplication factors shown are appropriate for sands having other fineness moduli. These values are intended as a guide in establishing the first trial mixtures. Further adjustments will be necessary.

here as consistency tests, and even the best are considered to give only limited comparisons between the actual workability of various concrete mixtures.

Consistency testing methods for use with the stiffer grades of concrete are (1) the flow test, (2) Thaulow drop table, (3) the remolding test, (4) the Vebe method, (5) the compacting factor method, and (6) the compacting ratio method.

Concrete Flow Test

In this test method, a truncated cone of concrete is cast on a flow table that is 30 in. in diameter (Fig. 4-1). The table is raised and dropped 15 times in 15 seconds and the average diameter of the dispersed concrete is then measured. Flow is expressed as the percent spread of the concrete measured from the base diameter.

This test method is not the best for measuring workability and it has largely been replaced by the other consistency measurement procedures.

Thaulow Drop Table

A 10-liter sample container is clamped to a drop table that drops about 0.4 in., four times with each revolution of a hand crank. The container is scribed inside at the 5-liter mark.

A standard slump cone is filled and compacted within the sample container, after which the funnel adapter is filled and the whole assembly is subjected to 15 drops. After strike off and removal of the cone, the number of drops required to deform the 5½-liter cone to a 5-liter cylinder (with a ½-liter topping) is a measure of the consistency of the concrete.

Powers' Remolding Test

The flow table just mentioned is used in conjunction with the apparatus shown in Figure 4-2 to determine concrete consistency by the remold-

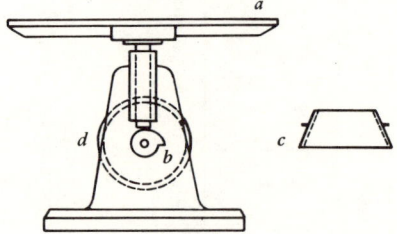

Figure 4-1 Flow table. (a) table, (b) drop cam, (c) truncated cone, (d) crank.

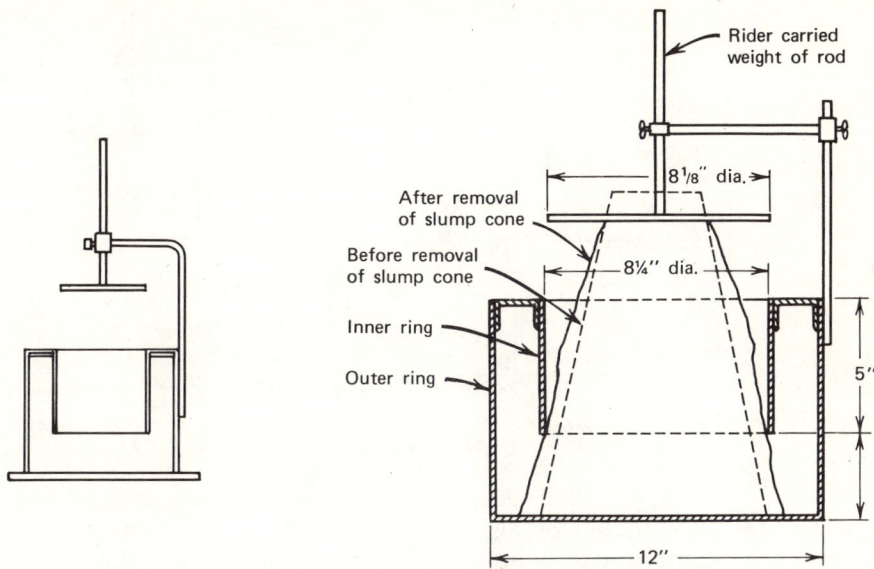

Figure 4–2 Ring and follower assembly for the remolding test.

ing test. The inner and outer ring assemblies in the figure are fastened to the flow table before running the test. A slump cone, from which the foot stirrups have been removed, is placed inside the ring assembly and rests on the flow table. Concrete is compacted in the slump cone in the standard way, the slump cone is removed, and the rider disk is lowered to the surface of the concrete. Then the flow table is dropped until the conical concrete pile is reduced to a cylindrical shape. This is determined by the rider falling to a predetermined level inside the inner ring. The total number of drops required to remold the concrete is recorded as the remolding effort.

This remolding test is apparently a fair measure of concrete workability, but it has not been used much in recent years. One of the reasons for this is that the equipment required for the test is too bulky for anything but a laboratory method. Second, and more important, this jolting method of remolding has been supplanted by the more modern "Vebe" method of remolding by vibration.

Vebe Test Method
The Vebe apparatus, shown in Figure 4-3, consists of a vibrator, a cylindrical container affixed to the vibrating table top, a slump cone and filling

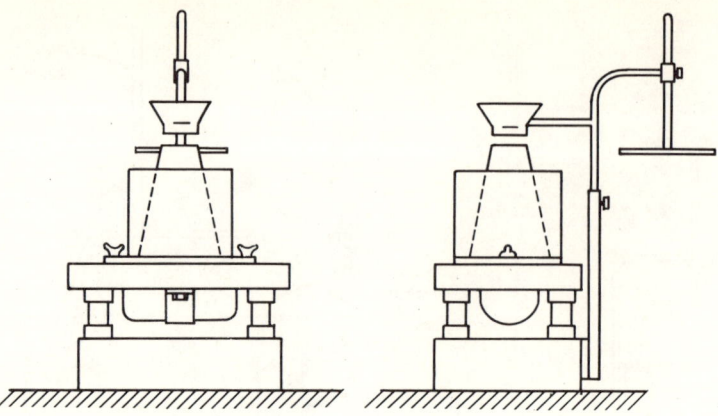

Figure 4-3 The Vebe apparatus.

funnel, and a glass or plastic rider quite similar to that used in the remolding test. Concrete is compacted in the slump cone, the cone is removed, the rider is placed atop the concrete, and the Vebe test result is determined as the number of seconds required for vibration to reduce the concrete to a cylinder in the bottom of the container.

The following are detailed laboratory procedures.

1. Secure the sample container to the vibrating table. Because of the violent vibration, nuts should be tightened with wrenches.
2. Center the premoistened slump cone in the container and fill it in three equal concrete lifts; each lift is tamped 25 times with a hemispherical-tipped rod.
3. Remove the slump cone. Measure the slump of the concrete, in inches, and record this value.
4. Put the glass or plastic plate of the follower on top of the concrete.
5. Start the vibrator and determine with a stopwatch the number of seconds between the start of vibration and the time at which the concrete attains a cylindrical shape. This latter time can be determined fairly well as the time when the entire bottom surface of the follower contacts the concrete and there are no large voids at the edges of the follower plate. Record this elapsed time. The elapsed time is sometimes described as "seconds Vebe" or "degrees Vebe."

The seconds Vebe is reasonably well related to the concrete workability, with lower workability being associated with the higher Vebe times. Vebe testing is usually confined to the laboratory or plant because the test equipment is not very portable. For accuracy in testing really stiff grades of no-slump concrete, this apparatus is unequaled. It is not sensitive to plastic concretes having slumps of 2 in. or greater.

Compacting Factor Method

In this procedure, a concrete charge is loosely compacted into a cylinder by dropping it from a controlled height. The density of the loosely compacted mass is obtained and divided by the fully compacted density of the same concrete to determine the "compacting factor" of the concrete, which is a reasonably good measure of the concrete's workability.

Figure 4-4 gives two views of the compacting factor apparatus. It consists of a frame holding two upper funnel-shaped hoppers (each equipped with bottom trap doors) and a lower cylinder. The procedure for making a compacting factor test is:

1. Using a scoop, gently place a sample of the concrete in the upper hopper. Fill the hopper so the concrete is about level with the top.
2. Open the trap door of the upper hopper, dropping the concrete into the lower hopper.
3. Open the trap door of the lower hopper, dropping the concrete into the cylinder.
4. Strike off the concrete without further consolidation of any sort. Weigh the mold and its contents. Record this weight and then discard the concrete. Calculate the weight of concrete filling the cylinder.
5. Refill the mold with concrete from the same batch, using vibration or very thorough tamping to fully consolidate the concrete. Strike off the concrete level with the top of the cylinder and weigh the mold and its contents. Calculate the weight of concrete filling the cylinder. The ratio of the weight of the partially compacted concrete divided by the weight of the fully compacted concrete is the compacting factor of the concrete.

Like the other procedures already described here, the compacting factor apparatus is not very portable, making the test primarily a laboratory or plant method. Although it is a good procedure for determining the consistency of concrete over a wide workability range, it can only freely handle the more workable grades of "no-slump" concrete. Really stiff, unworkable concrete may hang up in the hoppers and require gentle rodding to loosen it.

Compacting Ratio Method

A derivative of the compacting factor test method involving simpler equipment is being investigated at this time. In it the concrete is placed

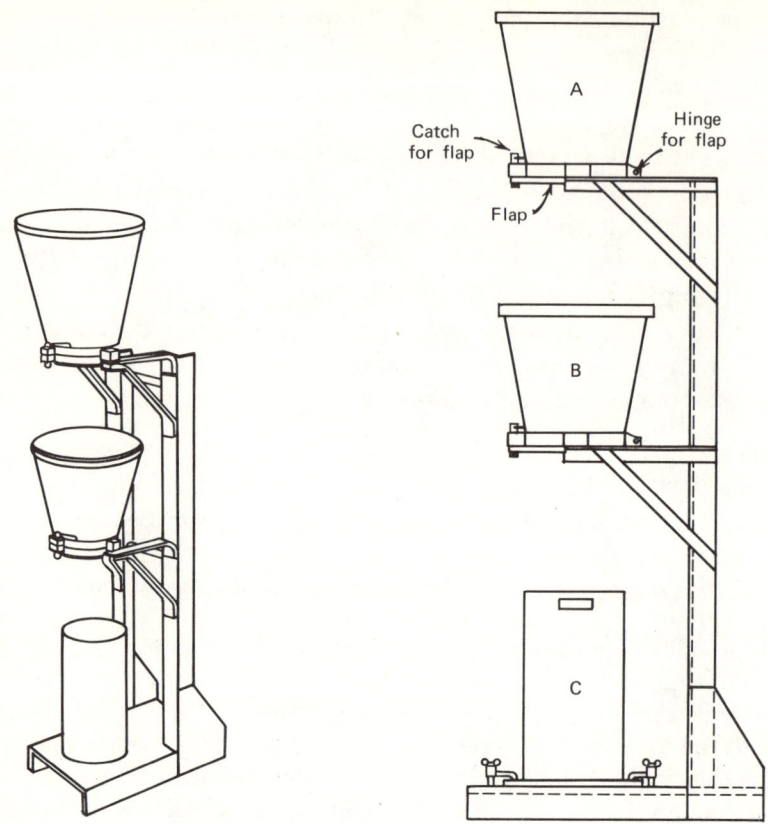

Figure 4-4   Two views of the compacting factor test apparatus.

in a container by gently dropping it from the top edge of the container. The density of the concrete lightly compacted in this fashion is compared with the density of identical concrete that has been thoroughly vibrated by an internal vibrator, and the compacting ratio thus established is taken as a measure of the workability of the concrete.

The compacting ratio procedure is a sort of field expedient for the compacting factor method involving less sophisticated test equipment.

Table 4-F compares consistencies using three of the testing methods just described and the slump cone. If none of these methods is available, compaction of the trial mixture under actual placing conditions in the field or laboratory will serve as a means of determining whether the consistency and workability are adequate.

Table 4–F
COMPARISON OF CONSISTENCY
MEASUREMENTS BY VARIOUS METHODS

Consistency Description	Slump (in.)	Vebe (sec)	Compacting Factor (average)	Drop Table Revolutions
Extremely dry	—	32 to 18	—	112 to 56
Very stiff	—	18 to 10	0.70	56 to 28
Stiff	0 to 1	10 to 5	0.75	28 to 14
Stiff plastic	1 to 2	5 to 3	0.85	14 to 7
Plastic	3 to 4	3 to 0	0.91	<7
Flowing	6 to 7	—	0.95	—

A SAMPLE MIX DESIGN PROBLEM

As an example, the specifications for a prestressed bridge girder, which will be exposed to severe weather with frequent cycles of freezing and thawing, require the concrete to have a compressive strength of 4600 psi at 28 days. The size of the girder and the spacing of the prestressing strands are such that a maximum size aggregate of 1½ in. can be used, and equipment for vibrating the form is available.

The laboratory data is as follows: Type I air-entrained cement will be used; its specific gravity is 3.15. The dry bulk specific gravity of the coarse aggregate is 2.68 and it has an absorption of 0.5%. The fine aggregate has a dry bulk specific gravity of 2.64, an absorption of 0.7%, and fineness modulus of 2.80. The dry-rodded weight of the coarse aggregate is 100 lb per cu ft.

In order to estimate the quantities of materials for the first trial batch of air-entrained concrete, a water-cement ratio must be selected. There are two factors that determine this: (1) the exposure conditions and (2) the strength required. Water-cement ratios for exposure conditions and types of structure are given in Table 4-A. Since, in this case, a girder is considered a thin section and exposure conditions will be severe, the recommended maximum water-cement ratio is 0.45 lb of water per lb of cement. By using Table 4-B, it is possible to determine what water-cement ratio would produce a strength of 4600 psi. Since 4600 is not one of the values listed, it is necessary to interpolate between 4800 psi, which requires a water-cement ratio of 0.35, and 4000 psi, with a water-cement ratio of 0.44. Since 4600 is ¼ the distance between 4800 and 4000, the quantity to be determined is ¼ of the difference between 0.35 and 0.44

(¼ of 0.09), or 0.023. To complete the interpolation, 0.023 is added to 0.35; and this sum, 0.373 (rounded down to 0.37), is the water-cement ratio for a strength of 4600 psi. Since the 0.37 value required for strength is lower than the 0.45 required for durability, 0.37 is the water-cement ratio to be used.

Next, Table 4-C can be used to determine the amount of mixing water required for concrete which, because of the use of heavy vibration equipment, will have a very stiff consistency. The water requirement for air-entrained concrete (very stiff, 1½-in. aggregate) is 216 lb (27 gal) per cubic yard. The air content for a stiff mix will be significantly lower than that obtained in a plastic mix using the same amount of air-entraining agent. Since a plastic mix would develop about 4.5% entrained air, it is possible to assume that the air content will be about 3% when the procedure suggested in the note below Table 4-C is used.

Since the water-cement ratio and the quantity of water required is now known, the amount of cement per cubic yard can be calculated by dividing the estimated water content, 216 lb, by the water-cement ratio of 0.37, resulting in 584 lb of cement per cubic yard.

The next step is to find out the quantity of coarse aggregate needed to produce a very stiff mix in which the maximum aggregate size is 1½ in. and the fineness modulus of the fine aggregate is 2.80. Table 4-D indicates that the volume of dry-rodded coarse aggregate needed per unit volume of concrete is 0.71—this is for concrete having about a 3- to 4-in. slump. Table 4-E indicates that for 1½ in. aggregate, a very stiff mix should have 125% of the volume given in Table 4-D. Thus, to calculate the volume of coarse aggregate needed in the stiff mix, 0.71 is multiplied by 1.25, and the resulting 0.89 cu ft is the volume of coarse aggregate per cubic foot of concrete. This value multiplied by 27 is the number of cubic feet of coarse aggregate in a cubic yard, 24.0 cu ft—dry rodded. Multiplying this number by 100 lb (the weight of 1 cu ft of dry-rodded coarse aggregate) yields a product of 2400 lb of coarse aggregate per cubic yard of concrete.

The amounts of cement, water, coarse aggregate, and entrained air needed in this mix are now known. The only unknown is the quantity of fine aggregate, which is determined by calculating the absolute volume occupied by each of the other ingredients, then totaling these quantities and subtracting this sum from the volume of a cubic yard (27 cu ft) to find the volume of fine aggregate needed.

To calculate the solid volume of cement, the previously determined weight (584 lb) is divided by the specific gravity of the cement (3.15) multiplied by the weight of 1 cu ft of water (62.4 lb). This gives an absolute volume of 2.97 cu ft of cement (584 ÷ 196.56).

The volume of water is calculated by dividing its weight (216 lb) by the weight of 1 cu ft of water (62.4 lb), which gives 3.47 cu ft of water.

To determine the solid volume of coarse aggregate used, the total calculated weight (2400 lb) is divided by the dry bulk specific gravity of the coarse aggregate (2.68) multiplied by the weight of 1 cu ft of water (62.4 lb), which gives 14.36 cu ft of coarse aggregate.

The entrained air needed is estimated as 3%; 3% of 27 cu ft is 0.81 cu ft.

The total volume of these four ingredients (cement, water, coarse aggregate, and air) is 21.61 cu ft (2.97+3.47+14.36+0.81). To determine the solid volume of fine aggregate needed in the mix, this sum (21.61) is subtracted from 27 cu ft; the difference (5.39 cu ft) is multiplied by the specific gravity of the fine aggregate (2.64) and multiplied again by the weight of 1 cu ft of water (62.4 lb) to convert the solid volume of sand to a weight of dry sand, giving 888 lb of fine aggregate. The batch weights for this sample mix have now been determined: 584 lb of cement, 216 lb of water, 888 lb of sand, and 2400 lb of coarse aggregate will be needed per cubic yard of concrete.

Batch Weights for Field Use

For the sake of convenience in making trial mixture computations in this sample problem, the aggregates have been assumed to be in a dry state. Under field conditions, however, they will generally be moist, and the quantities to be batched into the mixer must be adjusted accordingly.

With the batch weights determined in the example, it will be assumed that tests show the sand to contain 5% and the coarse aggregate 1% total moisture. Since the quantity of dry sand required is 888 lb, the amount of moist sand to be weighed out must be 888 ÷ (1−0.05) = 888 ÷ 0.95 = 935 lb. Similarly, the weight of moist coarse aggregate must be 2400 ÷ (1−0.01) = 2400 ÷ 0.99 = 2424 lb.

The free water on aggregates in excess of their absorption must be considered as part of the mixing water. Since the absorption of sand is given as 0.7%, the amount of free water that the sand contains is 5.0% minus 0.7% = 4.3%. The free water on the coarse aggregate is 1.0% minus 0.5% = 0.5%. Therefore, the mixing water contributed by the sand is 0.043 × 888 lb = 38 lb, and that contributed by the coarse aggregate is 0.005 × 2400 lb = 12 lb. The quantity of mixing water to be added, then, is 216 minus (38+12) = 166 lb.

Adjustment of Trial Mixture

In the notes to Table 4-C, it mentions that more water might be required than indicated, and, that in such cases, the cement factor should be increased to maintain the water-cement ratio, unless otherwise indicated by laboratory tests for strength. This adjustment will be illustrated by assuming that the concrete in the example was found to require 233 lb of water instead of 216 lb.

Since the water-cement ratio selected is 0.37, 630 lb of cement will be needed ($233 \div 0.37$), instead of 584 lb, when the water is raised to 233 lb.

The other ingredients in the batch then should be adjusted accordingly.

It was also pointed out that less water than indicated in Table 4-D may sometimes be required, but it was recommended that no adjustment be made in the cement factor, except as indicated by laboratory tests. Nevertheless, some adjustment in batch quantities is necessary to compensate for the loss of volume due to the reduced water. This is done by increasing the solid volume of sand in an amount equal to the volume of the reduction in water. For example, if it is assumed that 208 lb of water are required instead of 216 lb for the concrete of the example, then 208 lb is divided by 62.4 (the weight of 1 cu ft of water) in computing the volume of water in the batch, and the solid volume is 3.34 instead of 3.47 cu ft.

Amount of Entrained Air

The percentage of air in concrete can be measured directly with an air meter, or it can be computed from theoretical and measured unit weights according to the gravimetric method. For any particular set of conditions and materials, the amount of air entrained is roughly proportional to the quantity of air-entraining admixture used. Increasing the cement content or the fine fraction of the sand, decreasing slump, or raising the temperature of the concrete usually decreases the amount of air entrained for a given amount of admixture. The grading and particle shape of aggregate also have an effect on the amount of air entrained. The job mixture should not be adjusted for minor fluctuations in the water-cement ratio or air content.

SECTION TWO
DECORATIVE CONCRETES

CHAPTER 5
ARCHITECTURAL CONCRETE

Advancing technology has made the field of decorative concrete challenging, rewarding, and exciting. Dramatic developments have given designers a new dimension that combines rugged strength, attractive appearance, and easy upkeep. The variety of textures, forms, and patterns possible with concrete affords an aesthetic freedom not possible with other building materials.

The speed and efficiency of concrete construction usually means reduced labor costs. Lower maintenance after construction also reduces costs. Fire resistance is an important consideration in building, and concrete generally affords lower insurance rates because of its fire-resistant properties.

EXPOSED AGGREGATE

In 1938, the Portland Cement Association established at its laboratories a small outdoor exhibit of architectural concrete made by the aggregate-transfer method. That exhibit still stands as an example of the excellent durability of this type of concrete. Since World War II, exposed aggregate has enjoyed increasing popularity for several reasons:

1. Architects have realized its possibilities.

2. Better handling equipment is available.
3. Methods of construction have improved.
4. New materials are available.

New methods and new effects are being conceived regularly. Exposed aggregate is used effectively on patios, walls, plazas, sidewalks, building approaches, and anywhere else that the architect wants a colorful, textured effect.

Exposed aggregate surfaces that are made with good materials and good workmanship are durable in severe weather conditions. The use of air-entrained concrete is recommended, not only to resist freezing and thawing cycles but also to make the mix more workable.

Aggregates used for exposed aggregate surfaces vary in price from $10 a ton to $600 a ton for special glass aggregate. However, the cost can be kept down by using special aggregate only on the thin exterior layer. Natural aggregate is generally less expensive than manufactured aggregate. Quartz, granite, gravel, and marble are natural aggregates. Glass and ceramic or vitreous tiles are specially manufactured aggregate materials.

Size of the aggregate can vary from pea gravel to fist-size rocks, depending on the texture desired. Color can be uniform, a prepared mixture, or a random selection of hues, shades, and tones. Aggregate should be hard and sound. Flat, sliver-shaped particles or aggregate that is too smoothly polished or ball-shaped should be avoided; these aggregates may not bond properly, are easily dislodged during exposing operations, and may be loosened, once exposed, by freezing and thawing.

Aggregate should be cleaned prior to use, and a test panel should be made using a sample of the selected aggregate to insure that the chosen material will produce the desired finish.

For normal concrete, uniformly graded aggregate is used. For cast-in-place and decorative precast panels (where the aggregate will be exposed by a surface retarder, or by sandblasting or bushhammering), gap-graded aggregate is often used. *Gap-graded* means that some of the intermediate sizes are omitted. Often, two sizes are mixed—for example, 1½ to ¾ in. and ⅜ in. to No. 4 sieve leaving a gap between ¾- and ⅜-in. coarse aggregate, or ¾ in. to 1½ in. mixed only with masonry sand, eliminating most material between about ¾ in. and the No. 8 sieve.

The concrete mix used for precast decorative panels varies according to results desired, but a typical mix might be 1 part cement, 1 part sand, 3 parts coarse aggregate by weight and a water-cement ratio not higher

Figure 5-1 This precast component is being sandblasted to expose the attractive aggregate. Since a white cement matrix was used in the concrete, a white sand is being used for abrasive blasting so that there will not be any discoloration to its bright appearance.

than 0.35. Air-entraining agents are helpful for workability since this is a harsh mix and hard to handle. Specifications for strength of precast panels may require 28-day compressive strengths up to 7500 psi for 6-in. cubes, which corresponds to about 6200 psi for 6×12-in. cylinders.

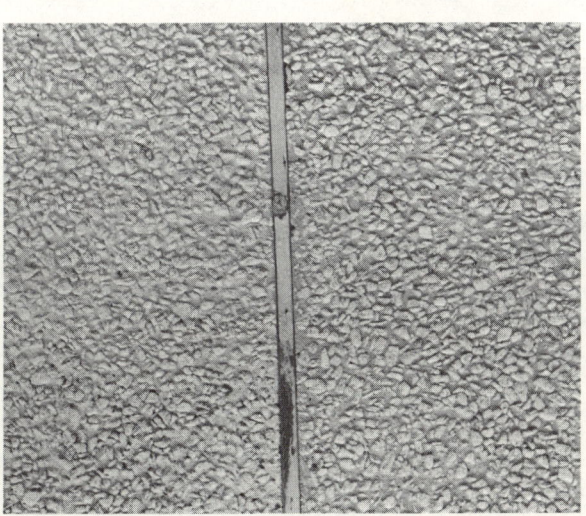

Figure 5-2 Notice the gap-graded aggregate used on this tilt-up panel.

Figure 5-3 Sandblast pattern achieved with the use of sheet rubber templets.

Precast Panel Production

Standard Face-Down Method. Steel, plywood, or concrete forms are painted with a chemical solution to retard the set of the concrete in contact with the form. The retarder is allowed to dry as required. A facing of rich white cement mix with special aggregate is deposited to a depth of 1 in. or more. This is vibrated to consolidate the mix and to make the aggregate come to the surface (which is the bottom in this case). Reinforcing steel and a back-up mix are placed. The back-up mix has a low slump because it is desirable to make it "pull" water from the facing mix. Air entrainment should be used to make the concrete more workable. The mix is placed to a depth of the form and leveled.

The next day, approximately 20 to 30 hours later, the panel is picked up and placed on edge. The face is brushed and washed to remove the retarded cement matrix. Several weeks later, a wash of muriatic acid may be used to brighten the face and remove any remaining film. When the panel is to be used in an area where air pollution or dirt are problems, the surface may be treated with a clear acrylic solution.

Patterns can be introduced by putting divider strips in the forms. Different colors of aggregate can be put between the divider strips.

Sand-Bedding. Another face-down method is the sand-bedding technique, used when boldness and massiveness are desired. Stones ranging from 1 to 8 in. in diameter as well as flagstone can be sand-bedded to produce precast stone-faced wall panels.

The sand-bedding concept involves the use of a sand layer in the bottom of the form—usually the same sand as in the concrete—to a depth equal to the desired exposure of aggregate. Depth of sand can vary, depending on the aggregate shape and the effect desired.

In one method, the sand is leveled and the aggregates are hand placed and forced into the sand until they contact the form. In the case of flagstone, the pieces are positioned on the bottom of the form in the desired orientation, and masonry sand is poured in all joints to a depth of about 1 in. The stone and sand are given a fine spray of water to help consolidate the sand, and a structural concrete mix is then cast on this prepared base. The structural concrete bonds to the stone and to a layer of the bedding sand, transferring a sand-textured background to the panel. When the concrete reaches the specified strength, it is positioned vertically and the loose sand is removed by spraying with water or an air jet.

Figure 5-4 Spreading a single layer of aggregate on a sand bed.

Figure 5-5 Exposed aggregate gives a unique and interesting design.

Figure 5-6 Tilt-up wall panels with unusually large exposed aggregate.

This technique produces a precast wall of uniform aggregate distribution and exposure in which about two-thirds of each stone is embedded in the concrete and one-third exposed. This depth of exposure, ranging from ½ in. to possibly 2 in. or more, is not feasible with chemical retarders or with sandblasting.

Good workability is required in the concrete for this method. The mix should be designed for a slump of 3 to 5 in. with air entrainment about 5%. No other admixtures are generally recommended. If early handling of the panels or removal from the forms is required, Type III portland cement should be used.

Another method involves placing a single layer of aggregate in the panel form and packing it as closely as possible. Then, fine, dry white sand is sprinkled over it and allowed to sift down around the pieces until the depth of the sand is one-third to one-half the aggregate size. Excess sand on top of the rocks is removed with a soft-bristled brush and an air blast, and the aggregate is moistened with a fog spray to settle the sand firmly into place. Next, a mortar of white portland cement and white sand is placed over the aggregate. A typical mortar mix is 1 part cement to 2½ parts well-graded sand, with sufficient water to make a creamy mix.

Reinforcement, with bolting and welding inserts attached, is then set into position and the backup of standard structural concrete is placed to fill the forms. The backup mix is usually of very low slump in order to absorb excess water from the facing mix. The mix should be designed for a minimum strength of 4000 psi at 28 days, very low slump, and air entrainment of about 5% to improve workability. After proper curing, the panels are raised and placed on edge, and any clinging sand is removed with a stream of water or an air jet.

To avoid a cold joint or a possible plane of separation, no appreciable time interval should occur between the placing of the facing mix and the backup concrete. Normal concreting procedures may require a short time between placing the two mixes, but extended delays should be avoided, especially in hot weather.

Face-Up Method

A concrete mix with a minimum cement content of 550 lb/cu yd and a maximum slump of 3 in. should be used. Aggregate to be exposed should be cleaned and wetted thoroughly before use to prevent mixing water being drawn from the concrete. Wetting also lubricates the aggregate as it is pushed into the surface of the slab. After wetting, however, the aggregate must be allowed to drain to prevent free water being carried into the concrete.

Immediately after the slab has been screeded and darbied, the selected aggregate should be scattered by hand—and evenly distributed so that the entire surface is completely covered with a single layer. Special care should be taken in covering the edges to insure a uniform appearance.

The initial embedding of the aggregate is usually done by patting with a darby or the flat side of a strike-off board. After the aggregate is embedded, and as soon as the concrete will support the weight of a man on kneeboards, the surface should be hand-floated with a magnesium float or darby. This operation should be performed thoroughly so that all aggregate is completely embedded just beneath the surface. The cement paste should completely surround and slightly cover all aggregate, with no holes or openings left in the surface. Following this floating, a reliable retarder may be sprayed or brushed on the surface following the manufacturer's recommendations. On small jobs a retarder probably will not be necessary; retarders are generally used on large jobs for better control of exposing conditions. When a retarder has been used, exposure of the aggregate can generally be delayed a little longer.

Proper timing is critical in exposing aggregate; a test panel should be made to determine the best time to begin. Exposing should be done as soon as the grout can be removed by simultaneous brushing and flushing with water without overexposing or dislodging the aggregate. If it is necesssary for men to move about on the surface during exposing operations, kneeboards must be used with special care. If possible, this should be avoided because of the risk of breaking aggregate bond.

Exposed aggregate slabs should be cured thoroughly. Care must be taken that the method of curing does not stain the surface. Straw, earth, and some types of building paper may cause staining.

Arbeton or Naturbetong Method

This new patented method was developed in Europe where it is called Naturbetong. Hardware cloth is wrapped around the rebar cage of a column, or the outside layer of bars in a wall is faced with hardware cloth. Forms are built for concreting about 3 in. from the hardware cloth. Dry aggregate is put between the mesh and the forms, after which a rich concrete mix is placed inside the rebar cage and vibrated to force the mortar from the concrete through the mesh and aggregate to the forms. The forms are removed early and the area is sandblasted while the concrete is "green." About 160 sq ft can be sandblasted in an hour by three workmen. The concrete must be kept damp until it is sealed. For this purpose, a curing compound incorporating a protective sealing coat can be applied after sandblasting.

The aggregate used for Arbeton should be ¾ to 1½ in. in size. Even if it is prewashed by the supplier, it should be washed again by the contractor to get rid of all the fines. It should be stockpiled at the job site to insure uniformity. From the stockpile, it is lifted in a mortar box to the workmen who shovel it between the rebar cages and the forms.

Cages must be precisely set to withstand the force of the fresh concrete. The cages can be preassembled and hoisted into place. Concreting forms are built in the regular manner, spaced about 3 in. from the rebar cage.

A rich mix consisting of 700 lb of cement per cu yd, giving 5000 psi strength and having a 6-in. slump is desirable. The temperature of the concrete at discharge should be 60 to 80°F. Aggregate and water should be heated during cold weather if necessary to maintain temperatures within this range.

Arbeton has not been used extensively in the United States. The word is an acronym from the word *architectural* and the French word for concrete, *beton*.

Bushhammering

Bushhammering is the process by which pneumatic or hand hammers are used to remove mortar and fracture aggregates at the surface of hardened concrete to produce an attractive, varicolored, textured surface. Most form marks and minor surface blemishes are removed.

A satisfactory bushhammered finish can be achieved on any good structural concrete. However, certain precautions should be taken. Form joints should be taped so they won't show. Care should be taken in selecting the aggregates, designing the mix, placing, and curing. In localities where colorful natural aggregates are available, their use in the concrete will improve the final appearance of the bushhammered surfaces.

The concrete mix should contain at least 570 lb of cement per cubic yard, and the water-cement ratio should not exceed 0.49. Slump should not be more than 4 in. and the concrete should be air-entrained. Attractice finishes are most easily achieved when the sand content is kept to a minimum.

Uniformity of aggregate distribution can be accomplished by using a

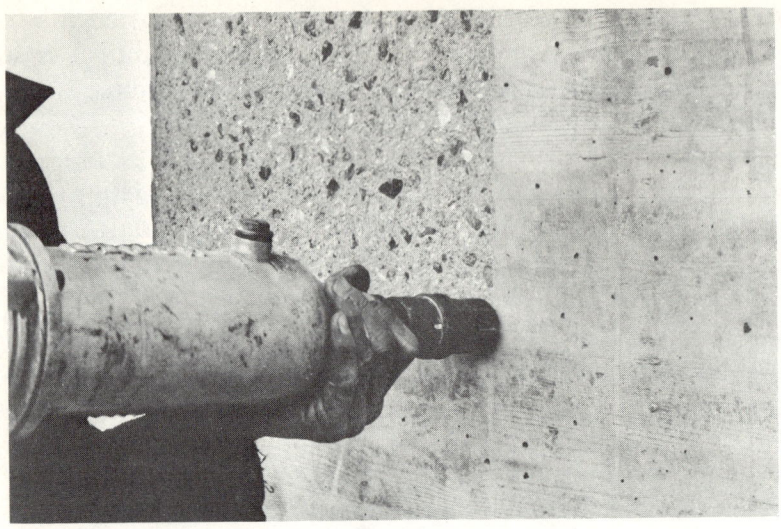
Figure 5-7   Pneumatic bushhammer in operation.

relatively stiff mixture of air-entrained concrete. Concrete should be deposited in level lifts to avoid segregation and sloping lift lines. Each lift should be vibrated properly to insure uniform consolidation of the concrete.

Concrete should not be bushhammered until it has reached a minimum of 80% of its 28-day strength. Bushhammering has been done as early as two weeks after casting, but a more uniform surface will be achieved after three or more weeks.

The only equipment needed is a hand hammer or power hammer. Power hammerheads are faced with a number of points. They are more efficient for large, flat surfaces. Hand hammers are used for small areas at corners and in other locations where a power tool cannot reach. It is advisable to stop power bushhammering about 2 in. from sharp corners to avoid damaging the edges.

A modification of the bushhammering technique is often used. Concrete is placed behind forms designed to produce a corrugated surface. After removal of the forms the fins are broken with a hammer, exposing the aggregate and giving the surface a rugged look. This is just one more method of utilizing the decorative potential of concrete.

Figure 5-8 Fractured vertical fins produce shadow effect on concrete walls.

PRECAST CONCRETE CURTAIN WALLS

A curtain wall is a prefabricated non-load-bearing unit suspended on the frame of a building. It serves both as a windshield and sunscreen.

The evolution of the curtain wall technique started with early man who covered a framework of poles with sod or woven grasses, or stretched pliable materials over widely-spaced supports to make a tent. The facing had the same characteristics as a curtain wall: it was a non-load-bearing exterior wall suspended on a structural frame.

In the case of a curtain wall, its own dead weight and wind loads are transferred to the frame at anchor points.

There are two main types of curtain wall construction: grid and panel.

In grid construction, a rectangular grid is formed by vertical and horizontal members making the framework for glass inserts. The grid is attached to the structure and is seen from the outside as part of the facade. The wind load and deadweight are transferred to the structure from the grid.

In panel construction, large jointless panels are made with a continuous outer surface. They are joined directly to the structural elements. Windows are cut out of the panels when it is structurally feasible. Window frames are rigidly connected.

Advantages

The advantages of curtain wall construction are many. One important feature is the opportunity for design virtuosity. A wide variety of materials can be employed and a vast number of decorative effects can be incorporated into the design.

The units are easy to manufacture and transport. Erection can be rapid and the structure can be ready for early occupancy. The interior will be protected at an early stage in construction so that fixtures can be installed sooner. Reduced labor costs and the efficient use of materials resulting from standardization are important. On-site construction can be faster with a potential for more uniform quality. All parts of the standardized facade can be removed and replaced easily.

Problems

Fastening techniques are an important consideration in curtain wall construction. Fasteners must be strong and firmly embedded. At the same time, they must be easy to erect and adjust, and their surfaces must be protected from weathering and rust. The minimum cover required on the weather surface is usually $1\frac{1}{2}$ in. to avoid corrosion.

During hurricanes, because wind causes suction as well as pressure, more windows are blown out than in. Fastenings must transfer the wind load to the structural frame. Excessive deflection caused by wind loads may rupture the seals between parts of a curtain wall and result in leaks. Construction employing materials capable of resisting bending action has a distinct advantage. Concrete is such a material.

Joints often cause problems. They must be designed to keep out water, dust, and air in order to preserve the insulating properties of the curtain wall. They must withstand work loads and be easily assembled and maintained.

Construction

Lightweight concrete with air entrainment is usually employed. Pumice or expanded clay and shale minerals used as aggregate make suitable lightweight concrete. The density of panels varies depending on the amount of handling and transportation they must withstand. The kinds of concrete that are practical are:

1. Natural aggregate concrete weighing 125 to 150 lb/cubic feet.
2. Lightweight concrete weighing 75 to 115 lb/cubic feet.
3. Foamed concrete weighing 30 to 50 lb/cubic feet.

Shrinkage increases with the lightness of the aggregate owing to the greater absorption of water, but most of the drying shrinkage will have taken place within 4 weeks.

Concrete panels are reinforced to withstand bending stresses from handling as well as from wind. Whether made on the site or in a precasting plant, they are cast in forms with all the inserts, anchor bolts, fastenings, and lifting eyes in place. Sometimes, lifting eyes are put in in such a way that they can be removed and reused.

FORM LINERS

An almost limitless range of textures and designs can be incorporated into concrete by the use of various form liners. Plastics, wood, rubber matting, plaster of Paris, concrete, and sheet metal are some of the materials used to make form liners.

Wood

Interesting finishes that resemble board siding can be made by arranging boards in a beveled or clapboard fashion. The forms must be treated with a parting agent or form oil before concrete is placed. When the concrete is cured, the form is removed and the panel resembles clapboard siding.

Sandblasting plywood gives a texture that resembles gnarled, weather-beaten wood. Rough-sawn boards placed in a variety of ways will give an interesting pattern.

Rubber Matting

Rubber matting works very well as a form liner and is available in a variety of patterns and textures. Many patterns are geometric or uniform in appearance while others present a carpetlike design. Most rubber mattings do not require an application of parting agents or bond breakers if the surface is moistened with water just before concrete is placed. Sometimes a slurry of cement grout of creamy consistency is placed on the matting just ahead of the concrete to aid in reducing or eliminating small air voids. However, most experienced users prefer to coat the rubber with a thin film of lightweight oil or a special parting agent. Rubber matting is reusable, so the added initial cost can be amortized.

Polyethylene Film
Dimpled surfaces are adapted to tilt-up panels as well as to smaller precast panels. A single layer of gravel or crushed rock of a fairly uniform size is spread on the casting slab within the edges of forms. A single sheet of lightweight polyethylene film is spread over the gravel and secured to the edge forms or extended over the edge. Concrete is placed on the plastic in the usual manner. The plastic comes away from the concrete easily when the concrete panel is removed, leaving a dense, glossy, and dimpled surface.

Sheet Metal
Intricate patterns or special designs can be achieved by the use of steel sheets embossed with a pattern. Some stock patterns are available, but special patterns can be stamped into steel sheets for custom-made designs.

Figure 5-9 Dimpled concrete panels created by covering gravel scattered on the bottom of the form with a thin plastic sheet.

Plaster of Paris

Plaster of Paris forms are made by first making a master mold of wood, clay, or other material. A creamy mix of plaster of Paris and water is poured in and allowed to harden. After the plaster of Paris has thoroughly hardened, the surface may be cut with a knife or other sharp instrument and scraped to a final shape, creating a smooth finish with sharp, uniform edges. Often a metal-edged templet can be made that will ensure uniformity of design in the plaster molds. Two coats of white shellac applied to the dry mold surface will make it waterproof and nonabsorbent as well as improve durability and produce a more uniform color in the concrete. The surface of the mold requires greasing with a nonstaining petroleum product similar to cup grease. The excess should be wiped off.

Fiberglass-Reinforced Plastic

Plastic forms reinforced with fiberglass have become increasingly popular. They are often the most economical type when repeated usage is desired. Great pattern flexibility is possible and large areas can be cast without joints or seams.

Fiberglass-reinforced plastic forms have a number of advantages and disadvantages.

Advantages:
1. They are lightweight, permitting easier handling.
2. The form surface has good resistance to abrasion, requires little cleaning, and is not subject to corrosion.
3. Fiberglass forms are not affected by weather because they do not rust or corrode and are not subject to rotting.
4. On many jobs, no parting compounds of any kind are necessary. If form oil is used, it should be a thin film of lightweight oil similar to diesel fuel oil.
5. Either internal or external vibration may be used without injury to the forms.
6. The use of fiberglass results in a concrete surface that is smooth and hard with a minmum of air voids and blemishes.

Disadvantages:
1. As a material, fiberglass has no salvage value.
2. In most instances, fiberglass forms cannot be readily built at the job site.

Since the largest cost in formed concrete is mold making, it is sensible to build forms that can be reused. The number of times a form can be reused will depend on how roughly it is handled. Some fiberglass forms have been used more than 60 times and still remained in near perfect condition.

Textured Plastic
A wide variety of designs and patterns may be produced in panels and large, tilt-up wall units by using textured plastic materials as form liners. Among these are vacuum-formed thermoplastics, fiberglass-reinforced plastics, or plastics formed into shape by heat and pressure.

Since plastic is quite rigid and its surface is smooth, concrete cast in plastic form liners is smooth and uniform in color. Moreover, since most plastics do not bond to concrete, no parting agents are needed.

The thermoplastic materials may be high-impact styrene, linear poly-

Figure 5–10 Panel cast against corrugated vinyl plastic form.

Figure 5-11 Textured fiberglass-reinforced plastic form for precast beam.

ethylene, or other similar materials. All are rigid at normal temperatures, but can be readily vacuum-molded at temperatures from 325 to 375°F.

Thermoplastic materials and vacuum-forming equipment are generally limited in size to 4×7- or 4×8-ft sheets; this must be considered in laying out a design for large panels. Rustication will conceal joints between plastic sheets when a number of sheets are needed for a pattern in a large concrete panel.

Concrete for Form Liners

A good structural concrete mixture is recommended using either standard concrete aggregates or lightweight aggregates. The concrete should be designed for a minimum strength of 4000 psi at 28 days, a slump of 4 in. with standard aggregates or 2½ in. with lightweight aggregates, and air

ARCHITECTURAL CONCRETE 85

Figure 5–12 Completed beam being placed in building.

Figure 5–13 Abstract design created by casting white concrete in sand molds.

86 DECORATIVE CONCRETES

entrainment of 5%. Other admixtures are not generally recommended. If conditions require early handling of the panels or early removal of the forms, Type III portland cement may be used.

Proper vibration is necessary to thoroughly compact and consolidate the concrete and to prevent air voids on the panel face. If a spud vibrator is used, extra care should be taken to prevent the spud from touching the form surface.

In general, all form liners that are designed with a deep pattern must provide for a draft or slope on the raised or depressed areas to allow the panel to be removed from the form after casting. A slope of 12 to 15% is generally adequate. Deeper designs normally require more draft than shallow patterns. After separation from the panel, the form should be cleaned immediately with clear water or a mild soap and water. Proper care of the forms will insure maximum reuse and reduced costs.

CHAPTER 6
STUCCO, PLASTER, MARBLECRETE, AND SGRAFFITO

Portland cement stucco or portland cement plaster is a facing material that can be applied to the surface of any building or structure. It is hard and durable and possesses the properties of quality concrete.

In essence, portland cement stucco and portland cement plaster are the same material. Stucco is the term generally associated with exterior use of the material, and plaster usually refers to interior use. In some areas the word *stucco* refers only to the finish coat. In most other areas, it refers to the entire thickness. When used for either exterior or interior application, the material can be applied in numerous textures on a wide variety of supporting bases.

The material is a combination of portland cement or masonry cement, sand and water, and frequently a plasticizing agent. Color pigments are also often used in the finish coat, which is usually a factory-prepared mix.

The end product has all the desirable properties of concrete. It is hard, strong, fire-resistant, weather-resistant, does not deteriorate after repeated wetting and drying, resists rot and fungus, and retains colors. It has proved to be a durable wall covering in both warm and very cold climates. Many structures that are built in cities in the far north have demonstrated the successful performance of this material in severe weather. It can be formed to any shape and can be finished with a wide variety of surface textures and colors. Moreover, it is economical and has a low maintenance cost.

MIXES

The proportions to be used for the first and second coats are 1 part, by volume, portland cement to not less than 3 nor more than 5 parts damp, loose sand. The variations in sand requirements are due to the wide variations found in sand gradation throughout the country. Unless one has experience with sand from a certain source, trial mixes must be used to determine the proper mix.

A good mix can be recognized by its workability, ease of troweling, adhesiveness to bases, and nonsagging on vertical surfaces. Batch-to-batch uniformity in proportioning mixes will help assure uniform suction and color.

Hydrated lime may be added to the mix as a plasticizer, but the amount used should not exceed 10% by weight or 25% by volume of the portland cement.

When other plasticizing agents such as asbestos fibers (shorts) or diatomaceous earth are used, the amount required will vary with the agent used and the degree of plasticity desired. The kind and amount of plasticizing agent necessary for proper workability should be determined in advance of starting the job. Use the smallest amount needed to obtain the desired plasticity.

When using masonry cement or plastic cement for stucco or plaster mixes, lime or other plasticizers must not be added, since these cements already contain plasticizers. Recommended proportions are 1 part, by volume, masonry cement or plastic cement and from 3 to 5 parts sand. The use of these cements simplifies job site proportioning and mixing because only sand and water need to be added.

For the finish coat, recommendations of manufacturers of finish coat stucco preparations should be closely followed. If the finish coat is job-mixed, truer colors and a more pleasing appearance will be obtained if a white portland cement or a waterproofed white portland cement and a fine-graded, light-colored sand are used. The proportions should be 1 part white portland cement, up to $\frac{1}{4}$ part hydrated lime, and between 2 and 3 parts sand, plus the desired mineral oxide pigment. Color pigment should always be weighed for each batch to help insure a uniform color from batch to batch. The sand should all pass the No. 16 sieve and should be well graded from coarse to fine.

Masonry cement is often used for both the base and finish in some areas where two-coat work is applied over masonry bases.

The materials for all coats should be thoroughly mixed. Dry materials should be mixed first to a uniform color before adding the water. A power mixer is recommended for uniform mixing and blending of materials. The material should be mixed for at least 5 min. after all materials are in the mixer.

Aggregates

The aggregate used in portland cement stucco and portland cement plaster can greatly affect the quality and performance of the finished product. It should be well graded, clean, and free from loam, clay, and vegetable matter since these foreign materials prevent the cement paste from properly binding the aggregate particles together.

A cubic foot of portland cement plaster contains approximately 0.97 cut ft of sand and 0.25 to 0.33 cu ft of cementitious material. The sand fills practically all the volume, and the cementitious material and water fill the voids (air spaces) between the sand particles.

Since the sand occupies such a large percentage of the volume, the

Figure 6-1 Power mixers should be used in blending materials uniformly.

importance of rigid sand requirements cannot be overemphasized. Obtaining a sand aggregate that is clean and sound is of course necessary, but the requirements go much further than this.

Void Content. The gradation of the sand used will affect the void content—the air space between the individual pieces of sand. Sand for stucco and plaster should be well graded with particles ranging from coarse (maximum size—$\frac{1}{8}$ in.) to fine. In a well-graded sand the smaller particles will fill in most of the spaces between the larger particles, and there will be a minimum of voids.

The voids must be filled with the cement paste—a combination of cement and water. The best quality cement paste is obtained when the least possible water is used that will result in a workable mix.

The amount of cement is determined by the stucco or plaster mix proportions, whereas the amount of mixing water will vary depending upon the amount of voids in the sand that must be filled with the cement paste. With poorly graded sand the amount of mixing water that is needed becomes excessive. This dilutes the paste and results in a poor-quality portland cement stucco or plaster that is likely to have excessive shrinkage.

Many times, the cause of cracking in stucco is attributed to too much cement in the mix; actually, in most cases, it is because of too much water due to poor sand gradation.

In order to obtain quality stucco or plaster, the aggregate that is used should meet the current requirements of *Aggregate for Masonry Mortar* (ASTM C 144), except that the gradation of the sand should meet the requirements of Table 6–A.

Table 6–A
AGGREGATE GRADATION

Sieve Size Number	Percentage[a] Passing Each Sieve
4	100
8	100
16	60–90
30	35–70
50	10–30
100	0–5

[a] The aggregate should not have more than 50% retained between any two consecutive sieves nor more than 25% between the No. 50 and No. 100 sieves.

A word of caution regarding gradation: If the gradation of the sand is such that it approaches the maximum percentage permitted by the specification for fine particles, it is nearing the trouble point. The percentage of sand passing each sieve should fall well within the allowable range for maximum and minimum size particles.

Vermiculite and perlite aggregates are sometimes used in portland cement plaster. Their use is important where weight and/or fireproofing are prime factors. Proportioning of these materials should be determined by laboratory tests or by making test panels.

BASES

All structures to receive stucco and plaster must be rigid and well braced. Likewise, all metal reinforcement and masonry bases must be rigid and sound. Otherwise, subsequent movements in the structure will probably cause cracks in the stucco or plaster.

Surfaces to which stucco or plaster are to be applied must contain no frost.

Figure 6–2 This is a sample of well-graded sand after separation into various sizes as indicated in Table 6-A.

STUCCO, PLASTER, MARBLECRETE, AND SGRAFFITO 93

Stucco and plaster can be applied to the following types of supporting bases: metal reinforcement, masonry, cast-in-place concrete, old stucco or plaster, wood frame, and steel frame.

Metal Reinforcement

Metal reinforcement must be used wherever portland cement stucco or plaster is applied on the following:

Wood frame of open or sheathed construction.

Steel frame structures.

Flashings.

Masonry surfaces providing unsatisfactory bond.

Chimneys and disintegrating masonry surfaces.

Unsound stucco and stucco surfaces that do not provide good bond.

Metal reinforcement should always be securely attached to the supporting structure, but there should be at least $\frac{1}{4}$-in. space between the supports and reinforcement. The latter must be stretched between supports

Figure 6-3 Expanded metal lath with dimples provides proper furring for embedment of the metal.

to eliminate any slack. Loosely attached reinforcement generally results in an uneven thickness of stucco or plaster, with thinner sections at the supports. These thinner sections are weak and frequently develop cracks.

Nails or other attachments for securing the reinforcement to the supports should be spaced close enough to provide good, rigid support. They should have a self-furring device to keep the reinforcement at least $\frac{1}{4}$ in. away from the supports unless the reinforcement is self-furring.

Metal reinforcement must for ma continuous network of metal over the entire surface. Laps at the ends and sides of large mesh reinforcement must be at least one full mesh and not less than 2 in. For small mesh reinforcement, the laps should be at least 1 in. Laps should be wired securely and end laps staggered. In open frame construction, joints and laps are made at supports.

Metal corner beads having solid metal noses should not be used on exterior surfaces that are exposed to the weather. It is virtually impossible for the stucco or plaster to get behind these beads to completely encase the metal and thereby prevent corrosion. However, corner reinforcment built up with a series of wires can be used for such corners where the spacing of the wires permits complete encasement of the metal.

Masonry

In applying portland cement stucco or plaster to any masonry surface, the question frequently arises as to whether or not good bond can be obtained.

Good bond with masonry surfaces is dependent on mechanical key and/or suction. If the surface texture is coarse or rough enough for the stucco or plaster to cling to the surface, it has good mechanical key. Suction is the property of a base to absorb moisure; it helps fresh plastering material stick to the base.

The masonry surface should be checked for good suction by spraying the surface with clean water and observing whether the moisture is sucked into the masonry. If the water droplets ball up as they would on wax paper, then it is obviously impossible to obtain good bond. If the water soaks in, it is evident that the surface is absorbent.

Some masonry surfaces may have too much suction and absorb moisture too quickly. Thus, the stucco or plaster quickly stiffens and becomes difficult to properly work and straighten out. Such suction can be controlled by spraying—not soaking—the masonry with several applications of water.

It is also important that suction be controlled uniformly over the entire masonry surface. Otherwise, some parts of the wall draw more moisture from the stucco or plaster than others, and the final finish may be spotty in color. It may not be necessary to dampen masonry bases, however, when the stucco or plaster is machine-applied.

Obtaining Good Bond. If it is doubtful that good bond can be obtained when applying the first coat on a concrete or masonry supporting base; it is sometimes desirable to use a dash bond coat. The mix for a dash bond coat is 1 part portland cement and from 1 to 2 parts sand.

A small straw broom or a long-fibered stiff brush is dipped in the mix, and the material is splattered on the wall with a quick "throwing" motion. The dash bond coat can also be applied by machine. The dash bond coat must not be troweled and must be allowed to set before applying the next coat. However, if there is a delay in applying the next coat, the dash bond coat should be kept moist for 48 hours, or until the next coat is applied.

Another method of obtaining bond on various masonry surfaces is by applying a bonding agent. These agents are relatively new and are in liquid form. Usually they are a water-based emulsion and produce a strong adhesion between the surface and the stucco or plaster coat. The

Figure 6-4 Method of applying a dash bond coat with a long-bristle brush.

surface must be structurally sound and rigid and it must be clean of all materials soluble in water such as water-base paints and glue.

If good bond cannot be obtained over the entire surface, either by mechanical key, suction, or bonding agent, then it is not advisable to apply stucco or plaster directly to that base. Poor or nonuniform bonding will likely cause cracking since the stucco or plaster surface will move with the supporting base in areas of good bond and independently of the supporting base in areas of no bond.

If bond cannot be obtained over the entire surface, a waterproof building paper or felt should be applied to the surface, and a metal reinforcement then nailed or mechanically anchored to the base through the paper or felt. Another method would be to attach a waterproof, paper-backed wire fabric lath to the masonry surface.

Similar precautions are needed for such surfaces as old brick softened by weathering, glass block, glazed masonry, metal, wood, and surfaces that cannot be cleaned thoroughly, such as masonry painted with substances other than portland-cement-based paints.

Metal reinforcement attached to masonry must be furred out at least $1/4$ in. to insure complete embedment of the metal. Nails are available that can be driven into concrete masonry with a hammer. Many forms of gun-driven nails are also available for attaching metal reinforcement to masonry.

Chimneys to be stuccoed should have metal reinforcement nailed to the masonry with the metal furred out not less than $1/4$ in. Waterproof paper or felt should not be used on chimneys because the high temperatures will deteriorate this material. Proper flashings should also be installed.

Concrete Masonry. When stucco or plaster is applied to new concrete masonry construction, there is usually a good bonding surface available. Concrete masonry usually has a fairly rough surface texture for good mechanical key and has good suction, which will help ensure good bond. Mortar joints should be struck flush on masonry work that is to be stuccoed or plastered.

The concrete masonry surface to be stuccoed or plastered should be carefully inspected to determine if paint, dust, grease, or any other substance is present that will be detrimental to obtaining good bond.

If any substances are on the surface that will adversely affect the bond, they should be removed if possible. Or the entire wall surface can be covered with waterproof building paper or felt and metal reinforce-

Figure 6-5 Stucco scratch coat being applied over concrete masonry.

ment furred out 1/4 in.; or it can be covered with waterproof paper-backed wire fabric lath.

An old surface such as paint, other than most portland-cement-based paints, usually has no suction for proper bond and can often be removed by sandblasting or chipping with bush-hammers or similar tools. The resulting surface then can be tested for suction.

When applying stucco or plaster over an old masonry surface, all loose or deteriorated material must be removed to provide a sound structural base. The surface should then be checked for proper suction or good mechanical key.

Clay Masonry. When applying stucco or plaster to clay masonry walls, precautions should be taken similar to those for concrete masonry. Surfaces of hard or medium clay tile, or clay brick (not glazed) usually have textures rough enough to provide good mechanical key and proper suction. On old brick walls, all loose or deteriorated portions should be removed until nothing but sound material remains.

Cast-in-Place Concrete

When applying stucco or plaster to new or old cast-in-place concrete, the surface must be carefully inspected to ensure good bond. Any portions of the surfaces that are smooth due to smooth formwork should be roughened by sandblasting, wire-brushing or chipping, or should be acid-etched. Newly placed concrete needs only a light sandblasting immediately after the forms are stripped.

Acid wash for etching the surface should consist of a solution of 1 part muriatic acid and 6 parts water. The surface should first be wet with water before applying the acid so that the acid will act on the surface only. Several applications may be needed. After treatment, the concrete must be thoroughly washed with water to remove all traces of the acid and then allowed to dry to restore suction.

Another cause of smooth or slick spots on new concrete work may be the presence of form oil. The oil should be removed by scrubbing with strong soap and water and then thoroughly rinsing with clean water. The surface then should be allowed to dry to restore suction. Sandblasting can be used to remove oil.

A bonding agent may also be used.

Old Stucco or Plaster

Old, sound portland cement stucco or portland cement plaster can be refinished (overcoated) by the application of a new finish coat of portland cement stucco or plaster. These surfaces should be thoroughly inspected, particularly for any coatings of paint. If necessary, the surface should be cleaned to provide good suction for bond. Sometimes bonding agents are used.

If the old stucco or plaster is unsound, it must be removed. In case it cannot be removed, the surface should be covered with waterproof, paper-backed wire fabric lath, or with a combination of waterproof building paper or felt and metal reinforcement. The wire fabric lath or metal reinforcement must be furred out $\frac{1}{4}$ in. from the old surface and securely fastened to the structure as a base for the three new coats of stucco or plaster.

Portland cement plaster and stucco should never be applied in direct contact with gypsum products such as plaster, lath, or block. A chemical reaction takes place between these materials that destroys the bond.

Wood Frame

Stucco should be applied over metal reinforcement on all wood frame structures. Wood frame structures fall into two categories: open frame construction and frame construction with sheathing.

In open frame wood construction the structural frame should be properly braced and rigid. A soft annealed steel wire, No. 18 gage or heavier, often called "line wire," is stretched across the faces of the studs in horizontal strands about 6 in. apart. The line wire should be stretched taut by first being nailed or stapled to every fifth stud, then secured to the intermediate studs, and stretched tightly by raising or lowering the attachments.

A waterproof building paper or felt is then nailed over these wire strands with edges lapped at least 2 in. The line wire behind the paper forms a rigid backing and minimizes sagging of the paper between studs when the stucco is applied; this results in a more uniform thickness of the scratch coat.

Hexagonal wire mesh (stucco netting), which is usually used with this construction, is placed over the paper surface and properly furred out to receive the stucco. As the stucco is applied by hand trowel or plaster gun, it is forced through the openings in the mesh against the paper backing, thus completely embedding the metal reinforcement. The resulting stucco facing is in reality a thin reinforced concrete section.

Other types of metal reinforcement can also be used over the line wire and paper. Some have waterproof paper backing can be used directly over the studs without first applying the line wire and paper.

In open frame wood construction, all metal reinforcement should be started at least one full stud away from any corner and should be bent around corners to avoid junctions at corners. When wood sheathing is used, the metal reinforcement should be returned at least 6 in. around corners.

Wood frame construction with sheathing eliminates the necessity of wrapping the horizontal strands of annealed line wire around the structure. Wood sheathing should be applied horizontally and securely fastened to each stud. Other accepted types of sheathing may also be used. The sheathing should be covered with waterproof building paper or felt, and then a metal reinforcement is applied over the paper. The reinforcement should be fastened to the studs and furred out ¼ in. from the paper. Another method is to attach a waterproof paper-backed wire fabric lath to the studs directly over the sheathing.

Metal reinforcement should be installed with the long dimension across the studs. Nails or staples for attaching reinforcement to wood should be galvanized or rust-resistant. Since portland cement will react chemically with aluminum, aluminum nails should not be used where they will contact fresh portland cement stucco or portland cement plaster.

Steel Frame

Studding or furring of steel channels, pressed steel members, steel reinforcing rods, or metal studs should be properly spaced and securely fastened to the structural steel frame by wire-tying, clipping, bolting, or welding.

If welded, they should be separated into smaller panels and the panels tied together with wire. This allows movement and prevents the accumulation of high stresses in the portland cement stucco or plaster membrane.

Where walls are constructed so as to restrain all movement, deformation of the stucco or plaster will likely occur. Control joints should be placed between panels to relieve any stresses.

Portland cement stucco is ideally suited for curtain walls with steel frame construction. Stucco with a sanded finish is being used more and more as an exterior facing material for this type of wall.

The development and use of perlite and vermiculite as plaster or stucco aggregates in spandrel, panel, and curtain walls have made possible considerable savings in space and weight. Lightweight portland cement plaster and stucco may be applied in any desired thickness to obtain the required fire-resistance ratings and to improve the structural properties of a frame.

APPLICATION

Stucco and plaster should be applied in three coats. The first coat is called the "scratch" coat, the second is the "brown" coat, and the final one is the "finish" coat. However, if the base for the stucco or plaster is masonry, cast-in-place concrete, old portland cement stucco, or plaster with reasonably true surfaces, and if no metal reinforcement is used, then two coats may be sufficient.

Hand Application

Where stucco and plaster are applied by hand, the first coat should be pushed through the mesh to insure that the metal reinforcement is completely embedded. Second and finish coats should also be applied with sufficient pressure to help obtain compaction and good bond.

The first coat should be approximately ½ in. thick so that the metal reinforcement is covered. Not more than 10% of the area of the reinforcement should be exposed. After application, the first coat should horizontally scored or scratched so that good mechanical bond can be obtained by the second coat.

The second or brown coat should be applied as soon as the scratch coat has set up enough to carry the weight of both coats. This usually means about 4 or 5 hours. If the scratch coat is applied in the morning, then the brown coat can usually be applied in the afternoon. Weather conditions

Figure 6-6 Stucco being troweled over metal reinforcement (stucco netting). The scratch coat must be forced through and behind the reinforcement for complete embedment.

Figure 6-7 A scarifier is being used to scratch the base coat. This forms horizontal ridges that provide maximum mechanical key for the second coat.

will affect this, of course. If the weather is hot, dry, or windy, the brown coat may be applied sooner. If the weather is cool, then the scratch coat will take a longer time to set, and the application of the brown coat must be delayed. When the scratch coat is applied on open frame wood construction over paper with line-wire backing, however, the brown coat should not be applied any sooner than 48 hours. This allows the scratch coat to become hard and provides a rigid base for the application of the brown coat.

If there is a delay in the application of the brown coat, the scratch coat will begin to dry out. This should be prevented by keeping the scratch coat moist with occasional fine sprays of water until the brown coat is applied, or for 48 hours if the delay is more than two days. Just prior to the application of the brown coat, the scratch coat should be evenly dampened, but not saturated, to obtain uniform suction.

Figure 6-8 Applying the second or "brown" coat.

The second coat is applied approximately ⅜ in. thick. A straightedge is used to bring it to a true plane surface. Deep scratches or surface defects in the brown coat should be removed since these tend to show through the finish coat.

A floated texture surface will provide good bond for the finish coat. All plasterers working on the same wall should use the same type of float to assure uniform texture and suction.

When applying the brown coat, an entire wall panel should be covered without stopping if possible. Stops should be made only at corners, pilasters, belt courses, doorways, downspouts, or wall openings where abrupt changes in the uniform appearance of the finish coat might not be obvious. Otherwise, the stopping and starting joints may be visible through the finish coat.

Horizontal stopping joints, if any, should be cut square and at least 6 in. below joints in the scratch coat.

The brown coat should be moist-cured for at least 48 hours and then allowed to dry for at least 5 days—longer if possible—before applying the finish coat. This permits the brown coat to attain a more uniform

Figure 6–9 Floating skill is demonstrated here as the plasterer finishes an overhead flat surface. Note the position and pressure of the hand on the float.

suction, which will produce a more uniform color and texture in the finish coat.

Just prior to the application of the finish coat, the brown coat should be uniformly dampened to attain uniform suction.

In some areas, only two coats are being applied over masonry bases. Masonry cement is often used for both the base coat and finish coat. The base coat is approximately 3/8 in. thick and the finish coat about 1/4 in. thick.

In most cases, however, the finish coat is the third coat; it is approximately 1/8 in. in thickness. This coat is frequently pigmented to obtain decorative colors. Although this coat can be job-mixed, a factory-prepared mix is recommended. Factory-prepared stucco mixes have greater uniformity of color, texture, and quality.

The finish coat can be hand-applied or it can be sprayed on.

To prevent damage to the surface and to the color of the finish coat, moist-curing should be delayed until the day after application. The finish coat should be fog-sprayed very lightly the first time—not saturated. Moist-curing should be continued for a full day.

Since the appearance of the entire wall depends upon the finish coat, considerable care should be exercised in its application. The finished job also reflects the skill and care that go into each preliminary step, such as the structural framing, the application of metal reinforcement, the preparation of the masonry bases, and the application of the various coats.

Machine Application

Machine application of stucco or plaster is now used in most sections of the country. The mix is sprayed from the machine nozzle against the prepared base. Machines demand new techniques and procedures if they are to be used to the best advantage.

Adequate and uniform mixing time in a mechanical mixer is needed for successful machine application. Uniform measuring and batching methods are also important. Sloppy (wet) mixes should be avoided, even though they go through the hose more easily. Overwatering may cause the material to separate and result in such on-the-wall problems as glazing, shrinkage cracking, slow take-up, drop-outs, lower strengths, efflorescence, and crazing.

Hose length from the machine to the working face should be straight and no longer than necessary. This keeps the pump pressure load at a

Figure 6–10 This is a combination mixer and pump. When plaster or stucco is machine-applied, the mixer and pump should be set up so as to allow the material to be discharged directly from the mixer into the hopper.

Figure 6–11 As one workman applies the second coat with the plaster gun, another man screeds or levels off the material to bring it to a true surface.

minimum. Water should be run through the pump and hose before the first batch because the dry surfaces of the equipment may cause stiffening of the mix.

When pumping unusually long distances, pipe of the same diameter as the hose should be used as much as possible because the mix passes more easily through pipe than hose.

If a hose stoppage should occur, the pump should be shut down immediately instead of trying to force the stoppage with the pump. Some machines have a reverse gear for sucking material back into the hopper under such circumstances. If this is not possible, the hose should be disconnected and cleaned with water.

Regular preventive maintenance and periodic inspection are necessary. The hose, pump, and nozzle must be absolutely clean since even a very small plaster build-up or constriction can cause a sand block to form. Loose hose connections and pump fittings will allow the cement paste to separate out of the mix; this also can cause sand blocks.

The entire system should be thoroughly cleaned before and after using. Cleaning includes filling the hopper with water and running it through the hose until clear. The material hose should be disconnected and a wet sponge rolled up and placed in the feed end. The hose should then be reconnected and the sponge forced through the entire length of the hose under water pressure.

When application begins, the nozzle should be about 12 in. from the surface. Except in the case of highly absorbent surfaces, it is not usually necessary to prewet the base because the mix itself will carry sufficient moisture.

The pattern and particle size of the spray may be varied by:

1. Altering the air pressure.
2. Changing the size of the hose orifice to one of a different diameter.
3. Varying the distance between the orifice and end of the air stem.
4. Increasing or decreasing the pump speed.
5. Changing the water content to change the mix consistency.

These controls are especially valuable in attaining either coarse or fine textures in the final color coat.

Machine application has many advantages: it eliminates the lap and joint marks sometimes found on hand-applied jobs; it produces a more uniform appearance in texture and color; and it enables the operator to use deeper and darker colors than can be obtained by hand application.

Color coats should be sprayed in two applications, the first to insure a complete coverage and the second to obtain desired texture.

For scratch or brown coats, the nozzle should be moved with a steady, even stroke, laying on the proper thickness with one pass and overlapping successive strokes. The angle of the nozzle to the wall should remain uniform. If the nozzle turns out toward the end of the strokes, the material will build up much more heavily directly in front of the operator.

Around door bucks or window frames, the nozzle should be moved up to within a few inches of the surface and pointed away from the area where overspray is to be avoided.

The thickness of the various coats should be the same as for hand application.

Plastering machine manufacturers publish instructions for guidance in

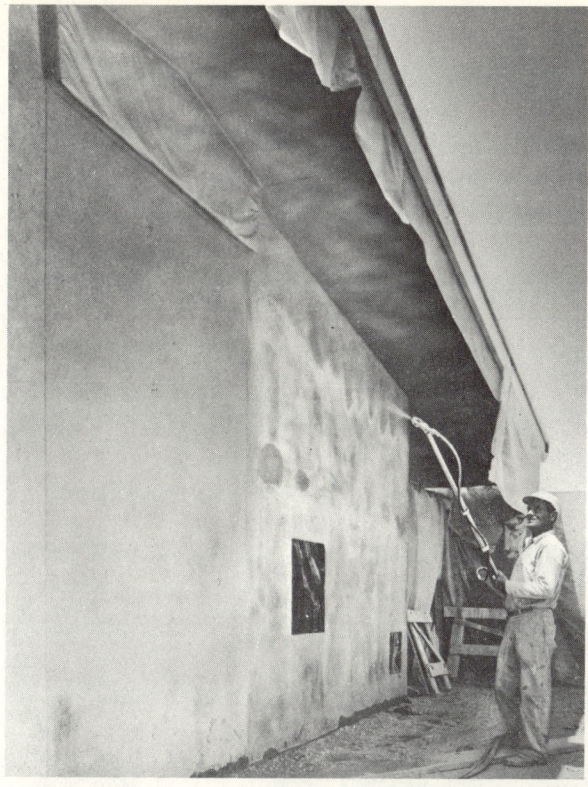

Figure 6-12 Machine application of colored final coat produces uniform texture and color. All openings or surfaces not to be coated must have protective shields.

the use use of their equipment. These should be carefully followed. Manufacturers of colored stucco finishes also issue machine application instructions.

As far as old stucco, masonry, or cast-in-place concrete bases are concerned, two machine coats will usually do a nice job of refinishing. First, the surface should be thoroughly cleaned and left dry. Sometimes a bonding agent is needed for good bond. Then the first coat should be sprayed on with a thin consistency until the entire surface is covered. It should then be allowed to set for 2 or more hours depending on the temperature, sun, wind, and relative humidity.

The second or final coat then should be sprayed over the first coat using a thicker consistency to obtain the desired texture. If necessary, the second coat may be deferred until the following day.

The operator should avoid spraying too heavily at any one place; instead he should use several successive light applications over the first coat. For uniform coloring, both the first and second coat should be sprayed with a finish coat mix that includes the color pigment desired.

SPECIAL EFFECTS

Marblecrete

Marblecrete is a stucco process in which two or three coats of weathertight portland cement stucco are applied first. Then marblecrete is

Figure 6–13 A straightedge is being placed at the corner to provide a screed to assure proper alignment of the wall and to form a good plumb corner. The operator keeps the hose nozzle close to the surface for better control.

formed by adding another step. The finish (bedding) coat is made with colored or white portland cement. Colorful aggregate chips are pneumatically or hand embedded in either a pure white or pastel-shaded finish coat to the desired textural depth. If a pastel finish coat is preferred, white portland cement should be mixed with color pigment. For best results, the aggregate should be "shot" into place with a rock dash gun.

A clear, nonyellowing glaze is often applied to the finish when it is installed in areas of heavy traffic or where a surface film from smog, smoke, or stains is likely to develop. This seals in the embedment aggregate more securely, keeps the white or pastel background true, and makes cleaning easier. Only an occasional hosing down will be necessary.

Marblecrete in areas of less traffic and cleaner atmosphere needs no sealant for protective purposes. Clear glaze, however, will intensify aggregate color.

Most popular aggregates for marblecrete are brightly colored or white marble chips, pebbles, quartz, and vitreous china. Other materials of suitable hardness are also used. Small particles of mica, glass, or aluminum flakes are often added to freshly finished marblecrete, lending increased sparkle to the surface.

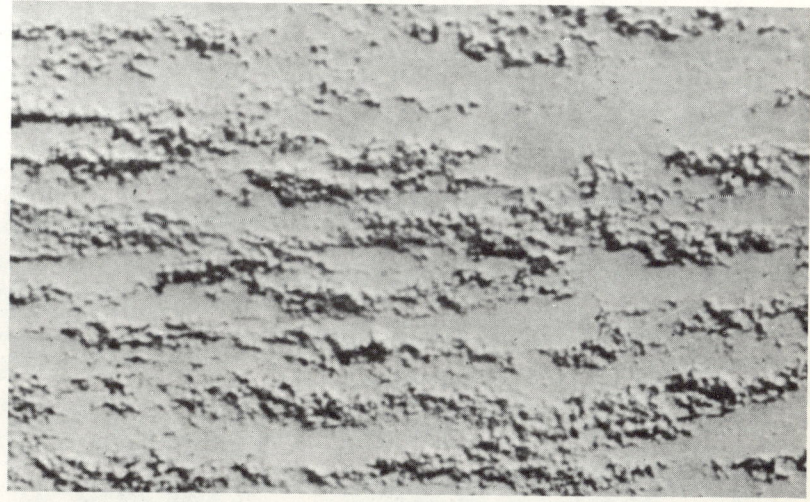

Figure 6–14 A stippled-troweled finish made by troweling a heavy finish coat fairly smooth, then stippling it with wire brush or whisk broom. High spots are then troweled smooth.

Figure 6–15 This is one type of machine used to apply a rock dash or marblecrete finish. Small pebbles, stones, or marble chips are placed in the hopper and are blown into the surface of the fresh stucco.

Simulated Stone or Brick

A simulated stone or old brick look is achieved with stucco at a much lower cost than real stone or brick. This method calls for a base of 3/8 in. standard stucco consisting of 3 parts sand, 1 part cement, and 2 parts water. For the old brick look, a red course is applied with a rough surface. The surface is scored in such a way that the red course is cut away exposing the gray. This has to be done while the concrete is soft. Then, stucco in yellow, blacks, white, and dark reds is applied with a sponge so it will be rough and uneven. It has to be done by a well-trained worker and takes practice to get genuine old brick effects.

Coral or stone are harder to simulate because some areas have to be built up and others dug out. This costs more than simulated brick but it is still less costly than real stone or coral.

Figure 6–16 Rock dash finishes having large variation in size are usually placed by hand. Large pieces are individually placed and the small pebbles and stones broadcast by hand or with a machine.

Sgraffito

Sgraffito is a method of creating colorful, artistic murals by scratching and exposing different colored layers of mortar. It has to be done rapidly before the different layers of stucco harden, so it is a difficult technique.

The design is planned in color; for example, four colors are used—green, red, blue, and black. Stucco is spread on the wall area in layers, color by color. Where the design shows an apple tree, for example, the stucco is scraped away to the green layer for the leaves and to the black layer for the trunk. Then the apples are scraped to the red layer and the sky to the blue. Precise artistry is demanded because once a layer has been dug out, it cannot be replaced. The artist must work fast and have the design carefully worked out before starting. The effect can be very dramatic even using just two layers of color. It is an inexpensive way of decorating a large surface very effectively.

CHAPTER 7
PIGMENTS, PAINTS, AND STAINS

INTEGRAL COLOR PIGMENTS

The increased use of concrete as an exposed building material has created a growing interest in the different ways of introducing color in the surface finish. One of the ways of doing this is by the use of integral color in the concrete mixture. This is a satisfactory way of incorporating decoration in concrete. If the materials are handled carefully, uniformity and lasting effects can be assured.

Materials

The most suitable pigments for coloring concrete are synthetic mineral oxides. Most of the colors are available in natural oxides, too, which are cheaper per pound but vary in purity from 10 to 90%, while the synthetics are about 98% pure. Brighter colors, more attractive shades, and higher color strengths can be achieved from synthetic oxides with less pigment, which often makes them cheaper to use. Only mineral pigments should be used, since others will fade or reduce the strength of the mortar or concrete.

A general guide to the selection of colors and coloring materials to obtain various effects follows:

Blue. Pure cobalt blue oxide is an entirely permanent color, but its prohibitive cost restricts its use to small areas of special ornamental items. Commercial grades of cobalt blue oxide are rather expensive, too, but are cheaper than the pure oxide. They have good stability of composition and permanency.

Green. Chromium oxide has durability and permanency if it is chemically pure. At the present time, research is being done on organic compounds in an effort to produce moderately priced blue and green pigments that will be entirely permanent in concrete.

Yellow, red, maroon, and brown. Best results are obtained by using synthetic iron oxides for brighter, deeper shades.

Tan and light brown. Cost can be saved by using natural iron oxides for light buffs, cream tints, light tans, light browns, and reddish brown. Some of the lighter shades of yellow and tan can be obtained by using yellow and brown sand without any pigment. When this is possible, it should be done.

Black. Black is obtained by using either synthetic black iron oxide—be careful to get a jet black rather than a blue-toned material—or carbon black. Carbon black is difficult to handle but it can be used successfully. Common lampblack should not be used.

Problems with Materials

Some of the pitfalls in color materials that should be avoided are:

1. Weak iron oxides that are drab and not permanent. Efflorescence may develop with them.
2. Iron oxides containing a high percentage of calcium sulfate. The sulfate is soluble and much of the color could disappear.
3. Prussian blue, which is a dark, brilliant color. Since it is soluble in alkalis, the color may disappear in concrete.
4. Chrome yellow will fade badly in concrete.
5. Chrome green, a blend of Prussian blue and chrome yellow, will fade.
6. Lampblack fades badly.
7. Bone black contains soluble salts that cause efflorescence.

Colors should not be mixed to get new shades. Let the manufacturer do the mixing. Most of the shades that one might want are available on the market.

How to Mix Color

The amount of pigment used should never exceed 10% by weight of the cement used. The amount of color expressed by a percentage is always based on the weight of cement only.

The strength of the color called for will dictate the amount of pigment used. Pastels will require very little pigment—about 1½% of cement by weight—while deeper colors may require as much as 7%. White portland cement should always be used for lighter shades, and it is often preferred in darker colors to insure clean, clear shades. The pigment and cement should always be proportioned by weight. The proportions must be accurate to maintain uniform color from batch to batch.

Mixing of color pigments and cement should always be done in a dry state, preferably in a ball mill. This is expensive, so there are color mixers available, usually cylindrical drums with rotating blades. Mixers can be made by attaching blades to the inside of a drum or cylinder and rotating it.

The color achieved from any pigment cannot be judged by seeing the dry pigment. Sample panels should be made and kept for five days in conditions similar to the actual work. They will look darker when wet than they will after they have thoroughly dried, so no choice of shade should be made until the samples are dry. If there is time, the samples should be allowed to spend some time in the sun to test color durability.

Additional air-entraining agents may be required when pigment is used because of its fineness.

Tests to Determine Quality

Most architects and builders depend upon the reputation of the manufacturer of pigments for assurance that the quality of the material is satisfactory. A complete examination would involve extensive physical and chemical tests. There are, however, a few simple tests that can easily be made that will often be of assistance in determining the suitability of pigments.

The finer a pigment is ground, the greater is its coloring ability and the less will be required to produce a given shade.

The ability to resist the action of lime can be tested by mixing a sample part of 20 parts of cement and 1 part of the pigment and observing the sample for a period of several days, meanwhile keeping the specimens moist. Any pronounced fading indicates that the pigment is not limeproof.

To test the durability of a color under the influence of light takes some time, unless a special artificial light is at hand. Pronounced fading of a colored mortar on exposure to sunlight for one month is evidence that the pigment is unsuitable.

PAINTING CONCRETE

Increased demand for decorative touches on concrete has brought about the development of paints made especially for use on concrete surfaces. Selection of the type of paint for a particular job depends on:

1. The condition of the concrete.
2. The service to which it will be subjected.
3. The effect desired.

Sunproof and limeproof pigments, water, and occasionally accelerators and waterproofing agents are added to a base mixture of portland cement and hydrated lime. This makes a paint that applies easily and bonds with:

1. Properly prepared concrete surfaces.
2. Concrete masonry.
3. Portland cement stucco.
4. Common brick.
5. Soft tile.
6. Limestone.
7. Clean absorptive surfaces.

Portland cement paint should not be used on:

1. Wood.
2. Metal.
3. Enameled brick.
4. Vitrified or glazed brick or tile.
5. Surfaces saturated with oil paints.

White portland cement is used in the manufacture of portland cement paint except when mixing some of the darker colors. The dry pigment and cement are carefully proportioned and ground together in a ball or pebble mill until thoroughly mixed and uniform. It is sold in dry form to be mixed with water. A wide range of colors can be obtained by blending the standard available colors.

The following table shows federal specifications for two types of portland cement paint with compositions as indicated.

Other provisions are:

1. The paint shall contain no organic binder.

2. Pigment shall be limeproof and shall be incorporated by a milling operation.
3. Under each type are two classes: Class A (without siliceous aggregate) for general use, and Class B (with siliceous aggregate) for first coat on open-textured surfaces.

Table 7–A
COMPOSITIONS FOR TWO TYPES OF PORTLAND CEMENT PAINT

Ingredient	Percentage by Weight			
	Type 1 Class A		Type 2 Class A	
	Maximum	Minimum	Maximum	Minimum
Portland cement	—	65	—	80
Hydrated lime[a]	25	—	10	—
Carbonates (calculated as carbon dioxide [CO_2])	3	—	3	—
Water repellents (calcium or aluminum stearate)	1	0.5	1	0.5
Hygroscopic salts (calcium or sodium chloride)	5	3	5	3
Titanium dioxide (TiO_2), zinc sulfide (ZnS), or mixture	5	3	5	3

[a] The total free (unhydrated) calcium oxide (CaO) and magnesium oxide (MgO) in the hydrated lime shall not exceed 8% by weight of the hydrated lime.

Type 2 paint should be used on exterior surfaces of walls where excessive moisture is present. Either Class A or B can be used for a second coat on open-textured surfaces, depending on the texture desired. Type, class, and color of paint should be included in the specifications for each job.

Pigments should be commercially pure oxides, limeproof, and finely ground. Varying grades of pigment are available, but only the best should be used in paint. The amount of pigment needed depends on the tinting strength of the pigment as well as its distribution throughout the paint.

White pigment such as titanium dioxide or zinc sulfide is desirable in white or light-colored paint to increase the opacity of the paint film. A larger amount than that indicated in the table may cause shrinkage and crazing.

Powdered forms of aluminum or calcium stearate should be milled with the paint powder or with the cement. Although the federal specification permits either calcium or sodium chloride (common salt), calcium chloride is preferred because it accelerates the hardening of the paint.

For painting coarse, open-textured masonry surfaces, a fine siliceous aggregate is incorporated in the cement paint. The aggregate should be mixed dry with the paint powder before water is added.

Water

The amount of mixing water needed depends on the fineness of the material in the paint powder and is best determined by trial. The mixed paint should have the consistency of rich cream, although a slightly thinner paint should be used for the first coat on open-textured concrete surfaces.

Generally, the water required will vary from $3\frac{1}{2}$ to $4\frac{1}{4}$ quarts for 10 lb of dry paint powder. Paint with a coarse filler, in which approximately 25% is retained on the No. 100 sieve, will require 2 to $2\frac{1}{2}$ quarts of water. The thinner consistency required for open-textured surfaces will call for about $\frac{1}{2}$ quart more water than these amounts.

Mixing

A good practice is to use a shallow pan 4 to 6 in. deep and add only about half the estimated amount of water at first, thoroughly mixing to a stiff paste. Add the rest of the water gradually, stirring until all the lumps are destroyed and the paint is uniform. A record of the total amount of water used should be made so that all subsequent batches will contain the same proportions.

The paint will have better workability if it is allowed to stand 30 to 45 min after it has been mixed. With frequent stirring, it should stay usable for 3 to 4 hours (less in hot weather), but it is good practice to mix only enough to last for 2 hours. Water may be added to white paint to keep it workable, but adding water to other pigmented paints will change their color. It will keep better if it is mixed in small batches or is kept in a covered container from which small amounts are poured as needed. Once painting has been started, it is essential to continue until an obvious stopping place has been reached to insure that no difference in colors can be noted.

Preparation of Surfaces

Before painting with portland cement paint, the surface must be prepared. It must be clean and free from:

1. Dust and dirt.

2. Efflorescence.
3. Form oil.
4. Organic-type paints.
5. Grease.

The dust and dirt can be removed with a stream of water from a hose. A brush may be needed in some areas. If large areas are coated with form oil, parting agents, or glaze from casting in plywood forms, the area can be sandblasted. Otherwise, steel brushes, abrasive stones, or lye solutions will be necessary. Efflorescence can be removed with a 5 to 10% solution of muriatic acid, but the resulting deposits must be carefully rinsed off. Particular care should be taken to protect any aluminum from the acid.

Painting is best delayed at least 3 weeks after curing on masonry and stucco. It should be delayed on cast-in-place concrete for several months. The advantages of waiting before painting are that form oil is allowed to weather off and efflorescence has time to appear. However, when necessary, painting can be started immediately after forms are removed or as soon as stucco or cement plaster has hardened.

Concrete cast against plastic, plywood, fiberboard, or steel has a smooth texture to which it is difficult for paint to adhere. Such surfaces should be washed with acid, sandblasted lightly, or rubbed with abrasive stone.

To avoid suction and to aid in curing the paint coat, the surface to be painted should be dampened. This can be done with a garden hose using a fine spray. It should be done several times, waiting a few minutes between each spraying for the moisture to soak in. For dense concrete, this should be repeated in ½ hour.

After the first coat of paint has hardened, the surface should be kept damp for 24 hours by spraying at intervals. Before the second coat is applied, the surface should be dampened again. Twenty-four hours should elapse between coats, and the surface should be kept damp at least 2 days after painting.

Portland cement paint should not be used in strong direct sunlight, high wind, or temperatures below 40°F. If it is necessary to paint on hot, windy days, the surface should be dampened before the paint hardens, which means as soon as possible after application. Painting should be done in the shade whenever possible. Frozen surfaces should be allowed to thaw before paint is applied.

Application

Portland cement paint should be applied with brushes having short, stiff bristles, such as fender brushes used under automobiles, or regular scrub brushes. This is also true of paint containing sand. For a smooth, sandy finish, a whitewash or Dutch calcimine brush is desirable.

The surface must be dampened to prevent loss of water from the paint into the concrete. The dampness also prevents the paint from "balling up" under the brush.

A dense, durable, weathertight surface will only be achieved if the painted area is kept damp for at least 48 hours. If the paint is not properly cured, it may chalk or dust.

Coverage

On very rough-textured masonry block, coverage of two coats may require as much as one gallon of paint every 35 sq ft. On moderately rough surfaces, two coats will require one gallon of paint for every 45 to 55 sq ft. A smooth surface will only require one gallon per 100 sq ft. It is questionable economy to stretch paint much farther than this.

Other Paints

Latex paints are also water-based paints. In this case, pigmented organic film is dispersed in water. Latex paints can be purchased in either exterior or interior grade; however, the interior paint will not wear well under exposure conditions. Latex paints are easy to apply by brush or roller and are available in all colors. Soap and water are all that is needed to clean up after painting.

Chalk should be removed from the surface before latex paint is applied since it will not adhere to a chalky surface. The paint itself tends to chalk when continuously exposed to weathering. Deep colors lighten by chalking in nonuniform patterns.

Dispersions of alkyd varnishes are available. They are susceptible to attack by alkalis, so they are usually used on interior partition walls that tend to remain fairly dry.

Other types of water-based paints are available. With careful use, following the manufacturer's directions, they will give good service.

Organic Paints. It is difficult to secure a good job with these paints because they form impermeable films that may be blistered off the concrete by moisture. If used, it should be on dry, well-seasoned concrete from which all oil, grease, and efflorescence is removed.

Old concrete should be wire brushed or sandblasted lightly. New concrete should be allowed to dry 8 to 10 weeks after moist-curing ends. A neutralizing wash should be given to the surface to prevent saponification of oils. Two to 3 lb of zinc sulfate crystals per gallon of water can be used. A small amount of pigment added to the mixture will make it easy to see where the wash has been applied. Allow it to dry 48 hours and brush off all protruding crystals.

A better method of preparing the surface for organic paint, as proven by laboratory tests, is to allow the concrete to dry for several weeks after curing and then to apply generously a solution of 3 oz zinc chloride and 5 oz orthophosphoric acid (85% phosphoric acid) per gallon of water. Let it dry for 24 to 48 hours and brush off the surface dust. Do not rewet the surface.

After the neutralizing wash has dried, a binding, suction-killing treatment of any of the following should be applied.

1. One or more coats of oil or varnish that is carrying some pigment.
2. One gal oil paint, ½ gal China wood oil spar varnish, and 1 qt turpentine or other thinner.
3. Other special priming paints.

The prime coats should be allowed to dry thoroughly, after which any of the organic paints can be applied, following the manufacturer's instructions.

Oil Paints. Oil paints are made by combining a pigment with an oil vehicle. Any of the following oils can be used.

1. Linseed oil.
2. Linseed oil and China wood oil.
3. Soybean oil.
4. Other drying oils.

Sometimes a fine silica sand is mixed with the final coat. These paints give a hard, flat finish, rough to the touch. They adhere well to concrete and other masonry. However, they are not recommended for floors.

There is a wide variety of oil paints and enamels made by grinding white lead in a vehicle such as linseed oil and adding other pigments to produce color. Modern paints are made from pigments other than white lead, such as aluminum or zinc, and from oils other than linseed. In some cases, pure metal rather than the oxides are used. Such paints made by reputable manufacturers are suitable for use on concrete, provided they are applied to properly prepared surfaces.

Special and Brand-Name Paints. Many of the newer paints and lacquers consist of resins, gums, synthetic plastics, and a wide range of other materials dissolved in hydrocarbon or other suitable solvents. These paints should be used in accordance with the directions of the manufacturer.

Painting Concrete Floors

Abrasion-resistant pigments are used when painting concrete floors, but even so, repainting will be necessary at intervals because there is no paint presently developed that will resist the wear to which a floor is subjected. A paint with a phenol resin base is as good as paint with a rubber base. The rubber base produces a paint that gives a hard, durable surface coating, resistant to water and alkali.

When painting a concrete floor, it is essential that the floor be perfectly dry. Do not try to paint basement floors where moisure penetrates from below because the moisture will cause the paint to peel and chip off. Painting should be done when the atmosphere is also dry.

A new concrete floor should be seasoned and neutralized with magnesium fluosilicate. This helps harden the floor while it neutralizes. The entire surface must be clean. To remove oil or grease, the floor should be scrubbed with gasoline, naptha, or a special solvent. This should be allowed to evaporate before painting.

Roughness should be removed with carborundum brick or a coarse emery. Smooth surfaces to which paint will not bond should be etched with 10% muriatic acid solution. First, wet the floor with the acid and scrub it with a fiber brush. Then, mop or flush it with water and let it dry thoroughly. Be sure to protect hands, shoes, and clothing from the acid.

Concrete floors should be given three coats for best service. The first coat should be a thin one for good penetration into the concrete. The subsequent coats should have the consistency of rich cream. Varnish or wax over the painted surface will help protect the paint.

CONCRETE STAINS

Another method of applying color to concrete surfaces, especially floors, patios, and walks, is by use of both pigmented and chemically reactive concrete stains. There are a number of good stains now on the market and, when used according to the manufacturer's directions, they are quite satisfactory. Color applied in this manner is not strong or bright but can be strengthened by coating with a wax of the same color. Such waxes are made and sold by all stain manufacturers with instructions for proper application. Rewaxing will be necessary, and frequency depends on the amount of traffic. Outdoor areas will require attention more frequently.

Stains should not be applied to concrete until it is at least six weeks old, and an even longer period of curing is desirable. Before staining, the concrete surface must be clean and free from all foreign substances such as curing compounds, hardeners, soap, and paint, for example. Acid should not be used to clean the concrete before staining.

Stains are generally applied in two coats with coverage varying with the porosity of the surface. A gallon of stain will provide two applications to about 200 sq ft. Coverage of the wax is 600 to 900 sq ft per gallon for a one-coat application.

WHITE CONCRETE

White concrete is made from white portland cement, which is used the same way as any other portland cement. It is more expensive than regular portland cement concrete, but the effects achieved can be so dramatic that in many cases it is worth the extra expense.

Some things to be considered when planning to build with white concrete are the effects of weather on buildings in the area, the effects of industrial gases on buildings, and what happens when dirt that is suspended in the atmosphere is washed over the structure by rain.

The aggregate to be used in white concrete should be carefully chosen and tested because many aggregates tend to discolor white concrete. In fact, light buff, yellow, or tan concrete can be made without pigment with the addition of yellow and brown sand to white cement. As the concrete weathers away, the aggregate will become exposed, therefore testing is necessary to see what effect exposure of the aggregate will have on the appearance of the structure.

The choice of aggregate for white concrete is a wide one in spite of the need for care. Among the natural decorative aggregates, marble offers a wide selection of colors from red, yellow, rose, pink, gray, green, white, blue, and black. Granite, which is durable as well as beautiful, is available in shades of pink, gray, black, and white. Quartz sparkles when embedded in concrete and can be obtained in three varieties: clear translucents, milk whites, and tinted crystals. Manufactured decorative aggregates are higher in cost than natural aggregates, but they are rich and lustrous with an intensity of color not readily achieved with natural aggregates.

White concrete requires particular care to prevent staining. Stains can be caused by form oil, rust from tools and equipment, or from curing compounds. Curing should be done with nonstaining materials, such as waterproof paper overlapped and sealed with nonstaining material.

SECTION THREE
REINFORCED CONCRETE, PRECASTING

CHAPTER 8
REINFORCED CONCRETE

Reinforced concrete is a combination of concrete and steel—a combination which gives the material good properties of both components, the compressive strength of concrete, and the tensile strength of steel.

In construction, steel must be placed where it is needed; otherwise, there will be portions of concrete improperly reinforced, possibly leading to failure. The tensile strength of concrete is only about 10% of its compressive strength. To design the concrete for tension, steel bars are added. Because concrete bonds to steel, and steel and concrete expand and contract to the same degree, the steel and concrete stick together and act together. Deformations on the steel bars increase this adhesion.

Plain bars bond to the concrete in which they are placed only by skin friction. This very limited amount of bond is not strong enough for the reinforcing bars to accomplish their job. Plain bars therefore are not used to reinforce concrete except where a bond is not required, or where bonding is undesirable, such as dowel pins at joints that must be free to move parallel to the dowels. These kinds of joints are commonly used in pavements to allow the slab to expand and contract and still transfer load from one slab to another.

Deformed bars develop bond through both skin friction and the mechanical action of the raised shoulder against the hardened concrete. A study of bond characteristics of deformed bars has shown that the strength of the concrete is more important than the surface condition of the bar as long as the height of deformation and unit weight of the bar meet ASTM standards. This means that rust on the reinforcing bars can be disregarded if it is not flaking.

REINFORCING STEEL—PURPOSE AND LOCATION IN CONCRETE

Concrete is strong in compression and shear, but weak in tension. Whenever large tensile forces are present, and for most structural uses, the concrete is reinforced with steel. In some cases, concrete can accommodate low tensile stresses, and so can be built without bars; it is then known as "plain concrete."

Quality control of both concrete and steel is important to assure the engineer or architect that the specified design requirements are met. The compressive strength of the concrete is checked on almost every important job by taking test samples as the concrete is being mixed.

Standard molds are used that form the concrete into test cylinders 6 in. in diameter and 12 in. high. After 7 or 28 days of curing under job site conditions, these cylinders are placed in a testing machine and crushed in compression while gauge readings are taken to determine the crushing load. For example, where 3000 psi concrete is specified, the cylinders should break at loads of about 90,000 to 100,000 lb after 28 days.

Shear is a more complex force than simple tension or compression. A common example of almost pure shear is illustrated by a bolted lap joint in steel plates, producing what is called single shear. In concrete this is best illustrated by a loaded column bracket.

If a set of books (Fig. 8-3) were carried horizontally, they would have

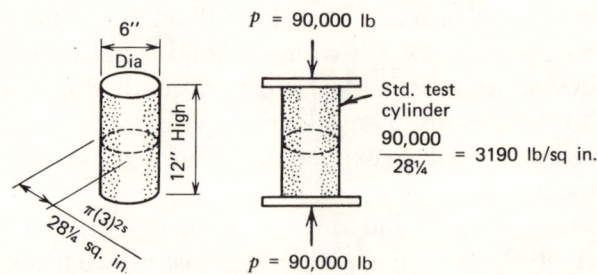

Figure 8–1 Test cylinders of plain concrete.

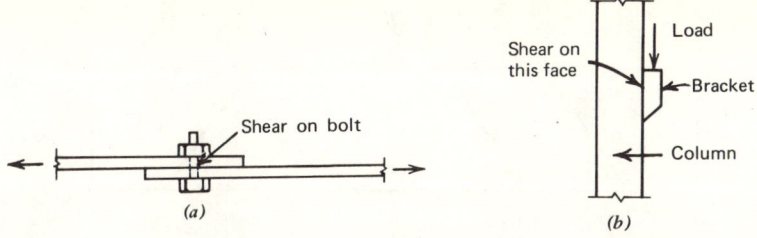

Figure 8–2 (*a*) Bolted joint-single shear. (*b*) Column bracket.

to be squeezed tightly together or they would slip and fall. In a beam, each imaginary vertical slice (analogous to a single book) with a load on top tends to slip down, due to what is called vertical shear.

One may think of a loaded reinforced concrete beam as a truss. High loads cause shear cracks to incline upward from the support. In Figure 8-4, the truss diagonal represents an inclined concrete strut between two inclined shear cracks. The strut is in compression. The vertical truss member represents a stirrup and is in tension.

In a loaded beam or slab, both vertical and horizontal shear are present and the net result of the two forces is called diagonal tension. A crack resulting from these forces would occur near the support and extend upward and outward at an angle of approximately 45°. These forces are also present in a floor around supporting columns. Punching shear is the vertical shear, or the portion of the shear in a vertical plane.

Tension in concrete can be caused by bending or shear as in beams, by drying shrinkage after concrete hardens, and by temperature changes. An example of bending tension is shown below. As a beam is loaded, tension cracks will appear at the bottom of the beam and would develop rapidly in an unreinforced concrete beam with the cracks beginning near midspan. Plain concrete would be brittle and would fail suddenly without warning shortly after the first crack appeared.

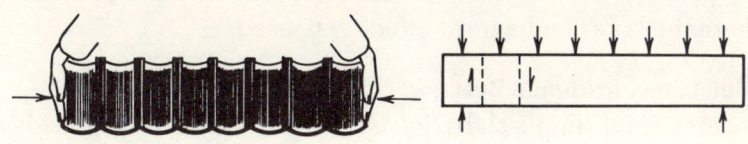

Figure 8–3

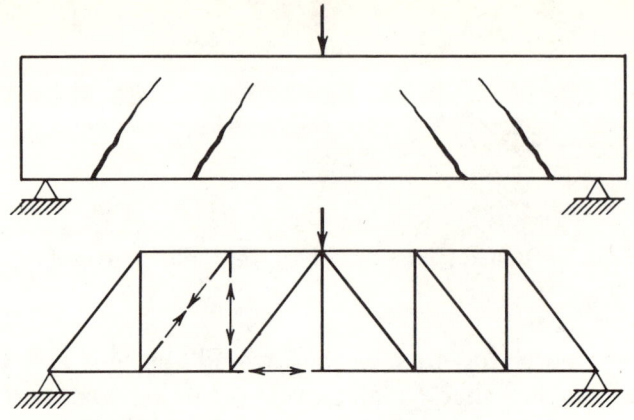

Figure 8–4

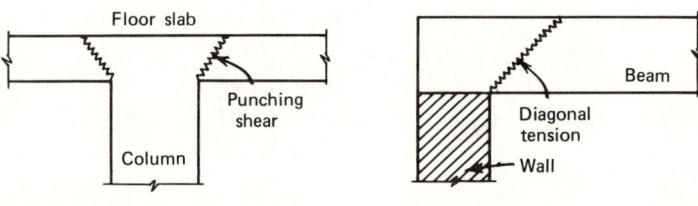

Figure 8–5

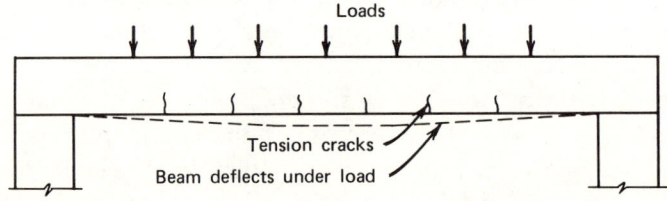

Figure 8–6

Steel provides reinforcement to the concrete where forces exist that the concrete cannot withstand; this is why it is called reinforcing steel. The following paragraphs will illustrate in a general way the location in concrete where the bars can be most effectively used.

Longitudinal Steel in Simple Beams
Simple beams (and simple slabs, joists, and girders) carrying loads assume the shape shown below.

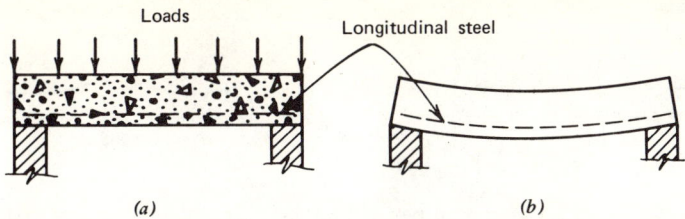

Figure 8-7 (a) Loaded member. (b) Shape it assumes.

The top of the beam is in compression and the bottom is in tension, so longitudinal steel is located near the bottom to resist the tensile stresses. The concrete will crack part way up the beam at midspan where the steel is highly stressed, but the steel will control the cracking tendency.

When a beam deflects under a load, shear stresses are also present. To resist the diagonal tension, small U- or W-shaped bars called stirrups are used and are placed vertically across the beam.

Since shear is usually at a maximum near the support and decreases toward midspan, the stirrups are more closely spaced near the support and spaced increasingly farther apart toward midspan. Since the concrete is able to take some shear, stirrups can often be omitted near midspan.

Steel in Continuous Beams

Continuous beams (or slabs) that extend over more than one span deflect downward between supports and are thrust upward over the supports. This requires tension steel in the *bottom between supports* and in the *top over the supports*.

Cantilever beams (or slabs) loaded at the top deflect as shown below and, in the cantilever (or overhang), tension bars must be placed in the *top*. They must be carried back into the main span. If the cantilever does not extend back to a main span, it must be securely anchored by bending or hooking into an outside beam or column.

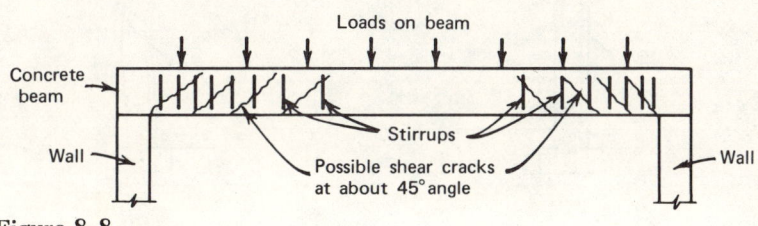

Figure 8-8

REINFORCED CONCRETE

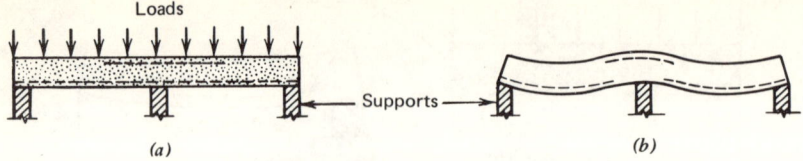

Figure 8-9 (*a*) Loaded member. (*b*) Shape it assumes.

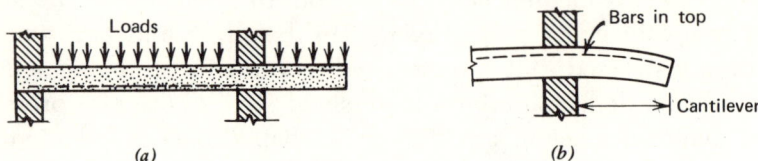

Figure 8-10 (*a*) Loaded member. (*b*) Shape it assumes.

Walls

Walls, braced at the top and bottom by floor slabs and loaded with horizontal pressures on the outside face, deflect inward. This requires bars near the inside face where there is tension. Sometimes, when specified by the engineer, bars may be required around the outside top corner. Basement walls of houses can be built without reinforcing bars.

Cantilever retaining walls deflect as shown and require main bars on the side toward the earth. Bars should be placed where called for on the placing plans. The proper location of bars in such structures is very important.

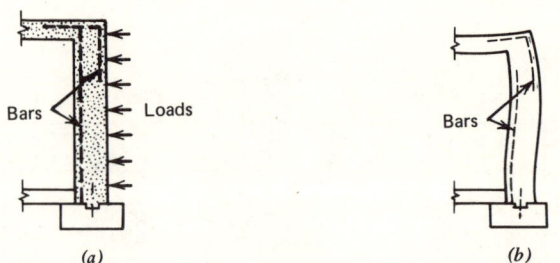

Figure 8-11 (*a*) Loaded member. (*b*) Shape it assumes.

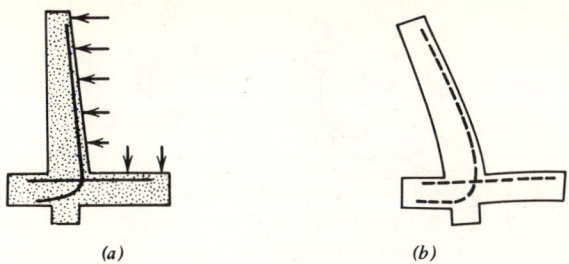

Figure 8-12 (a) Loaded member. (b) Shape it assumes.

Footings

Combined footings carry column loads at two or more points, supported by a uniform soil pressure upward over the entire area between columns and at the projecting end. The forces acting on a footing are roughly in a reverse direction to the forces acting on a normal floor beam or slab. For this reason, these footings need the bars placed in an inverted position as illustrated by straight bars (A) and truss bars (B) in Figure 8-13. Truss bars will be more fully covered later.

Bottom cross bars (C) and (D) are also provided to take care of the curling effect. Sometimes stirrups or ties are used in combined footings and, when used, they are placed in the inverted position (hooks down) or are completely closed on four sides and hooked at a corner.

Square or rectangular footings have a concentrated column load downward in the center and soil pressure upward all over the base. They tend to curl up in all directions toward the corners. Bars are placed in two directions at right angles to each other and located a prescribed minimum distance from the bottom of the footing.

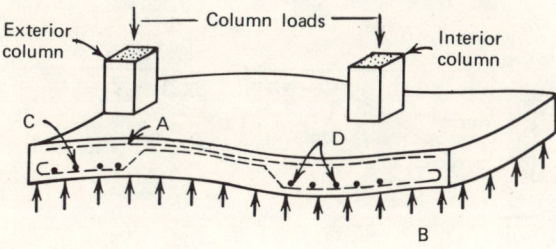

Figure 8-13

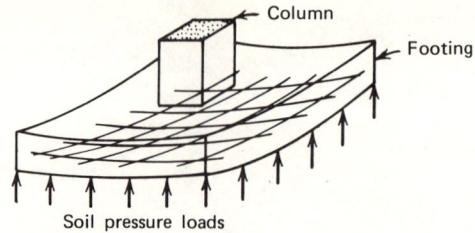

Figure 8–14

Reentrant or Inside Corners

This condition is common at the intersection of two walls or grade beams or at the intersection of a wall with a beam or slab. Bars are in tension and should never be bent around an inside corner, since they would tend to pull straight and could break out the concrete. Instead, they should extend past the corner from each direction to the far face and be hooked if necessary for sufficient anchorage. (See Fig. 8-15). This bar arrangement is shown by the engineer on his structural drawings and by the fabricator on the placing plans for the ironworker's use.

For stair landings, as for inside corners, a similar condition exists. The illustration below shows the wrong and the right way. The bars are in tension and should continue across this point and be bent into the stair and landing slabs as shown.

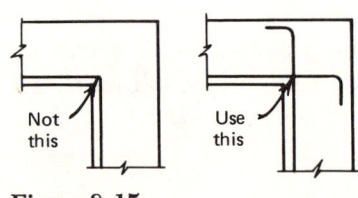

Figure 8–15

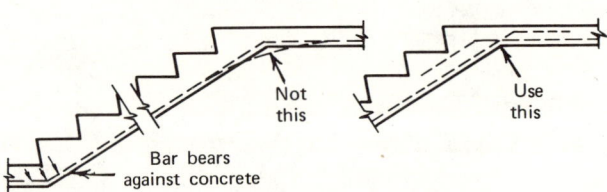

Figure 8–16 Stair landing steel placement.

136   REINFORCED CONCRETE, PRECASTING

Shrinkage and Temperature Tension

Much concrete is lightly reinforced to resist drying shrinkage and to hold tight the cracks caused by seasonal temperature changes. The best example of this is a concrete highway. In continuously reinforced concrete pavement, longitudinal bars eliminate all need for cross-joints. Initial drying shrinkage, cold or cool nights, and the first winter temperature cause cracking about every 6 ft. The bars hold these cracks so tightly closed, in summer and winter, that the pavement behaves essentially like an unbroken slab.

Roof slabs of buildings are subject to wide temperature changes as well as concrete shrinkage. Floor slabs, although not usually subjected to wide temperature changes, are subject to shrinkage. It is general practice to provide reinforcement in structural slabs, at right angles to the main stress-carrying steel, to offset these effects. This reinforcement is referred to as temperature steel and may consist of either tied bars or welded wire fabric. This steel also serves to distribute concentrated loads.

In reinforced concrete walls, temperature bars are used at right angles to the main steel for the same purpose. Often, in wall construction, temperature bars are placed both horizontally and vertically, even though that face of the wall does not carry any tension forces other than temperature changes.

Compression Reinforcement—Columns

While the main purpose of steel in concrete is to resist tension forces, it can also be used advantageously to resist compression forces.

The most common use of steel for compression forces is in columns. If concrete alone were used, the size of column required would be so large as to be impractical and would add too much to the weight of a structure and crowd the floor. Since steel bars are about 20 times stronger in compression than an equivalent area of concrete, they are used to carry part of the column load. The concrete and the steel work together and the resulting column is much smaller in size and lighter in weight.

Column vertical bars are in compression, and if not restrained in some manner, the compression load would allow the bars to buckle and break out of the concrete. Column ties are spaced to prevent this. They are in tension, holding the vertical bars in position. Column spirals act in the same manner as column ties, but in addition they restrain the concrete somewhat like a stove pipe restrains a column of sand inside it. Spiral turns are more closely spaced to accomplish this purpose. A spiral rod should never be cut without splicing as directed by the engineer.

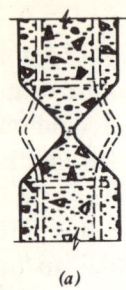 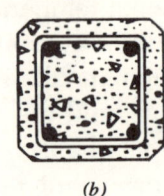 

(a)   (b)   (c)

Figure 8–17   (a) Buckled column bars. (b) Column ties. (c) Column spirals.

Compression Reinforcement—Beams

In some beams, where compression stresses are high or where the size needs to be restricted, it may be necessary to use bars in the top of the beam. These bars take the compressive load from some of the concrete, very much the same as the column vertical bars. When used in this manner, the bars are always restrained by closed stirrups, which may be in one or two pieces as shown.

In continuous beams (those which have more than two supports) the horizontal compression loading in concrete is high next to the support. For this reason, bottom bars are sometimes extended and lapped with those of an adjacent span.

REINFORCING BARS

Grades of bars

Bars are furnished in several grades, according to the engineer's requirement. Each grade has different strength properties. There are parts of a job where one grade may be used to better advantage than another, and the grades are identified to assure proper use where each is required.

Each grade has what is known as yield strength and ultimate strength. Yield strength is the load beyond which the steel will stretch and will not return to its original length, much the same as overstretching a spring. Ultimate strength is the load limit at which the steel breaks.

Reinforcing bars are furnished in several grades that vary in yield strength, ultimate strength, percentage of elongation, bend test requirements, and chemical composition. The grade of steel is an important factor in the design of concrete members. The engineer will state in his specifications what grades he wants furnished for the various parts of the reinforced concrete structure.

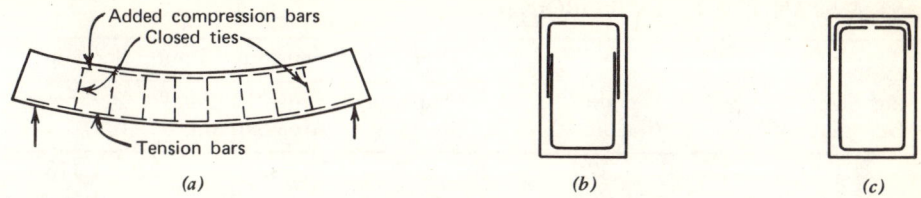

Figure 8-18 (a) Deflection of a loaded member. (b) Two piece tie. (c) Cap stirrup.

To obtain uniformity throughout the United States, the American Society for Testing and Materials (ASTM) has prepared standard specifications for these steels. They are:

1. A615-billet steel bars for concrete reinforcement (rolled from new billets).
   a. Grade 40 (40,000 psi minimum yield strength)
   b. Grade 60 (60,000 psi minimum yield strength)
   c. Grade 75 (75,000 psi minimum yield strength)
2. A616-rail steel bars for concrete reinforcement (rolled from used railroad track).
   a. Grade 50
   b. Grade 60
3. A617-axle steel bars for concrete reinforcement (rolled from used freight car axles)
   a. Grade 40
   b. Grade 60

Bar Sizes

Deformed bars are always designated by number. There are eleven standard sizes—No. 3 to No. 11, and No. 14 and No. 18. The number denotes the nominal diameter of the bar in eighths of an inch. For example, a No. 5 bar has an approximate diameter of ⅝ in.; a No. 9 bar, 1⅛ in. The nominal dimensions of a deformed bar (not including deformations) are equivalent to those of a plain round bar having the same weight per foot as the deformed bar. The approximate diameter may be helpful in identifying a bar size when the tag is missing or the roll marking is not clear.

Bar Identification

With the various grades and sizes available, it was found necessary to provide some means of easy identification. ASTM specifications require that each bar producer shall roll onto the bar (a) a letter or symbol to

Table 8–A
ASTM STANDARD REINFORCING BARS

Bar Size Designation	Weight (lbs per ft)	Nominal Dimensions—Round Sections		
		Diameter (in.)	Cross Sectional Area (sq. in.)	Perimeter (in.)
#3	.376	.375	.11	1.178
#4	.668	.500	.20	1.571
#5	1.043	.625	.31	1.963
#6	1.502	.750	.44	2.356
#7	2.044	.875	.60	2.749
#8	2.670	1.000	.79	3.142
#9	3.400	1.128	1.00	3.544
#10	4.303	1.270	1.27	3.990
#11	5.313	1.410	1.56	4.430
#14	7.65	1.693	2.25	5.32
#18	13.60	2.257	4.00	7.09

show the producer's mill, (*b*) a number corresponding to the size number of the bar, (*c*) a symbol or marking to indicate the type of steel, and (*d*) a marking as shown below to designate the grade.

Fabrication

Reinforcing bars, whether plain or deformed, are furnished by the mill to the fabricator in mill lengths and by the fabricator to the job, either (1) cut-to-length or (2) cut-and-bent. Fabricated bars are furnished to the job by the supplier in one of these two ways depending on local practice. Either way, the supplier shop-details all bars, assigns them reference

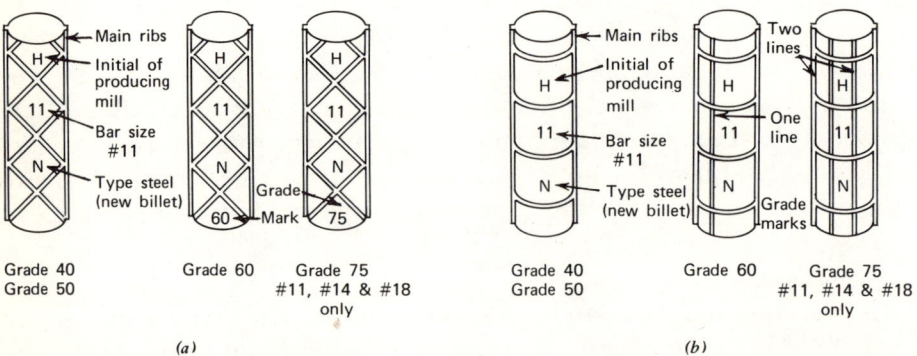

Figure 8–19 (*a*) Number system-grade marks. (*b*) Continuous line system-grade marks.

marks, makes list, cuts to certain tolerances, bundles, tags, and delivers the bars to the job site complete with approved detailed placing plans, material list, and bending diagrams. The two ways differ only in the amount of fabrication before delivery: (1) cut-to-length bars are straight length cut according to a list, each size and length bundled separately; and (2) cut-and-bent bars are similar except that, in addition to the foregoing, the supplier does the bending, so that the bars are completely fabricated and ready to set.

Stocking Steel on the Job
Steel should be stockpiled on the job so that it is separated by size and length of bar. If the bars are prebent, each shape should be stored together. All steel should be kept out of mud and water. This is usually done by placing heavy timbers on the ground and putting the steel on top of the timbers.

Stockpiling of steel follows the same common-sense rules as good housekeeping. Steel should be placed so that the size and length of the bar wanted can be quickly found, easily removed from the pile, and readily transported to the job.

Trusses, Stirrups, and Ties
Truss bars may be called for in beams, joists, slabs, girders, and occasionally in mats or walls and may be bent as shown in the sketch (with or without end hooks). They are usually placed with the low center part in the bottom of the member between supports, and with the high ends over supports and into the next span.

Stirrups may be called for in beams, girders, and some footings and are placed in variously spaced groups in a vertical position across the beam near each end support. They are usually bent with a 90° hook rather than the 135° hook shown here.

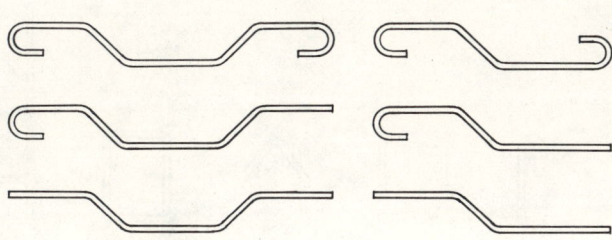

Figure 8–20   Various types of truss bars.

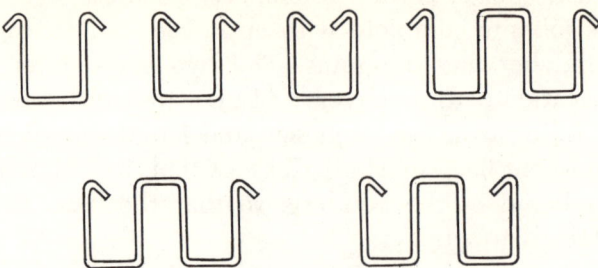

Figure 8–21   Various types of stirrups.

Ties are used in tied columns and in beams or girders where compression (top) steel is used. They are bent as shown and are ordinarily of small size, No. 3 or possibly No. 4, and they are wrapped around a group of bars to form a cage. This assembling is usually done at the site before placing steel.

Spirals
Used in spirally reinforced columns, piers, and piles, they are made of heavy wire bent into the form of a coil spring to enclose the vertical bars for assembly into a cylindrical cage. Spirals usually arrive on the job preformed and are delivered in a flattened or collapsed state with spacers to hold them together. They are generally assembled on the job with the vertical bars and placed as a unit.

Radial Bending
For curved surfaces such as tanks, chimneys, silos, or covers, bars are bent into an arc. When bars are supplied completely fabricated, this radial bending is done by the fabricator. If the bar is bent only a little—for

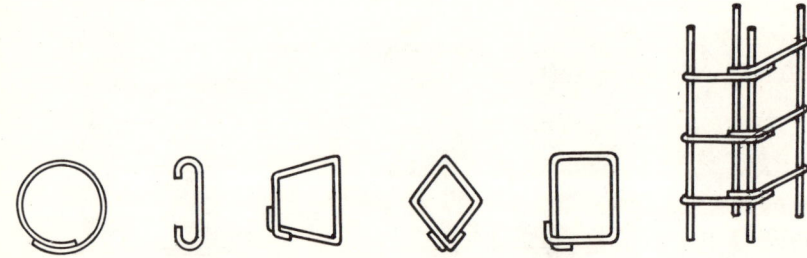

Figure 8–22   Various types of ties.

142   REINFORCED CONCRETE, PRECASTING

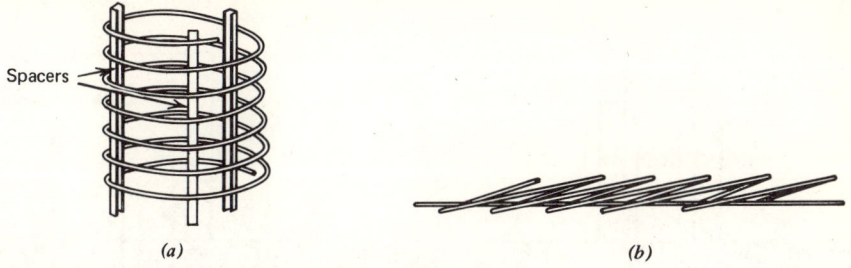

Figure 8–23 Spirals. (*a*) Expanded. (*b*) Collapsed.

example, on a 25-ft radius for a No. 5 bar—it will be delivered straight and sprung into place on the job.

Off-Set Column Bars

Off-set column bars are bent so that the upper part that projects above the floor will come inside of the reinforcing cage of the column above. Such prefabrication is done when the off-set is 3 in. or less and the bend is a flat one sloping about 2 in. per ft with the top of the bend just below the top of the slab. When the off-set exceeds about 3 in., the lower verticals stop below the floor and separate dowels are used extending 20 bar diameters (or the distance specified on the placing drawings) into the column below and the same distance into the column above.

Hooks

Hooks are used for the anchorage of bars and concrete where there is no possibility of extending the bar to develop anchorage by bond alone.

Types of Bends

Type 3 bend, for example, describes a truss bar with a circular hook at each end. Since these types are standard nationwide, the fabricator will know exactly what is desired. When no value is given for a dimension, that item will be omitted in bending the bar. For example, if distance G is omitted, there would be no hook on that end. How and where each of these shapes is used in reinforced concrete will be discussed later.

PLACING DRAWINGS AND BAR LISTS

Drawings are a very important part of any building project. They include all phases of the work necessary for the finished building. In general they are:

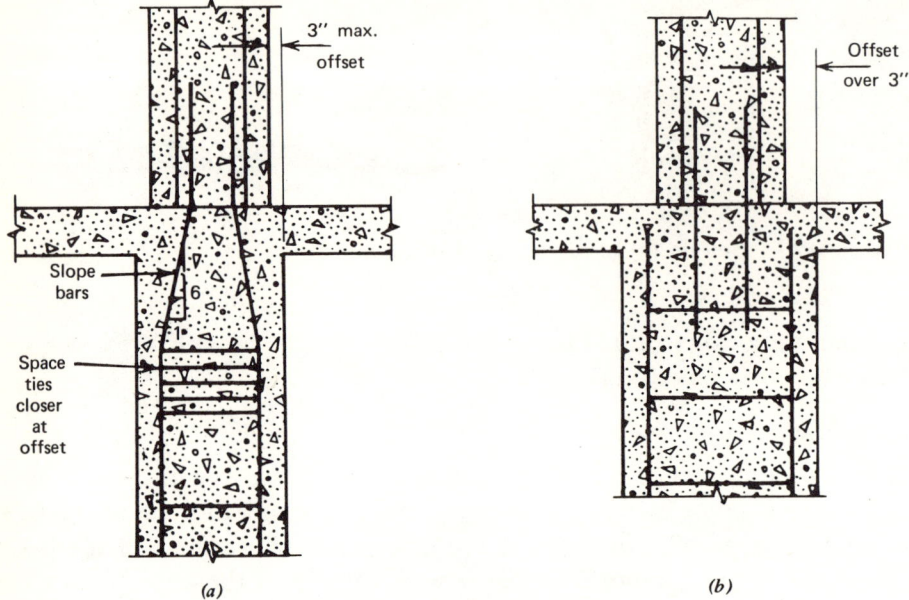

Figure 8-24 (a) Off-set column bars. (b) Separate dowels.

1. Site drawings that show the location of the building on the property; utility lines; drainage and drainage structures; outside walks, drives, and curbs; and elevations of natural ground and finished grades.
2. Structural drawings that show all plans, details, and elevations necessary to construct the building frame, with complete working dimensions and elevations.
3. Architectural drawings that show general outlines of the finished appearance of the building in elevation; plans and sections completely dimensioned; the relationship of the various kinds of materals, such as concrete, steel, brick, and wood and stone, to each other; plans showing room arrangements with sections and elevations to illustrate details; finishes such as plaster or tile walls; and ceilings, floor surfaces, and fixtures; for example. In short, architectural drawings are those on which all other drawings, such as structural, mechanical, and electrical, are based.
4. Mechanical drawings that show generally all piping; heating and air conditioning ducts; and mechanical equipment.
5. Electrical drawings that show all electric wiring, control panels, motors, and electric pumps, for example.

Placing drawings do not as a rule show complete building dimensions since all of the framework is presumed to be in place before the ironworker starts. Where necessary, structural and architectural drawings should be consulted for dimensions and for any other information that might be needed to supplement the placing drawings.

The bar fabricator uses the structural and architectural plans to prepare bar-placing drawings. At times it may be necessary to refer to other drawings for additional information.

A bar-placing drawing contains only the information required by the fabricator to prepare bar lists and bending details and by the ironworker to put the bars in place on the job. Placing drawings do not as a rule show dimensions.

An example of a bar-placing plan is shown in Figure 8-27. It is a portion of a plan for concrete joist construction.

The exterior solid and parallel dashed lines on the plan show the floor bearing on a wall. The supporting columns are indicated by the small rectangles, each numbered. The beams are shown by dashed parallel lines running from column to column (shown solid at the stair opening along the exterior wall). Each beam is given its reference number, prefixed 1B for first-floor beams, followed by a serial number.

The joist forms are shown in solid lines. (It is standard practice to show them this way rather than dashed as for beams.) The joists are also given reference numbers, prefixed 1J for first-floor joists, followed by serial numbers. The metal forms are shown tapered near the supports, although they are sometimes square. Holes and openings are indicated by rectangles of heavy lines crossed with light diagonal lines to indicate that there is nothing in that space. Midway between the tapered ends of the forms in the two longer spans, there are double horizontal lines showing a distribution rib.

Beams and joists are numbered as described. The bars to be placed in them will be found in Table 8-B, a typical joist schedule. It is more convenient to show bar placing for beams and joists in this way since space does not permit showing all information directly on the floor plan. Bars that cannot be readily scheduled, such as temperature, distribution rib, and slab bars, are shown directly on the plan. A placing drawing shows typical details and cross sections to properly locate the bars, as indicated in Figure 8-28.

Although the detailer will put as much information as possible on plans and in schedules, many conditions occur that cannot be clearly shown in this manner, and sectional details such as shown below will be added. On the plan there will generally be symbols like that shown in Figure 8-27.

REINFORCED CONCRETE 145

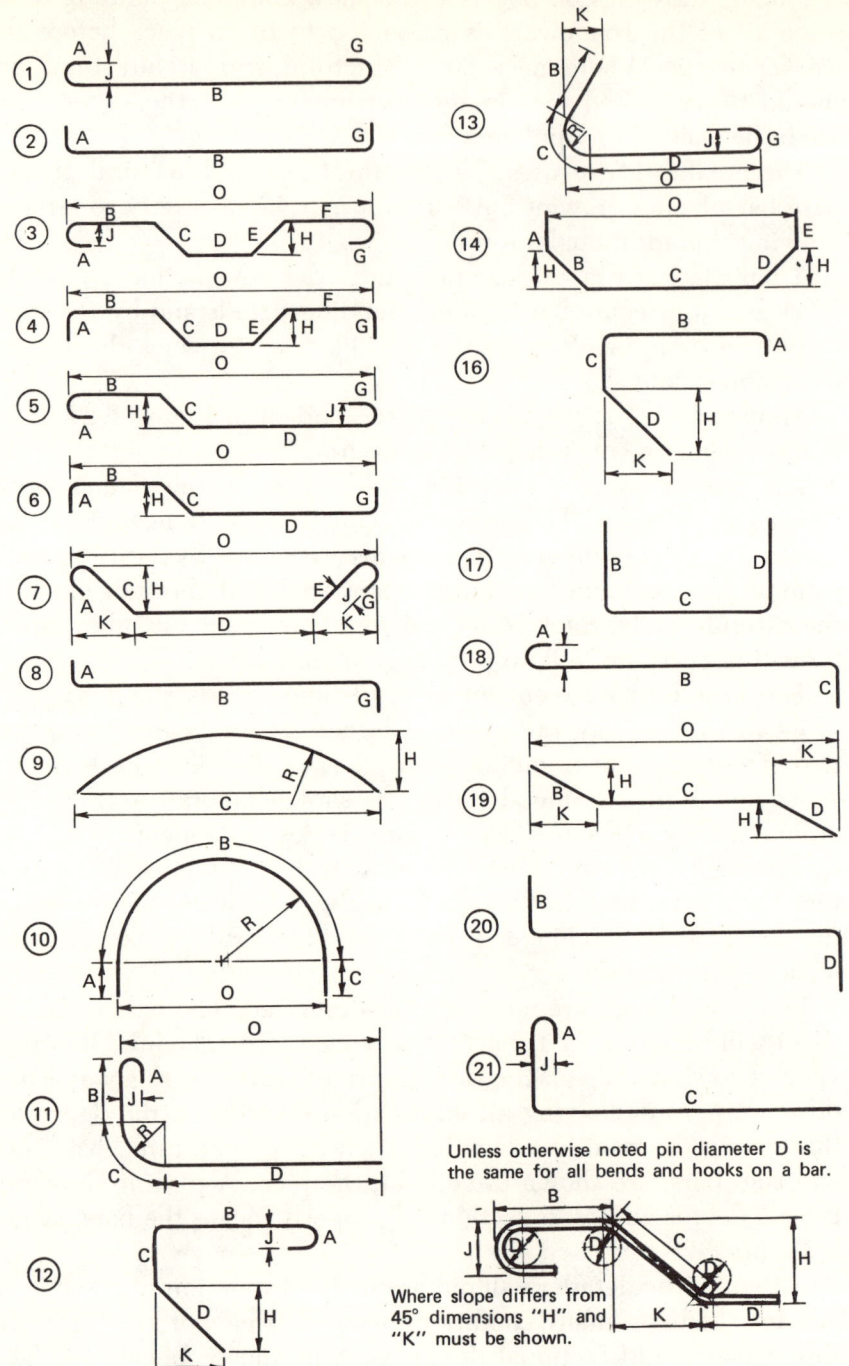

146 REINFORCED CONCRETE, PRECASTING

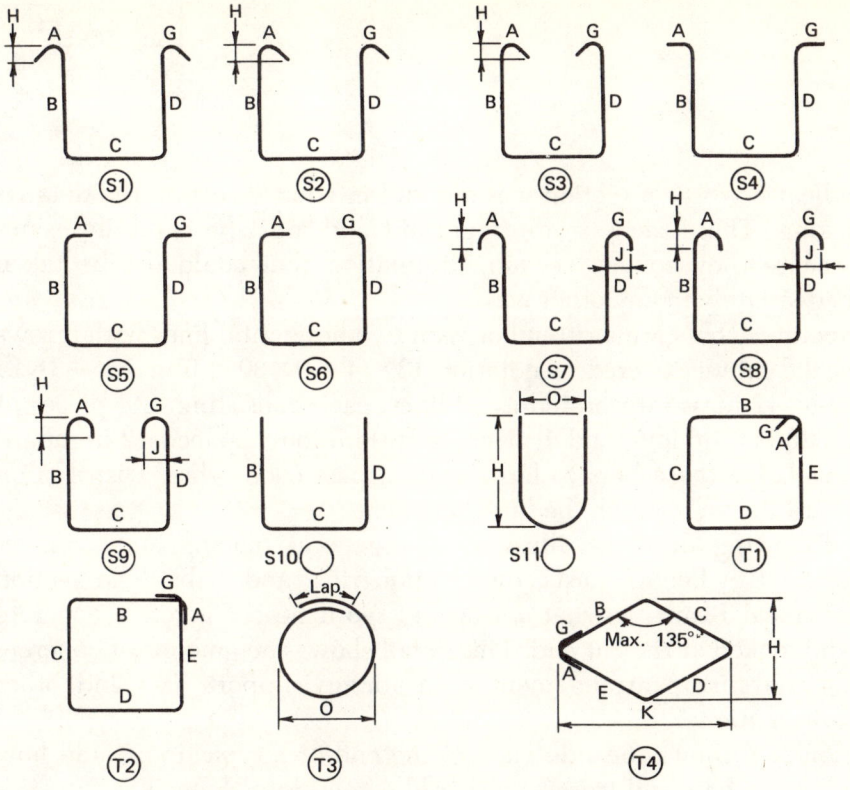

Notes: 1. All dimensions are out to out of bar except "A" and "G" on standard 180° and 135° hooks.
2. "J" dimension on 180° hooks to be shown only where necessary to restrict hook size, otherwise standard hooks are to be used.
3. Where "J" is now shown, "J" will be kept equal to or less than "H". Where "J" can exceed "H", it should be shown.
4. "H" dimension on stirrups to be shown where necessary to fit within concrete.
5. Where bars are to be bent more accurately than standard bending tolerances, bending dimensions which require closer working should have limits indicated.
6. Figures in circles show types.

Figure 8–25 Typical bar bends.

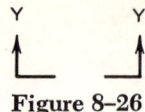
Figure 8-26

This indicates where a section was cut and in what direction the detailer was looking. These sections are important and must be carefully examined because they convey certain information that could not be taken care of adequately in any other way.

Temperature bars are written between two horizontal lines with arrows to show the extent covered. A notation 13×(4-#3×30 ft-0 in. + 1-#3×14 ft-0 in.) @ 12 in. is interpreted as 13 lines, each consisting of 4 pieces of #3 bars 30 ft-0 in. long and 1 piece 14 ft-0 in. long, spaced 12 in. apart. A note calls for these bars to lap 1 ft-0 in. past each other. Distribution rib bars are shown in much the same manner.

The following drawings illustrate the general placing and arrangement of bars in beams, joists, distribution ribs, and slabs. The section titled "Typical Beam" is what one would see if he cut across a concrete beam and looked at the cut end. This detail shows the amount of concrete covering the stirrup arrangement with stirrup support bars and other useful information.

Section X-X shows the side view of the end of a typical joist and how far the bottom bars and truss bars should extend into the wall.

Section Y-Y is a cut across the exterior wall and several joists. The position of the straight and the truss bar in a joist is shown. The vertical dotted line is the bent-up portion of the truss bar. This section also shows the clearance under the bars and the location of temperature and tie bars.

A schedule is a useful method for presenting details of bars for a group of similar items, such as a group of beams, joists, and columns. The bars can be described more clearly and compactly in a schedule. The schedule is used in connection with typical details and sections to be sure bars are correctly positioned in the forms. Schedules can be arranged in different forms, according to the individual detailer.

Following across the top line for joist 1J1, the column headed "No." (for number alike) means that there are 20 such joists on this floor. "B" gives the width and "D" the total depth from the bottom of the joist to the top of the slab. Both of these dimensions are always given in inches.

Under the heading "Straight," the three columns headed "No.," "Size," "Length" show one piece of #7 bar 24 ft-0 in. long.

Under the heading "Bent," the five columns headed "No.," "Size,"

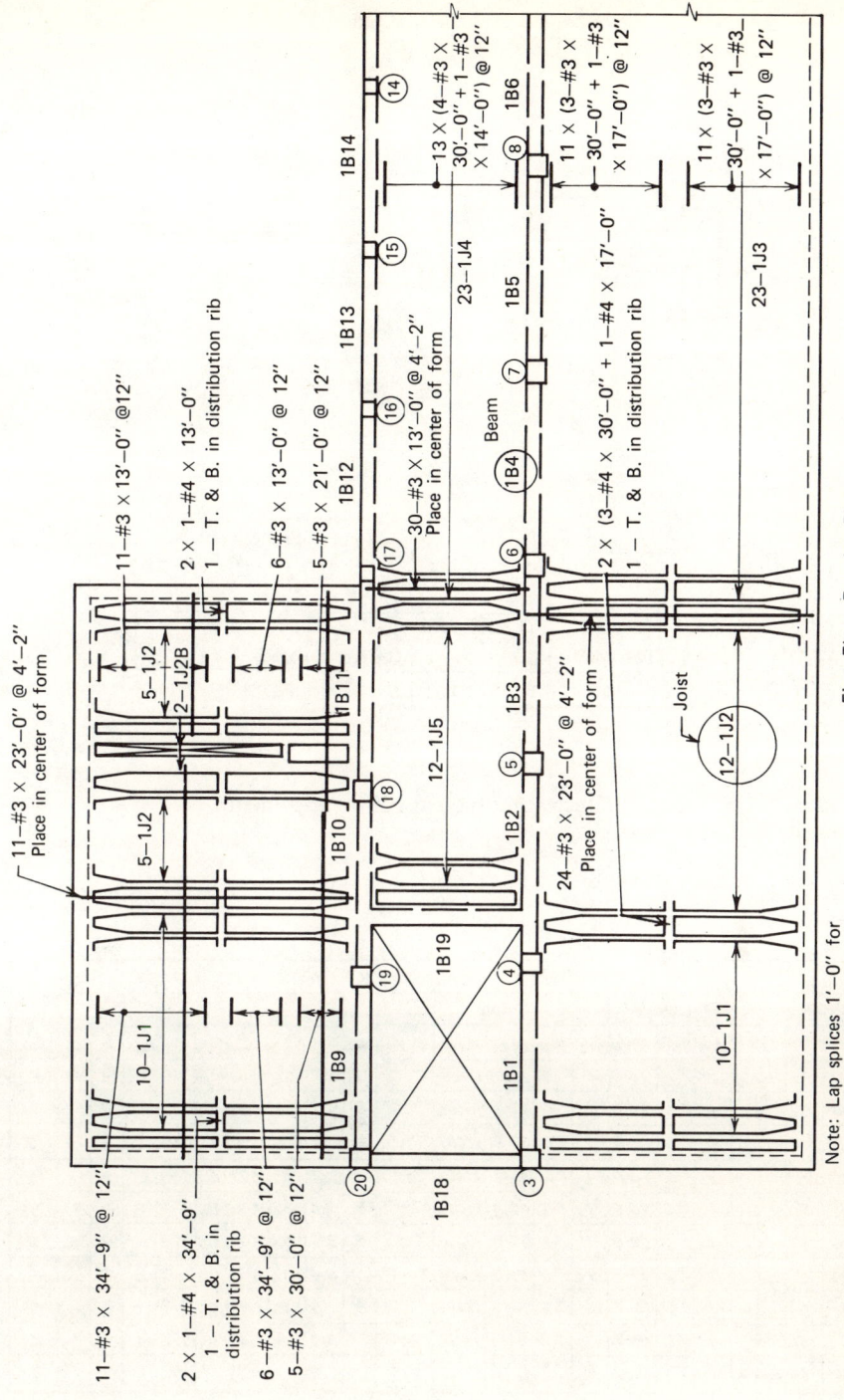

Figure 8-27 Typical placing drawing.

REINFORCED CONCRETE 149

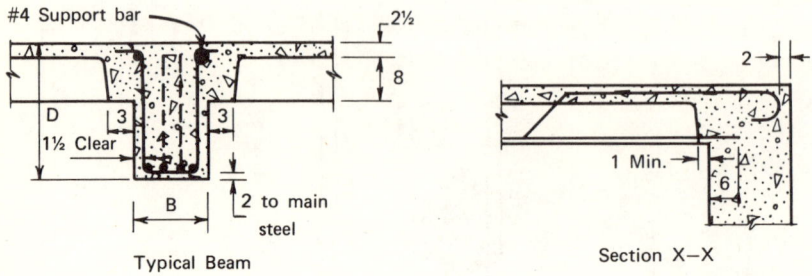

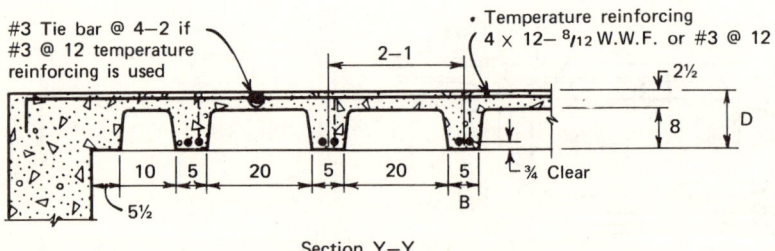

Figure 8-28 Section illustrating typical bar arrangement.

Table 8-B

	JOIST			STRAIGHT			BENT											3/4
MARK	No	B	D	No	SIZE	LGTH	No	SIZE	LGTH	MARK	TYPE	Hook		H		Hook	H	O	CHAIRS
1J1	20	5	8+2½	1	#7	24-0	1	#6	26-10	1J600	∼	8	3-7	1-0½	16-6	1-0½	3-4	8	9	24-11	6
1J2	58	5		1	#6	24-0	1	#6	45-7	1J603	∼	8	3-7	1-0½	15-2	1-0½	24-1		9		6
1J2A	2	Var		2	#6	24-0	2	#6	32-2	1J607	∼	8	3-7	1-0½	15-2	1-0½	10-8		9		12
1J2B	2	Var		2	#6	24-0	1	#6	45-7	1J603	∼	8	3-7	1-0½	15-2	1-0½	24-1		9		12
1J2C	1	5		1	#6	20-2	1	#6	22-8	1J606	∼	8	2-6	1-0½	13-8	1-0½	3-1	8	9	20-9	5
1J3	26	5		1	#6	24-0	1	#6	32-2	1J607	∼	8	3-7	1-0½	15-2	1-0½	10-8		9		6
1J4	23	5		1	#6	13-9	1	#6	21-4	1J604	∼		8-4	1-0½	8-4	1-0½	1-11	8	9		4
1J4A	1	5		1	#6	13-9	1	#6	16-4	1J605	∼	8	1-11	1-0½	9-1	1-0½	2-7		9		4
1J4B	5	5		1	#6	13-9	1	#6	21-4	1J601	∼		8-4	1-0½	8-4	1-0½	2-7		9		4
1J5	31	5	8+2½	1	#5	13-9															4

"Length," "Mark," and "Type" show one piece of #6 bar 26 ft-10 in. long, marked 1J600, and a sketch of the bent bar. In this schedule the marks 600 and 603, for example, signify #6 bars, but they could have been marked 1J1 and 1J2, for example, to signify the joist number. The shape of the bent bar, which is being shown in the schedule, may help the ironworker in locating the proper bar from a stockpile.

The next group of columns, with the sketch of a truss bar across the group, shows under each respective part of the sketch that there is an 8 in., 180° hook followed by a 3 ft-7 in. horizontal on top, then a downward slant of 12½ in. (1 ft-0-½ in.), a horizontal bottom dimension of 16 ft-6 in., an upward slant of 12½ in., and a horizontal top dimension of 3 ft-4 in., and closed with an 8 in., 180° hook. Column "H" shows that the height, out-to-out, is 9 in.; column "O" shows that the overall length (out-to-out of the bar) is 24 ft-11 in.

Beams are detailed in the same manner in a beam schedule that shows stirrup shape and spacing as well as the size and number of support bars.

Bar Lists

A bar list is just what the name implies—a list of reinforcing bars making up a bill of materials. It is prepared by the fabricator to cover a particular part of the structure. Bars are classified as to straight, heavy bending, and light bending and are grouped by sizes and lengths. The largest size is grouped first, then the next smaller size, and so on. Each size has the longest length listed first, graduating down to the shortest.

An explanation of the different bending classification might be of interest to the ironworker and inspector.

1. *Heavy bending.* Includes bar sizes #4 through #11, which are bent at not more than six points, radius bent to one radius, and bending not otherwise defined.
2. *Light bending.* Includes all #3 bars and all stirrups and column ties; and all bars #4 through #11 that are bent at more than six points, bent in more than one plane, radius bent with more than one radius in any one bar, or a combination of radius and other bending (radius bending being defined as all bends having a radius of 10 in. or more to inside of bar).
3. All bending of #14 and #18 bars is considered special fabrication.

A bar list usually includes the bend types and the bending dimensions for all bent bars. Special bending diagrams are added where the standard bend types do not apply.

ABC STEEL PRODUCTS CO.
CHICAGO, ILL.

Project Blue Warehouse Add'n.
Customer Jones Const. Co.
Location Jonesville, Illinois
Mat'l. For 1st Floor Beams & Cols.

Grade As noted
Order No. 4838
Drg. No. 23
Sheet 1 of 2
Date 2-1-65 Rev. 9-21-67
Made By D.R.S. Ch'k By C.R.W.

For typical bend types refer to _____

Item	Grade	No. Pieces	Size	Length	Mark	Type	A	B	C	D	E	F	G	H	J	K	R	O
1	STRAIGHT																	
2	60	4	7	22-0														
3	60	4	7	17-6														
4																		
5	60	2	5	28-3														
6	60	2	5	17-6														
7																		
8	HEAVY BENDING																	
9	60	2	9	38-6	IB901	3	1-3	10-0	2-3	12-4	2-3	9-2	1-3	1-7			34-8	
10	60	2	9	35-7	IB902	3		9-2	2-3	12-9	2-3	9-2		1-7				
11																		
12	60	2	8	23-6	IB801	1	11	22-7										
13																		
14	60	2	7	25-2	IB703	3	10	2-3	2-8½	9-10	2-8½	6-10		1-11				
15																		
16	60	2	6	26-2	IB601	3	8	5-7	2-7	8-6	2-7	5-7	8	1-10			23-4	
17																		
18	LIGHT BENDING																	
19	60	22	4	5-4	S401	S4	3½	1-11	11	1-11			3½					
20	60	34	4	5-0	S402	S5	3½	1-9	11	1-9			3½					
21																		
22	60	26	3	6-2	S301	S4	3	2-6	8	2-6			3					
23	60	24	3	5-10	S302	T2	3	2-0	8	2-0	8		3					
24																		
25																		
26																		

SPIRALS (COLD DRAWN WIRE)

	Grade	No. pcs	Size	Length	Mark	Diam.	Pitch	Turns	Spacs.
1	A82	2	½	10-6	C10	22	2	66	3
2	A82	2	½	10-0	C11	20	2½	53	2
3									
4	A82	2	⅜	10-0	C15	16	1½	83	2
5	A82	2	⅜	10-0	C16	16	1¾	72	2

Figure 8-29 Sample bar list.

The heading of each bar list generally includes the following information.

1. Name of project.
2. Customer name.
3. Job location.
4. Part of job.
5. Order number (helpful in relating the bar list items to those shown on truck bills).
6. Placing drawing reference number.
7. Grades of steel.

The bar list (Fig. 8-29) shown is only an example, since each bar fabricator has his own form of list. Sometimes, one form is used for straight bars and another for bent bars.

Bar lists serve several purposes. The fabricator uses them for shearing, bending, tagging, shipping, and invoicing. The ironworker foreman and his crew use them for checking quantities on each delivery, sorting bars in the storage area, hoisting of proper bars to the placing area, and for the actual placing of the bars in the forms. The inspector finds these bar lists useful in his checking of grades of steel, lengths, and bending. The column "O" is always used when there are hooks on both ends or when it is necessary to restrict the bar and keep this dimension within prescribed tolerance. It is not used when the out-to-out dimension does not have to fit into a restricted position.

Placing drawings and bar lists are being prepared by computers. The EDP (Electronic Data Processing) procedure saves a great deal of time and manpower as well as being neater and more accurate.

Each fabricator uses his own system, but the differences are usually just in details. The basic system, whether done by hand or performed by a computer, is the same throughout the industry.

BAR SUPPORTS

A bar support is used to firmly hold the bars at the required clearance from the forms before and during the placing of concrete. It must be sufficiently strong and properly spaced to provide this support under normal construction conditions. Bar supports may consist of heavy steel wire or precast concrete blocks.

Wire bar supports are classified according to the degree of protection they provide to minimize rust spots or blemishes on the surface of the concrete. The five classes and their intended degree of protection are:

Class A. Bright Basic—which has no protection against rusting.

Class B. Pregalvanized—which has minimal protection against rusting.

Class C. Plastic Protected—for use in situations of moderate exposure or situations requiring light grinding or sandblasting of the concrete surface.

Class D. Stainless Steel Protected—also for use in situations of moderate exposure or situations requiring light grinding or sandblasting of the concrete surface.

Class E. Special Stainless—for use in situations of moderately severe exposure or situations requiring heavy grinding or severe sandblasting of the concrete surface.

Sometimes several classes of bar supports are used on the same job, and the class for each part of the job must be used as directed on the placing drawing.

Figure 8-30 contains a general description of the available types of standard bar supports.

TOLERANCES IN PLACEMENT

It is important that bars be placed and held in position as shown on the placing drawings. The strength of any concrete member can be affected by the improper positioning of the reinforcing bars. For example, lowering the top bars or raising the bottom bars by ½ in. more than specified in a 6-in. deep slab could reduce its load-carrying capacity by 20%.

A tolerance of ±¼ in. is a generally accepted practice in the positioning of bars above forms in the bottom of beams, joists, and slabs. Wire bar supports, spaced according to the Concrete Reinforcing Steel Institute's *Specifications for Placing Bar Supports*, should be sufficient to locate the bars where they are intended to be.

The exact spacing of lengthwise bars is generally not as critical to the strength of the structure; therefore, the placing tolerance is not as close as that shown for cover. Where bars are hooked at one noncontinuous end only, the tolerance in placing should be +0, −½ in.; where hooked at both ends, it should be ±½ in.

The same tolerances apply to truss bars at the hooked end. At ends that extend into an adjacent span, there is no overtolerance needed, but a

ABC STEEL PRODUCTS CO.
CHICAGO, ILL.

Grade As noted
Order No. 4838
Drg. No. 23
Sheet 1 of 2
Date 2-1-65 Rev. 9-21-67
Made By D.R.S. Ch'k By C.R.W.

Project Blue Warehouse Add'n.
Customer Jones Const. C˚
Location Jonesville, Illinois
Mat'l. For 1st Floor Beams & Cols.

For typical bend types refer to

Item	Grade	No. Pieces	Size	Length	Mark	Type	A	B	C	D	E	F	G	H	J	K	R	O
1			STRAIGHT															
2	60	4	7	22-0														
3	60	4	7	17-6														
4																		
5	60	2	5	28-3														
6	60	2	5	17-6														
7			HEAVY BENDING															
8																		
9	60	2	9	38-6	IB901	3	1- 3	10- 0	2- 3	12- 4	2- 3	9- 2	1- 3	1- 7				34- 8
10	60	2	9	35-7	IB902	3		9- 2	2- 3	12- 9	2- 3	9- 2		1- 7				
11																		
12	60	2	8	23-6	IB801	1	11	22- 7										
13																		
14	60	2	7	25-2	IB703	3	10	2- 3	2-8½	9-10	2-8½	6-10		1-11				
15																		
16	60	2	6	26-2	IB601	3	8	5- 7	2- 7	8- 6	2- 7	5- 7	8	1-10				23- 4
17			LIGHT BENDING															
18																		
19	60	22	4	5-4	S401	S4	3½	1-11	11	1-11		3½						
20	60	34	4	5-0	S402	S5	3½	1- 9	11	1- 9		3½						
21																		
22	60	26	3	6-2	S301	S4	3	2- 6	8	2- 6		3						
23	60	24	3	5-10	S302	T2	3	2- 0	8	2- 0	8	3						
24																		
25																		
26																		

SPIRALS (COLD DRAWN WIRE)

	Grade	No. pcs.	Size	Length	Mark	Diam.	Pitch	Turns	Spaces
1	A82	2	½	10-6	C10	22	2	66	3
2	A82	2	½	10-0	C11	20	2½	53	2
3										
4	A82	2	⅜	10-0	C15	16	1½	83	2
5	A82	2	⅜	10-0	C16	16	1¾	72	2

Figure 8–29. Sample bar list.

Symbol	Bar Support Illustration	Type of Support	Standard Sizes
SB	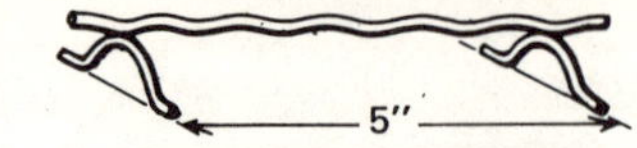 ⟵5"⟶	Slab Bolster	¾, 1, 1½, and 2 inch heights in 5 ft. and 10 ft. lengths
SBR*	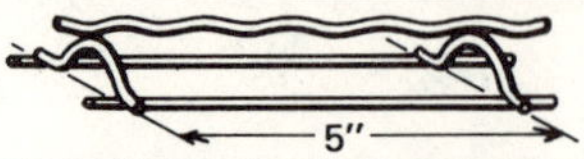 ⟵5"⟶	Slab Bolster with Runners	Same as SB
BB	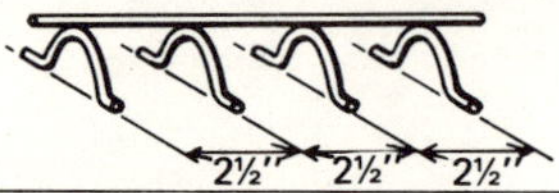 2½" 2½" 2½"	Beam Bolster	1", 1½", 2"; over 2" to 5" heights in increments of ¼"; in lengths of 5 ft.
UBB*	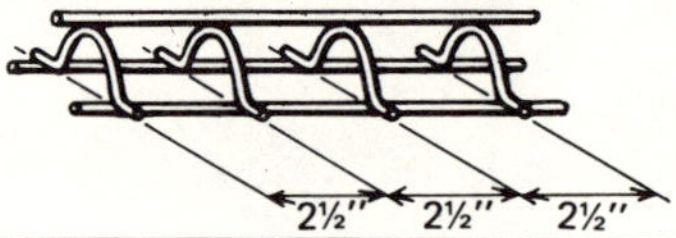 2½" 2½" 2½"	Upper Beam Bolster	Same as BB
BC	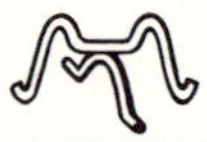	Individual Bar Chair	¾, 1, 1½, and 1¾" heights
JC	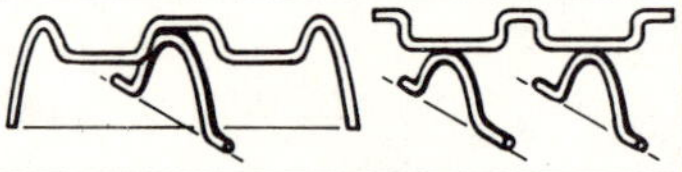	Joist Chair	4, 5, and 6 inch widths and ¾, 1, and 1½ inch heights
HC	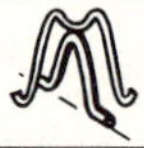	Individual High Chair	2 to 15 inch heights in increments of ¼ in
CHC	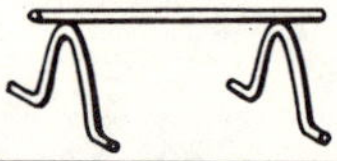	Continuous High Chair	Same as HC in 5 and 10 foot lengths
UCHC*	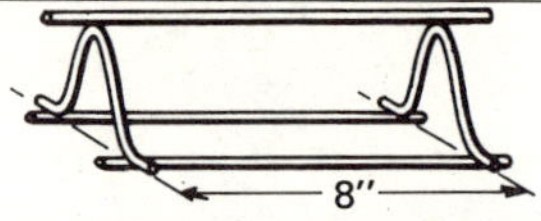 ⟵8"⟶	Upper Continuous High Chair	Same as CHC

*Available in Class A only, except on special order.

Figure 8–30 Standard types and sizes of wire bar supports.

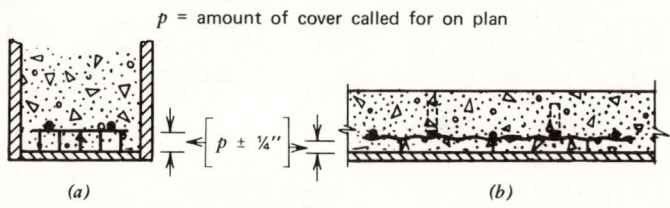

Figure 8–31 (a) Beam. (b) Slab.

tolerance of -2 in. should be used in placing. A tolerance of ± 2 in. in the lengthwise positioning of the "bend-up" or "bend-down" points on a truss bar should be sufficient.

It should be emphasized that where longitudinal bars are required to be hooked around or anchored to other bars, these tolerances will not always apply. Great care must be used to comply with these special situations.

These reasonable tolerances are allowed to permit economical bar placing and still meet the engineer's design requirements. They do not apply to the fabrication of the bars, but apply only to the workmanship required in positioning the bars.

Tolerances in Spacing Bars in Walls and Solid Slabs
In long runs of slabs or walls, the exact spacing of bars specified on plans is not critical. It is satisfactory practice to place the number of bars called for in any given length or panel at approximately the spacing required, but no exact spacing measurement is necessary for each bar.

Bar supports that are corrugated on 1-in. centers can generally provide the means for guidance in the spacing of slab bars. Where, for example, bars are specified at $6\frac{1}{2}$-in. spacing, they may be spaced alternately at 6 in. and 7 in. to satisfy the requirement.

A reasonable tolerance in spacing of bars would be ± 1 in. Greater tolerance may be allowed where the rebars must be shifted for inserts and openings, for example.

Lateral Spacing of Bars in Joists, Beams, and Girders
For buildings, ACI specifies the minimum amount of cover over stirrups, ties, and other bars, as well as the minimum clear space between bars. The American Association of State Highway and Transportation Officials (AASHTO) does the same for bridges.

The engineer determines that such spacings can be maintained and will call for bars to be positioned over each other in multiple layers, should it be necessary. Side cover should be maintained within a tolerance of $\frac{1}{4}$ in.

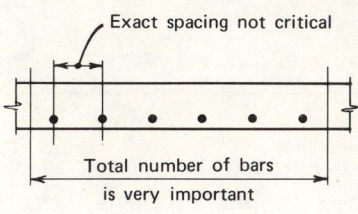

Figure 8–32

REINFORCED CONCRETE 157

A certain amount of wiring of bars to bar supports and to stirrups and ties will be required to keep the longitudinal bars reasonably straight, parallel, and the proper distance apart. Beyond that, additional tying is unnecessary.

Where beams or girders, with many bars, bear upon columns that are heavily reinforced, it is sometimes worthwhile to make a full-sized template for these intersections to hold the column vertical bars in such a position that the beam bars will later pass through them satisfactorily. When these bars cross at right angles, it is perfectly proper for them to be in contact with each other.

Tolerances in Height of Top Bars

A variation in the height of top bars changes the strength of the beam, and every effort should be made to hold them to within $\frac{1}{4}$ in. of the height called for. Top bars often interfere with other bars at right angles to the top bars, or with other facilities buried in the slab, such as underfloor ducts and conduits. Usually, this fact is foreseen by the designer and provisions are made on his drawings. It is not always physically possible to place top bars to a tolerance of $\frac{1}{4}$ in. In such cases, it is usually easier for the engineer to work out a relocation for the bars than to move the mechanical equipment. Whenever top bars cannot be placed within $\frac{1}{4}$ in. of where they are called for, the ironworker foreman should advise the inspector. If it cannot be remedied, the engineer is obliged to relocate the bars and, if necessary, authorize an extra order to increase the size or number of bars.

Tolerance in Stirrup Spacing

Stirrups are spaced by the ironworker as indicated on the placing drawings. A variation of 1 in. in any stirrup space is acceptable, but this should be adjusted in the next space or two in such a way that the end stirrups in each group having the same spacing are not more than 1 in. either way from the position called for.

OTHER ASPECTS OF PLACING

Tying

Not only is it necessary that bars be placed as called for, but they also should be securely wired to bar supports, stirrups, and ties and held against displacement.

Checking

Since bars sometimes get heavy abuse on the job, the ironworker foreman inspects and checks the steel before concrete is placed to see that it is all in position. If bars are displaced, they must be properly relocated before the concrete is placed around them.

Important

In all cases, follow the placing drawings. Drawings are checked by the fabricator and checked again by the engineer. If there is any doubt about the correctness of any item, or if the drawings are not clearly understood, the ironworker should consult his foreman. There must be no departure from the placing drawings without proper authorization.

CONCRETE COVER OR PROTECTION

The engineer determines the amount of concrete protection for each part of the job. He bases his design on these amounts, taking into consideration the requirements of building codes, fire hazards, possibility of corrosion, and exposure to weather.

Where not specified, the following minimum standard covers* (outside of bar to face of concrete) should be observed.

1. Three inches at sides where concrete is cast against earth and on bottom of footings or other principal structural members where concrete is deposited on ground (Fig. 8-33).
2. Two inches for bars larger than #5, where concrete surfaces, after the removal of forms, would be exposed to the weather or be in contact with the ground, and 1½ in for #5 bars and smaller (Fig. 8-34).
3. One and one-half inches over spirals and ties in columns (Fig. 8-35).
4. One and one-half inches to nearest bars on the top, bottom, and sides of beams and girders (Fig. 8-36).
5. Three-fourths inch on top, bottom, and sides of joists and on top and bottom of slabs where concrete surfaces are not exposed directly to ground or weather (Figs. 8-37 and 8-38).
6. Three-fourths inch from the faces of all walls not exposed directly to ground or weather (Fig. 8-34).
7. At least equal to the bar diameter, except for slabs and joists not exposed directly to ground or weather.
8. Maximum aggregate size should be no more than three-fourths (75%) of the *least* spacing dimension.

* From "ACI Standard Building Code Requirements for Reinforced Concrete, ACI 318-63," Section 808.

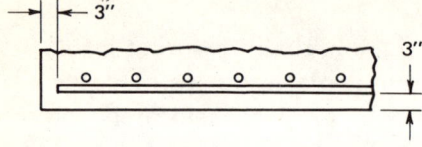

Figure 8-33 Footings.

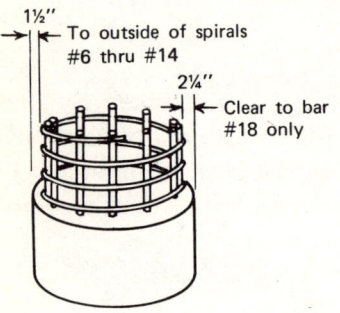

Figure 8-34 Columns.

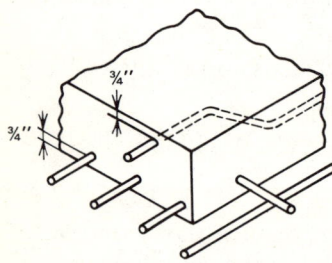

Figure 8-35 Slabs.

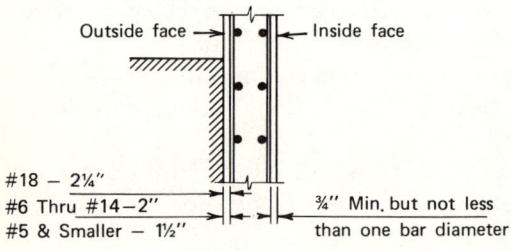

Figure 8-36 Walls.

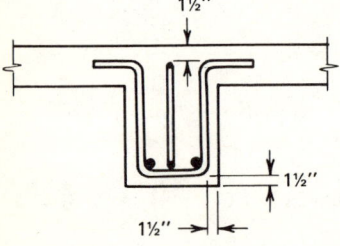

Figure 8-37 Beams.

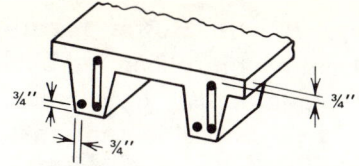

Figure 8–38 Joists.

SPLICES

Reinforced concrete structures are generally designed so that the separate parts act as a single unit. This makes it necessary to properly locate construction joints and provide continuity in the steel reinforcement through the joints. Just as it is impossible to cast all the concrete at the same time, it is also impossible to provide full length continuous bars in most structures. This is because of manufacturing, fabrication, or transportation limitations. Since splices in reinforcement cannot be avoided, it is necessary for the engineer to provide for properly designed splices.

Some knowledge of bar length limitations may be helpful.

1. Bars are ordinarily stocked in 60-ft lengths.
2. Sizes #14 and #18 are not ordinarily stocked but may be obtained in lengths up to 60 ft.
3. Lengths over 60 ft in all sizes, although not ordinarily stocked, may usually be obtained from suppliers by special arrangement.

Trucking regulations in the various states and cities often determine the maximum length and width of bars. Sometimes, longer lengths can be shipped by rail or water. Bending dimensions often govern the length that can be transported.

Practical construction limitations on bar lengths must also be considered. Except for slabs on grade, long lengths of horizontal bars projecting far beyond required construction joints are generally undesirable. A normal lap splice at or near the joint is preferable. Vertical bar lengths in columns and walls are most severely restricted. In multistory construc-

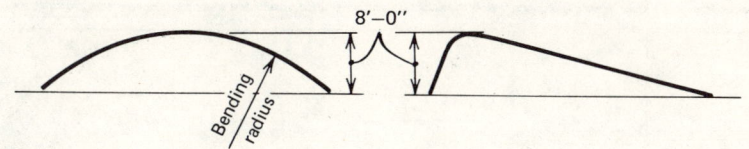

Figure 8–39 Maximum dimensions of bent bars for truck delivery.

REINFORCED CONCRETE 161

tion, usual practice is to use bars that are one story high. However, with heavily reinforced columns and staggered location of splices, vertical bars two stories long are sometimes used. In high walls, vertical bars extending full height are sometimes difficult to hold in position and may need to be spliced at one or more places, either to suit the contractor's concrete operations or the location of construction joints as determined by the engineer.

Splices are made at construction joints, such as between footings and columns or walls, between columns below and above a floor, or between walls and floors. Bars projecting through the joint are either lapped with other bars or connected directly by welding or mechanical means.

Bars are also spliced, usually by lapping, when used as horizontal bars in walls, temperature bars in slabs, or vertical bars extending through horizontal construction joints in high walls and piers, for example.

Although not considered desirable, it is sometimes necessary to splice main bars in beams and girders because of length limitations. This is usually done at locations where the forces on the bars are below their maximum.

The location and kind of splices are shown on the placing drawings. No substitution in type or location should be made without proper authority and specific instructions as to application. The three general types of splices are (*a*) lapped, (*b*) welded, and (*c*) mechanical.

Lapped Splices

In most cases, lapped splices are the most economical. The length of lap varies with the concrete strength, the yield strength of the steel, and the bar size. Laps are always shown on placing drawings and will be found either in the details or in the general notes. Some typical lapped splice details follow.

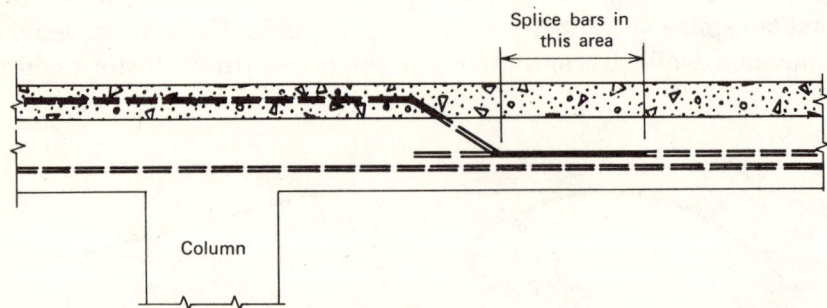

Figure 8–40 Detail showing tension splice in beam bars.

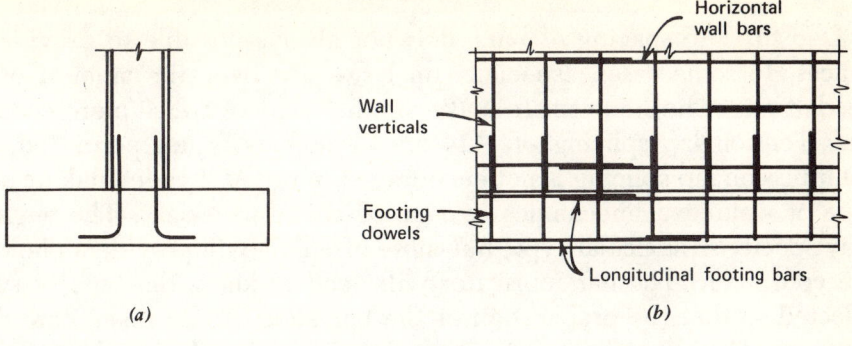

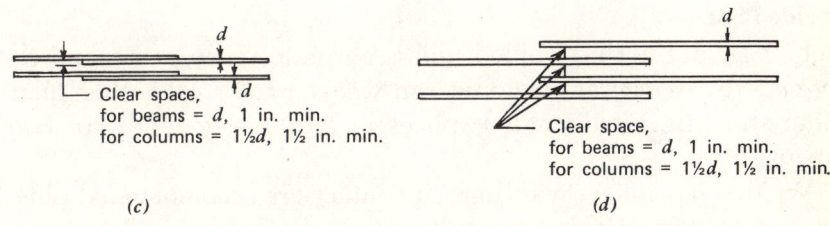

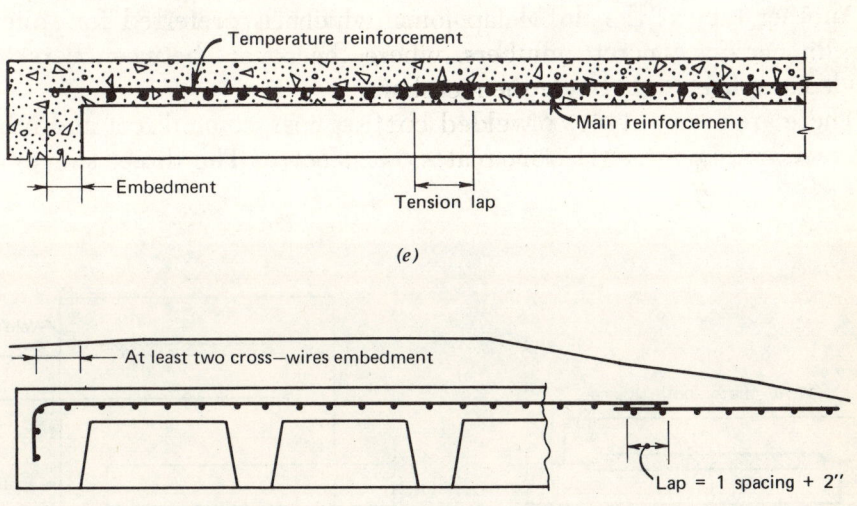

Figure 8–41 (*a*) Footing dowels-lap splice to wall verticals. (*b*) Lap splice-horizontal wall bars, longitudinal footing bars and dowels. (*c*) Lap splice-contact. (*d*) Lap splice-spaced. (*e*) Slab with shrinkage and temperature reinforcement. (*f*) Joist construction-temperature reinforcement welded wire fabric.

Due to close spacing of bars, it is not always possible to provide lap splices. In some cases, especially on large size bars, the amount of lap needed might be sufficient to make another type of splice more economical. Tension lap splicing of #14 and #18 bars is not permitted, and compression lap splicing is not encouraged in the ACI code, making some type of welded or mechanically coupled splice necessary. The engineer may specify a particular type, but more often, he will provide a choice to the contractor. The bar fabricator will need to know the type of splice selected so that the preparation of the bar ends can be taken care of in his shop. Also, he will need to show details on the placing drawings for use by the ironworkers.

Welded Splices
Rebars should not be welded unless absolutely necessary, and then only by experts. Inaccurate welding can weaken the steel rather than give it more strength. Field-welded splices of bars are generally of two types: lap and butt.

Welded lap splices with bars in contact are recommended only for #5 bars or smaller. Welded lap splices using a "backup" plate may also be used. In either case, it is necessary to consider the bar eccentricity in the design.

Another type is the double lap joint, which is preferred for splicing bars in narrow concrete members, where the offset between bars in a single lap splice might buckle the member.

There are several types of welded butt splices, some direct and others indirect, employing angle iron, plates, or sleeves. The direct splices are

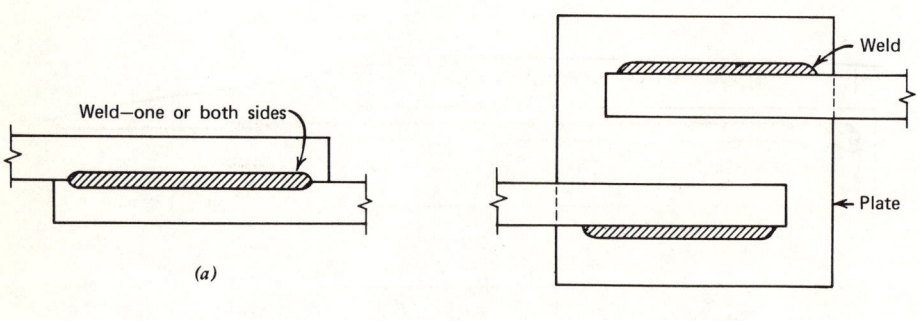

Figure 8–42 (a) Welded contact lap splice. (b) Welded lap splice with plate.

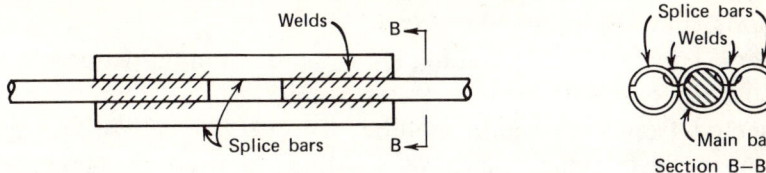

Figure 8–43 Welded double lap joint.

illustrated in Figure 8-44 and require special cutting or sawing of the bar ends. Types (*a*) and (*b*) are normally used in horizontal position and types (*c*) and (*d*) in vertical position. The double-bevel groove weld (*d*) is commonly used for column verticals. An indirect butt splice with backup plate is shown in Figure 8-45. An angle may be used for backup instead of the plate, as indicated in the figure.

The properties of reinforcing bars vary considerably. To produce good welds it is necessary to have (*a*) mill test reports on the bars including

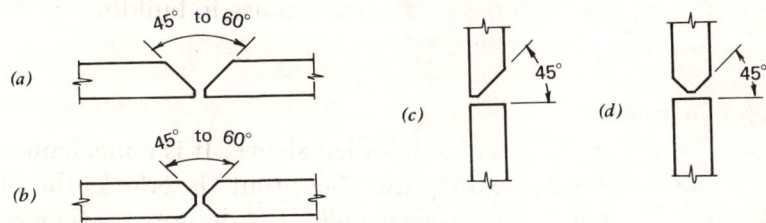

Figure 8–44 (*a*) Single-vee groove weld. (*b*) Double-vee groove weld. (*c*) Single-bevel groove weld. (*d*) Double-bevel groove weld.

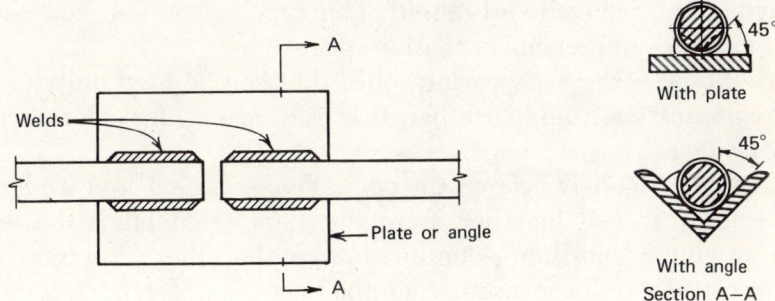

Figure 8–45 Double groove weld with back-up plate or angle.

the chemical properties, (b) a proper selection of electrodes based on these reports, and (c) qualified welders.

In precast concrete work, welded splices are common. Concrete members are usually cast with bars or plates projecting, and these projections are connected from member to member by butt- or lap-welded splices.

Another process, called thermite welding, is used in making butt-welded joints in large size bars, particularly #14 and #18. The ends of the bars should be reasonably square. Shear cutting, torch cutting, and saw cutting are all satisfactory. Two bar ends are placed in a mold with a gap of ⅜ in. to ½ in. between. The mold has a large chamber (the crucible) and a pouring gate leading to the gap between the bars. The mold is held in position by a set of removable jigs clamped to the bars. The crucible is filled with a thermite welding compound topped with a small amount of starting compound. When ignited, a high heat is produced, and the welding compound becomes molten and flows through the gap to preheat and fuse the bar ends to produce a good strong weld. After about 10 min. the jigs are removed and reused; the molds and excess metal are knocked off the bars with a hammer and are discarded.

Molds and jigs are available for welding bars in vertical, horizontal, or inclined positions, although the most common use in buildings is for butt-welding of vertical column bars.

Mechanical Couplings

One type in common use is the metal-filled sleeve. It is a mechanical butt splice in which the filler metal or "metallic grout" interlocks the grooves on the inside surface of the splice sleeve with the deformations on the bar.

A crucible is attached to the side of the sleeve by a spigot that is inserted into a tap hole. The crucible contains a filler metal that is ignited, becomes molten, and flows into the sleeve and around the bars, completely filling the space. The clamps holding the sleeve in position and the crucible are removed and reused. This type of coupling may be provided for either compression or tension splices.

Another type is the end-bearing splice that can be used only for transferring compression from bar to bar. It is used mainly for column vertical bars with square ends.

This detail (Fig. 8-47) shows the open, flanged sleeve and wedge that, when securely locked together, form the clamp that holds the vertical bars in an aligned position—one bearing on the other. Provision is also made for use of a reducer insert when the bars are different sizes.

The details above show the components of the clamp, plus two pieces

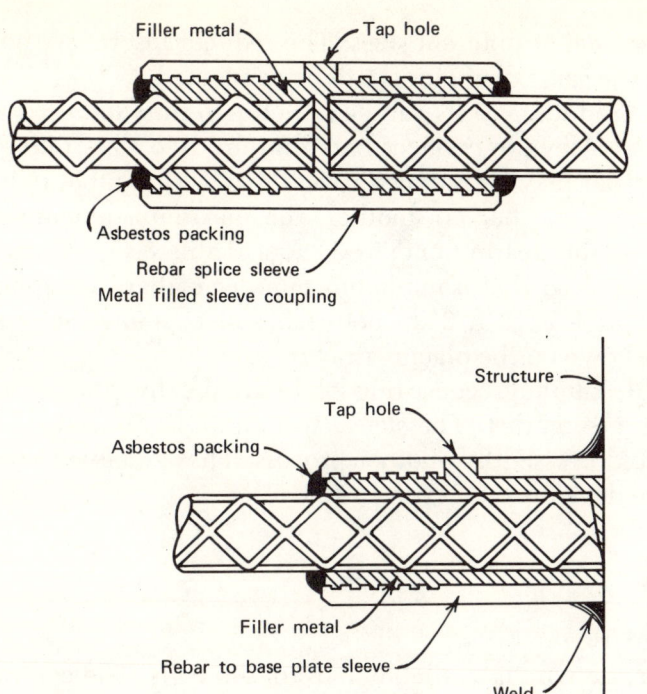

Figure 8-46 Coupling for splicing to structural steel members.

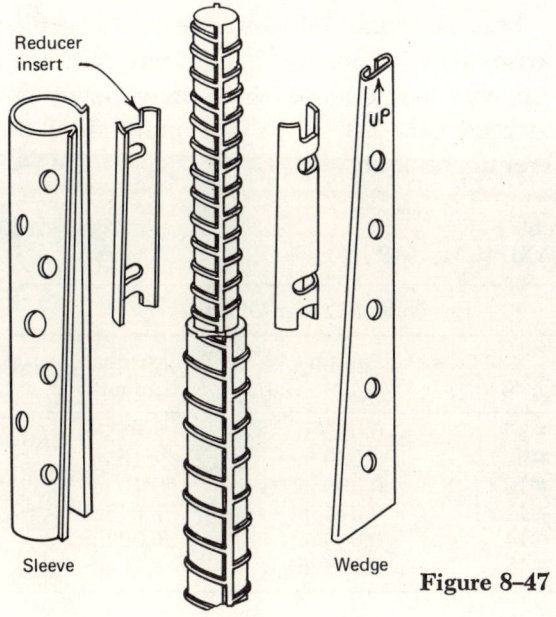

Figure 8-47

REINFORCED CONCRETE 167

of reinforcing steel of different sizes. The reducer inserts are not required when bars of the same diameter are to be connected.

Two sets of reducer inserts can be nested together and used along with the wedge and sleeve to connect bars that are two sizes different. As an example, a 14 bar may be reduced to a 10 bar with double reducers.

When placing one bar on another, the maximum amount of gap allowed between the bearing ends is 3° (see Table 8-C). Vertical column bars are arranged so that a minimum number of bars are spliced at one location with at least 2 ft-6 in. between splice points. Splices must be staggered as shown on the placing drawings.

Mechanical couplings consisting of hydraulically pressed sleeves are fairly new on the market. The sleeve for one process needs to be heated but can be cold pressed by another process. These enable the full tensile strength to be developed on even an 18 bar.

TIE WIRES

Carrying Tie Wire

In most areas, tie wire is available in 3- to 4-lb coils. The coils are readily placed in the tie wire holder or reel designed for this purpose. These reels are suspended from the belt for easy use.

Size of Tie Wire

Wire used for tying reinforcing bars is usually #16 gauge black, soft-annealed wire. These are some cases when a heavier gauge wire may be used. No. 15 and #14 gauge tie wire (or double #16) may be used when tying bars in heavily reinforced caissons or walls to maintain the proper position of the horizontal reinforcement.

Table 8–C
MAXIMUM GAP

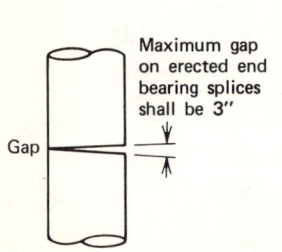

Bar Size	Equivalent Linear Offsets	
	Maximum Allowable Gap	Approximate Maximum Gap
#8	0.052 in.	3/64 in.
#9	0.059 in.	1/16 in.
#10	0.066 in.	1/16 in.
#11	0.074 in.	5/64 in.
#14	0.089 in.	3/32 in.
#18	0.118 in.	1/8 in.

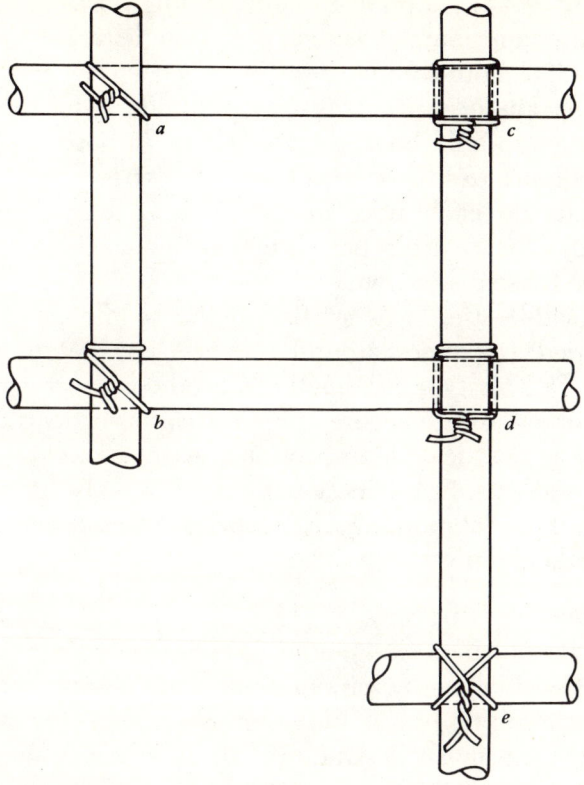

Figure 8-48

Types of Ties

In Figure 8-48, the snap or single tie, A, is normally used in flat horizontal work. This is a very simple tie and is wrapped once around the two crossing bars in a diagonal manner with two ends on top. The ends then are twisted together until they are very tight against the bars. The loose ends are cut and then flattened, to prevent them from snagging clothing and from protruding through the concrete.

The snap or single tie may be made stronger by doubling the wire rather than using a heavier gauge wire. Then the tie would be called a double snap tie or a single tie-double wire.

Wrap and snap tie, B, is normally used when tying wall reinforcement, holding the bars securely in position so that the horizontal bars do not shift during the construction progress or concreting.

The tie is made by wrapping the wire 1½ times around the vertical

REINFORCED CONCRETE 169

bar and then diagonally around the intersecting horizontal bar, completing the tie in the same manner as a snap tie (single tie).

Saddle tie, C, is more complicated than the two ties just described but is favored in certain localities. It is used particularly for the tying of footing or other mats to hold hooked ends of bars in position; it is also used for securing column ties to vertical bars. The wires pass halfway around one of the bars on each sides of the crossing bar, then are brought squarely around the crossing bar, and then up and around the first bar where they are twisted as shown.

Wrap and saddle tie, D, is similar to the saddle tie except that the wire is wrapped 1½ times around the first bar before proceeding as described for C. This type is sometimes used to secure heavy mats that are lifted by crane and to secure column ties to verticals where there is a tendency for a great deal of strain on the ties.

The figure eight tie, E, is occasionally used in walls instead of the wall tie in B, but it is not particularly recommended because of the time required to make the tie.

General Principles of Tying Bars

The proper tying of bars is essential in order to maintain their position during work that is done by various trades and during concrete placing. It is not necessary to tie bars at every intersection. Tying adds nothing to the strength of the finished structure. In most cases, every 4th or 5th intersection is all that is necessary. Ends of finished ties must be kept clear of the face of the concrete.

When tying bars in slabs that are being assembled in place, the spacing of ties should be governed by the bar sizes. Usually, snap ties are used.

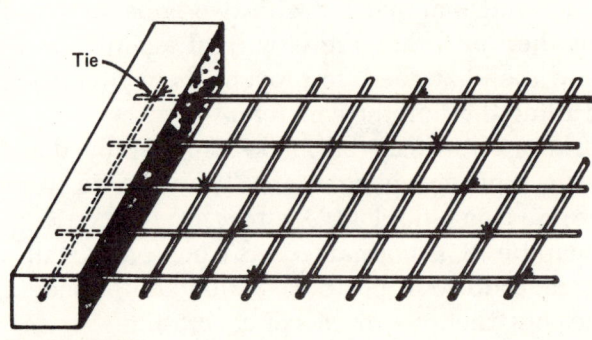

Figure 8–49

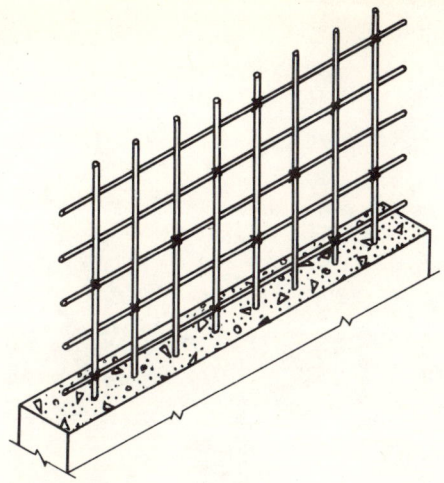

Figure 8–50

Wall bars that are assembled in place should be tied sufficiently to prevent shifting as the concrete is being placed. The wall or wrap and snap tie is generally used, but in many cases the snap tie is adequate.

For preassembled mats, a sufficient number of bar intersections should be tied to make the mats rigid enough for handling. When snap ties are used, it will make the mat more rigid if the direction of the ties is alternated.

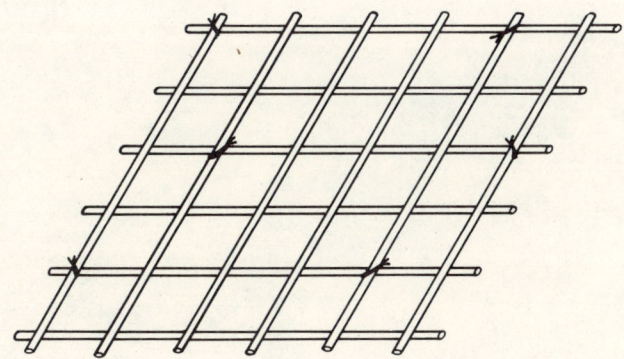

Figure 8–51 Snap ties alternated for added rigidity.

REINFORCED CONCRETE 171

CHAPTER 9
PLACEMENT OF STEEL

PLACING BARS IN FOOTINGS, WALLS, AND COLUMNS

Individual Square or Rectangular Footings

Individual footings are shown on the foundation placing drawings. The bars are usually called for in a footing schedule on the same drawing. The schedule indicates the size, length, mark number (if bent), and the number of bars each way for the various footings. Sometimes, the schedule may be found on another drawing that is specifically for schedules. Occasionally, where only a few footings are involved or where they are quite typical, the bars may be shown on the plan. The spacing of the bars may be given, but, if not, they should be evenly spaced.

Bars are frequently assembled into mats prior to installation. All of the bars to be located in one direction are spaced out evenly across saw horses. Then, the bars to be located in the opposite direction are placed on the first bars and spaced evenly. The two layers then are securely tied together at every second or third intersection around the exterior. The mat is tagged with the footing number and stockpiled ready for use or delivered to the excavated pit into which it will be placed. The pre-assembled mats are installed in the pit at the proper height.

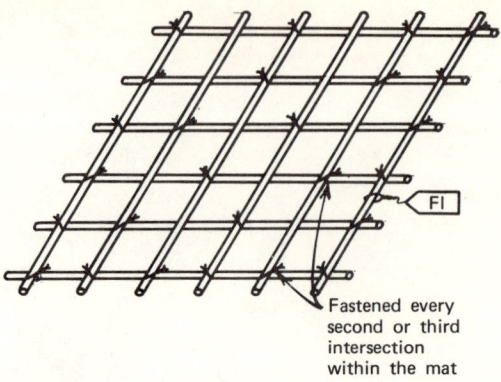

Fastened every second or third intersection within the mat

Figure 9–1

In large footings that require longer and heavier bars, the mats may need to be assembled in place. If the design permits, two (or perhaps three) bars in one direction are supported on concrete blocks (Fig. 9-2). All the cross bars then can be placed at right angles on top of these support bars, properly located, spaced, and tied to them. Although the tying need not produce a sufficiently stiff mat to allow hoisting and carrying around the job, it is necessary to tie each crossing bar to the support bar to keep an even spacing. Finally, the remainder of the bars are placed on top of those already installed, spaced out equally, but located to allow for the two or three support bars already spaced in the first layer. In the last layer, each bar needs to be tied at only two or possibly three intersections, depending on its length.

Foundation Mats

This type of construction usually consists of a thick concrete slab with bars in two layers, one near the bottom of the slab and one near the top. The bottom layer is usually supported on concrete blocks or on continuous high chairs with runner wire. To support the top layer, special supports designed to rest upon the lower layer are used, or a chair with a sand plate or support wires could be used resting on the ground. Also, dowel blocks that rest on the ground may be used to support both top and bottom layers of bars. These supports are spaced about 4 to 5 ft on center each way. There are many methods possible; only a few are shown. Selection of the method used will depend considerably on the type of ground condition, the size and spacing of the bars, the placing

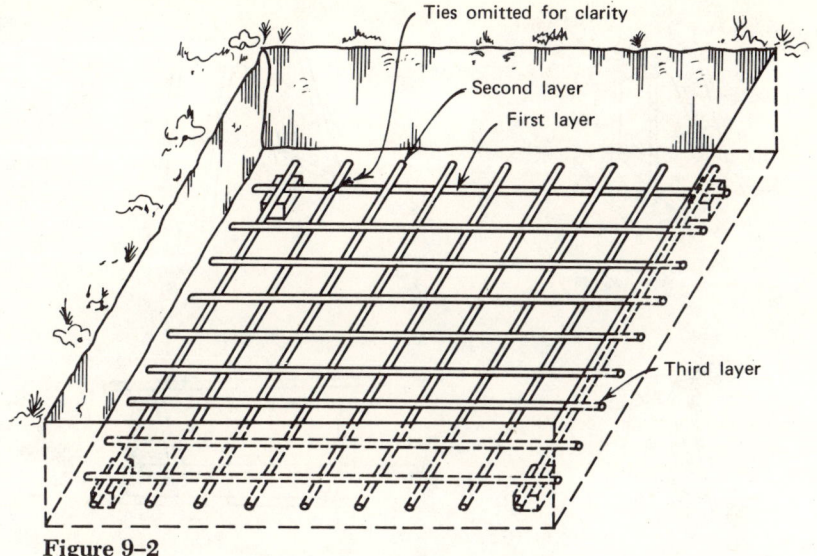

Figure 9-2

contractor's choice, and the bar fabricator's ability to furnish manufactured supports (Fig. 9-3).

In some localities, precast concrete blocks are used as bar supports. They are supplied in three styles: (1) plain, (2) with wires, and (3) doweled.

1. Plain blocks are used to support bars off the grade.
2. Concrete blocks with wires are supplied with two 16-gauge tie wires cast in the center. Concrete blocks with wires are used against vertical forms or in positions necessary to support the block by tying to the bars.
3. Doweled concrete blocks are cast with a hole in the center, approximately 2¼ in. deep and large enough to insert a #4 bar with a 90° bend at the top that is used to support bars above the concrete block. At the same time, the concrete block can be used to support bars off the grade by placing them on either side of the dowel bar (Fig. 9-4).

In general, maximum spacings for various conditions of usage and specifications for the placing of wire bar supports should be followed when using precast concrete block supports.

The styles and generally available sizes of concrete blocks are illustrated in Figure 9-4.

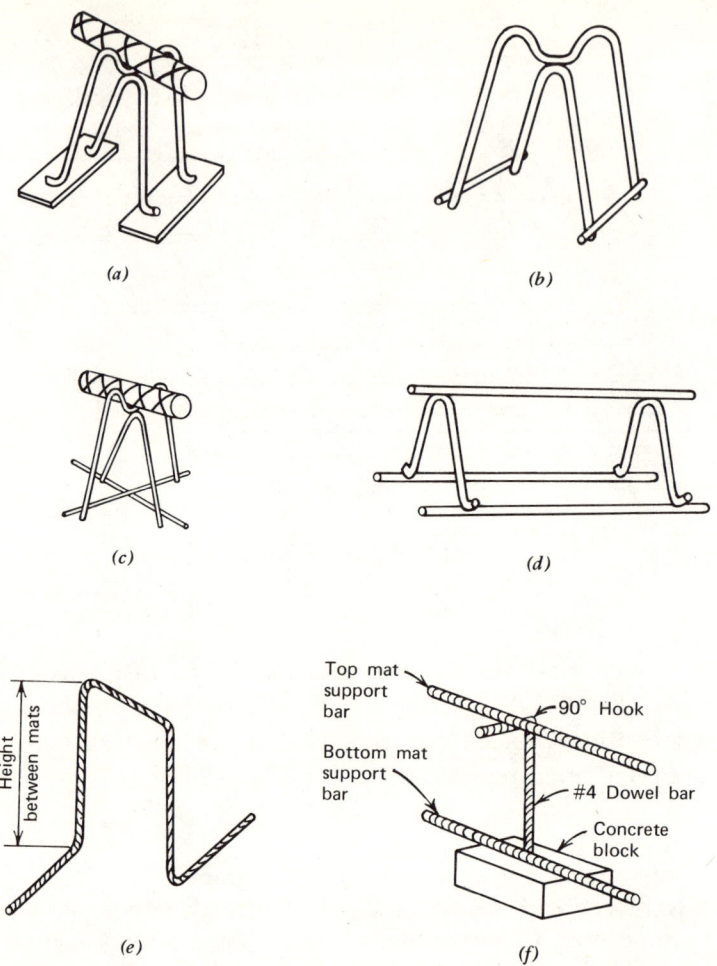

Figure 9-3 (a) HCP—High Chairs with Plates. (b) FHC—High Chair for Fill, (c) SHC—Special High Chairs. (d) UCHC—Upper Continuous High Chairs, (e) Bent Bar Chair ("Standee"), (f) Dowel block.

Continuous Wall Footings

Reinforcement usually includes two or more lengthwise bars, with crossing horizontal bars at a designated spacing and with dowels projecting into the walls if required. These bars are assembled in place in the trench in the same way as described for individual footings (Fig. 9-13). At corners, where two wall footings meet, the lengthwise bars extend a specified distance into the intersecting footing or are bent at right angles. The

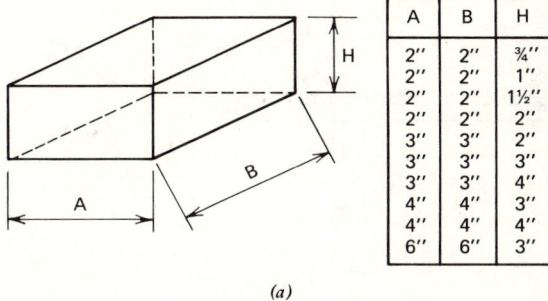

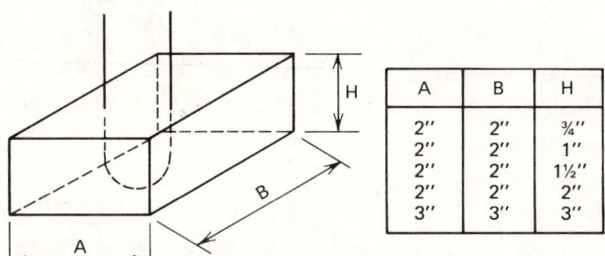

Figure 9-4 (a) Plain precast concrete blocks. (b) Precast concrete blocks with wires.

engineer specifies the method he wants. The bar fabricator shows the engineer's requirements on his placing drawings. The ironworker must follow these drawings exactly.

Pile Caps
Pile caps are usually square or rectangular but may be of different shapes, depending upon the number and arrangement of the piles. Bar placing in pile caps is similar to placing bars in individual footings. The piles are already driven and cut off at the proper elevation, so the mat rests on the piles.

Bars may not always be arranged in mats crossing each other at right angles but may be grouped in bands over pile caps. When in bands, the

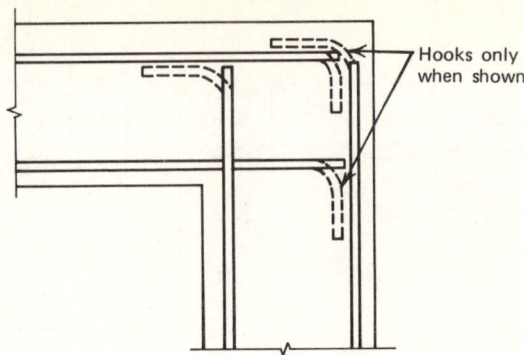

Figure 9-5 Typical corner of wall footing (top view).

bars usually cannot be preassembled but must be placed directly over the piles.

Piles are frequently used in cantilever footings, combined footings, raft foundations, foundation walls, or cantilevered retaining walls. The practice of placing bars for these elements is much the same as the practice described for individual pile caps, but the bars are placed directly and are not preassembled.

Column Dowels

Columns rest on footings, grade beams, piers, or caissons. Dowels are required to connect a concrete column to its support, although in some arrangements, precast columns do not need dowels.

Dowels must be placed accurately, not only with respect to the locations of the column bars above, but also vertically to assure that they project the specified distance above the footing. Dowels are spliced to the column vertical bars above either by lapping, welding, or the use of mechanical coupling devices.

Dowels, which lap to column verticals, may be located either alongside the vertical bars at the side of the tie or spiral or inside of the vertical bars. If located inside of the vertical bars, the center of the dowels should be about $3\frac{1}{2}$ to 4 in. inside the face of the column above.

The best accepted practice of positioning dowels is to use a template generally made of boards with holes drilled in it. Sometimes, a tie or a single turn of spiral is nailed to the guide boards. The dowels can then be located and carefully placed along with the footing mat bars. Where hooks are used on dowels, additional stability is obtained by tying the

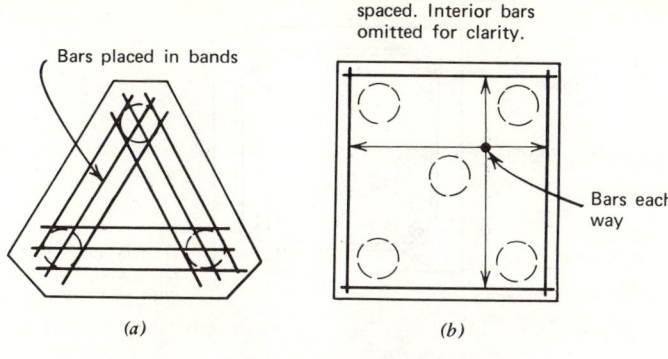

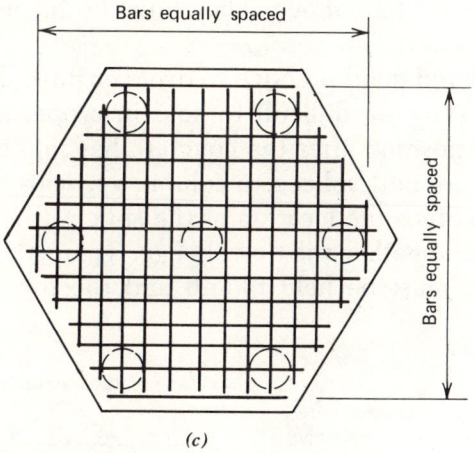

Figure 9-6 (*a*) 3-Pile footing. (*b*) 5-Pile footing. (*c*) 7-Pile footing.

hooked end to the footing mat. If welding or mechanical couplings are used, the dowels must be set to come directly under the column verticals above. The template method is also used here, and the dimensions of the tie or spiral turn must be the same as required for the column above.

Dowels are frequently provided with 90° hooks. These hooks provide a means of support where they rest on the footing mat. A dowel cage is made by assembling the dowels, using two sets of ties arranged so that they will be above the top of the footing. This cage is then wired to the footing bars so the entire assembly is held quite firmly in place. When the dowels are located at the side of the column verticals, the ties used are

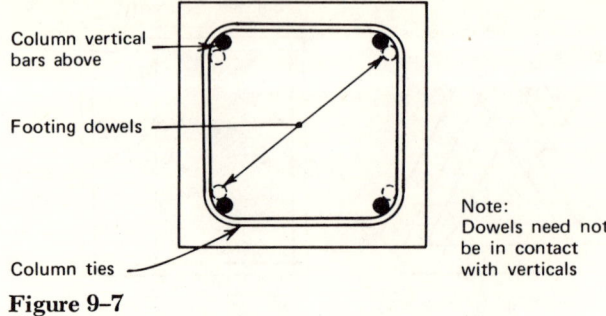

Figure 9-7

the same as in the column above. They must be furnished with the footing bars and dowels.

It is not considered good practice to drive or push dowels into position in wet concrete. They are difficult to hold in proper alignment, and corrections in dowel position after the concrete has set can be costly.

Where dowels are butt spliced to column verticals, certain precautions need to be taken. Since the location of the splices is staggered, care must be used in putting together the dowel cage. Templates should always be used. The dowels must be held plumb and the top ends held to exact

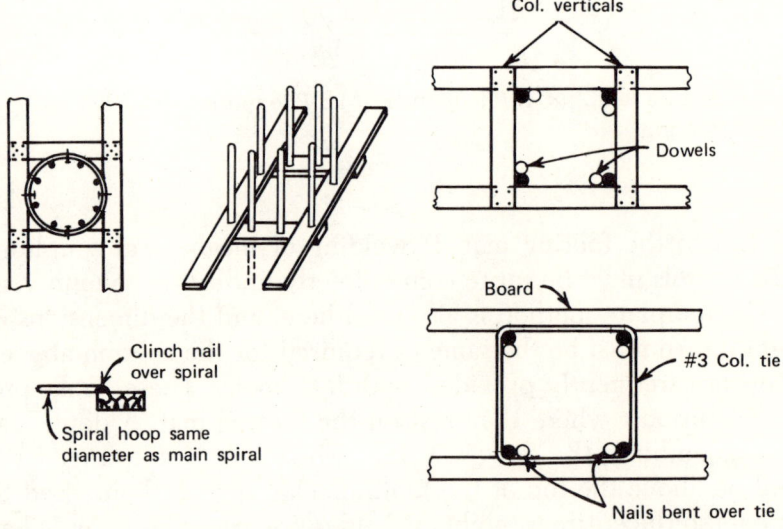

Figure 9-8 Templates for positioning dowels and verticals.

elevations. Placing drawings should be studied carefully to make sure of the arrangement of staggered splices.

In columns using interlocking double or triple spirals, the dowel arrangement is rather complicated. It is not necessary that dowels be in contact with the spirals. Figure 9-9 shows how dowels are best spaced in triple-spiraled columns to permit subsequent placement of the spirals.

Rectangular columns must be shown correctly oriented on the engineer's drawings; for example, E-W direction, long face; N-S direction, short face. If columns are located a little off center from the main grid lines, this must also be clearly shown and carefully followed in the field. Exterior side and corner columns, and often columns at corridors and elevator or stair openings, may be located by the faces instead of center lines. The field engineer should mark the grid lines established by his stakes and clearly instruct the ironworker if columns are located by one face or are off center, so that the marks can be distinguished from grid lines that are centered on the columns.

The templates must be carefully set and dowels placed from these lines. Misplaced dowels can cause great difficulty in properly aligning the columns above. Watch for any dowels required for grade beams abutting the sides of footings or for dowels required to extend into foundation walls, partitions, pit walls, or other construction units to be built later. The detailer should show these on the footing plan or schedule. When this cannot be done, it should be called for on grade beam, wall, or other placing drawings with appropriate reference notes placed on the footing plan. After all bars are placed, a further check should be made against these related drawings. The omission of dowels can cause serious construction problems that are expensive to correct.

Sometimes, in spite of all the care taken, dowels are misplaced. If the discrepancy is not very great, the solution may be to bend the dowel over with a long piece of pipe or a hickey (not a sledge hammer). It may be permissible to reduce the amount of cover over the dowel right at the top

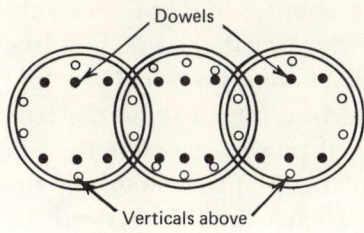

Figure 9–9

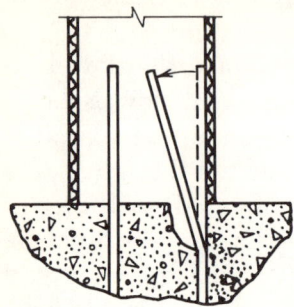

Figure 9-10

of the footing, especially if the floor slab will cover this point later. A bend of about 20° or 3 in. per ft with the vertical is about the maximum acceptable. It may be necessary to dig or chip out concrete in front of the dowel in order to move it over. Such field bending of partially embedded dowels can be done only with the engineer's approval.

Combined or Cantilever Footings
Bars are placed in combined or cantilever footings much the same as in individual footings. The bottom mat, either preassembled or assembled in the excavation, is supported on concrete blocks. The top mat then is assembled and placed, being held in position by special supports.

A combined or cantilever footing is shown in Figure 9-11.

Mat or Raft Slab Foundations
Sometimes, a single slab of concrete may be used underneath all of the columns or walls of a structure, taking the place of individual column footings or continuous wall footings. The slab thickness varies widely, depending upon soil conditions and the load to be supported, but it is usually in the range of 1 to 5 ft. The thickness could run up to 10 or 15 ft.

Generally, a mat slab is reinforced with a mat of reinforcing bars in two directions, the bottom steel being located about 3 in. above the subgrade (or mud slab) with a similar mat of top steel located an inch or two below the top of the slab. The placing drawings may show bottom mat bars on one plan and top mat bars on another in order to show the bars more clearly. The plans will give the number of pieces, size, length, and spacing of these bars. Often the bars may not run the full length of the slab. In these cases the beginning and end of each line of bars will be

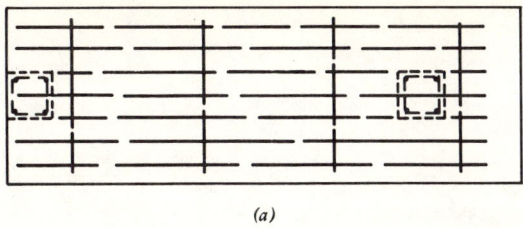

(a)

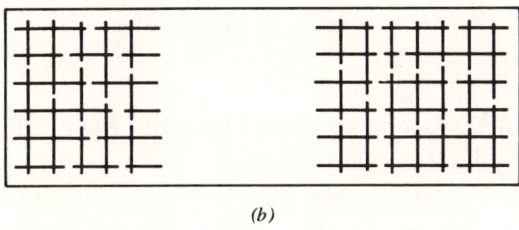

(b)

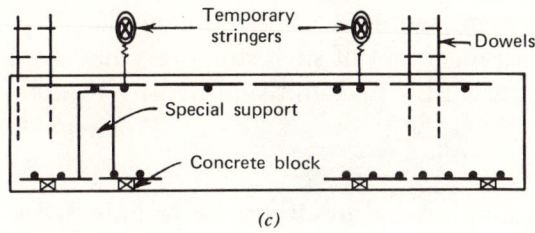

(c)

Figure 9-11 (a) Plan of top of footing. (b) Plan of bottom of footing. (c) Side elevation.

shown on the drawings. Where bars are lap spliced to make up a continuous line of bars, the number of overlapping bars of each length are shown.

After the subgrade is leveled, the lower mat of bars is placed in the excavation. Concrete blocks spaced about 5 ft apart in each direction, or other specified supports, are used to keep the bottom layer of bars at the proper height above subgrade. Because of irregularities of the subgrade, it may be necessary for the engineer to use an instrument to arrive at the correct level of bars. A second layer of cross bars rests directly on the first layer, and the two are wired together for stability. It is not necessary to wire every bar intersection or even alternate intersections. Only

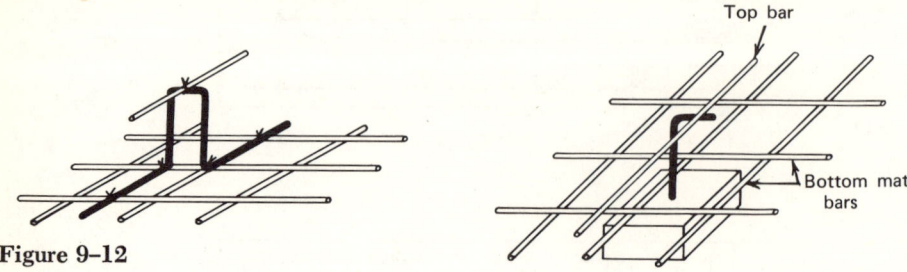

Figure 9–12

enough tying should be done to insure that the bars will remain exactly in place until covered with concrete.

For support of top bars in slabs of 4 ft or less, individual bar supports made of wire or bent bars are obtainable. For slabs over 4 ft thick, the contractor and the bar placing subcontractor will reach an agreement on a method for supporting the mats and will make arrangements with the bar fabricator for items required. Unless the engineer's plans are specific as to the type of arrangement of such supports, they are not supplied by the fabricator but are estimated and supplied by the placing contractor.

Walls

Basement walls are reinforced in either one or both faces. The reinforcement usually consists of vertical and horizontal bars that form a mat or curtain. The large, closely spaced bars may be either vertical or horizontal, depending upon the design. They are usually placed nearer to the surface or face of the wall. The smaller and more widely spaced tie bars (for temperature) cross the main bar at right angles and are farther from the face of the wall. In some thin walls, a single curtain of smaller-size, widely spaced temperature bars may be used both vertically and horizontally in the center of the wall.

After the continuous wall footing has been placed and the concrete hardened, with vertical dowels projecting, forms are erected and securely braced for only one face of the wall. The projecting dowels ordinarily are of the same size and spacing as the vertical bars. Vertical bars may be wired to the dowels. One horizontal bar is wired to the verticals to keep them plumb and at correct spacings. Then the other horizontals are placed. The horizontal and vertical bars are wired securely to each other at sufficiently frequent intervals to make a rigid mat. Tying is required at every second or third intersection, depending upon the size and spacing of bars, but with no less than three ties to any one bar, and not more than 4 to 6 ft apart in either direction.

The entire mat must be secured at the top to keep it in a vertical plane. There are various ways to do this. For example:

1. For walls not exposed or walls covered by other material, nails can be driven into formwork near the top of the bars, projecting to the required amount of cover, with wire looped around the nail head and the vertical bar.
2. Short lengths of slab, beam bolsters, or individual bar chairs can be used to space the mat from the form. The spacer is stapled to the forms, and the mat is wired to the spacer. Special arrangements must be made with the bar fabricator to obtain these, since they are not supplied as part of normal bar support requirements.

If a second mat of bars is called for, the first step is to complete the mat closest to the wall form already in place. Then place bars in the second mat in the same way, matching the second line of dowels and keeping the two mats the proper distance apart.

There are also various methods for holding the two mats in position.

1. The first mat may be secured as previously described for a single mat. The two mats can be held apart by using Z-shaped spreader bars tied

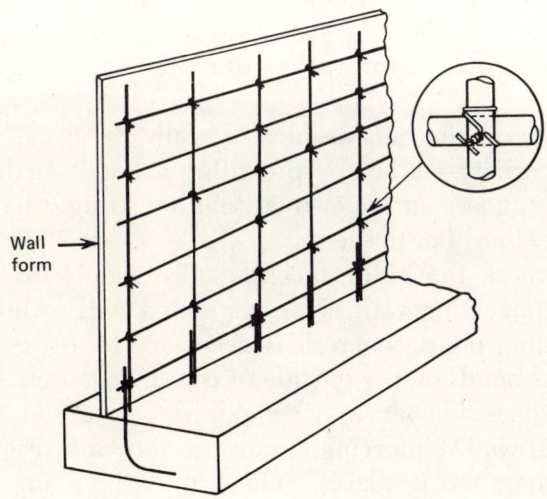

Figure 9–13

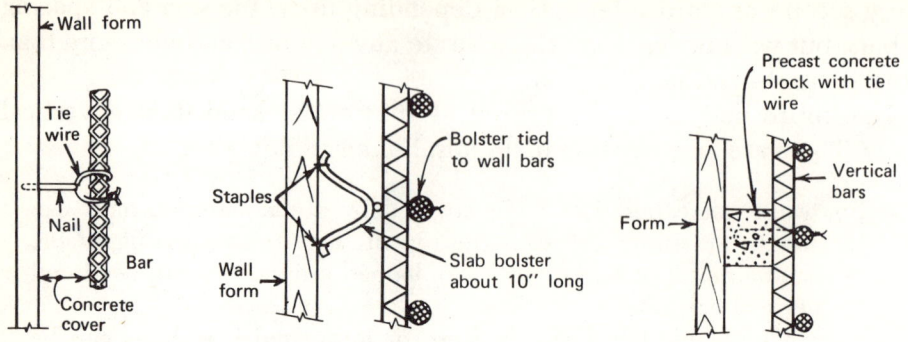

Figure 9-14 Types of wall mat spacers.

to both mats and using a short length of bar support wired to the second mat for clearance for the form.
2. For exposed or architectural concrete, a prefabricated wire spreader is obtainable, which may be tied or snapped onto the two mats and which extends to the wall forms. These act as both a spacer and spreader, with no wires or staples projecting from the concrete after removal of forms.

As a word of caution, the bar placer should make sure the wall curtains have the required clearance before the second form is erected.

Vertical bars often project out of the top of the wall and extend either vertically into a wall above or bend horizontally into a slab. When extending into a slab, they are usually prebent. Job bending with a hickey is sometimes required when the slab dowels are left straight to facilitate form removal. However, approval by the engineer is required for such job bending.

Horizontal bars at the outside face of walls are usually called for to extend around corners—either by providing a bend on the end of each of two bars meeting at a corner or by extending straight bars and splicing a short corner or elbow bar to them.

Horizontal bars at the inside face of walls usually extend as straight bars into the intersecting wall or are provided with a hook. For tanks, pits, and swimming pools, where it is necessary to assure strong, watertight corners, the bends on the outside of corner bars may be lengthened, and the ends of the inside bars may be hooked.

Several typical wall reinforcing details are shown in Figure 9-18.

After all the bars are in place, a check must be made to see that all dowels and anchors are installed before concrete is placed.

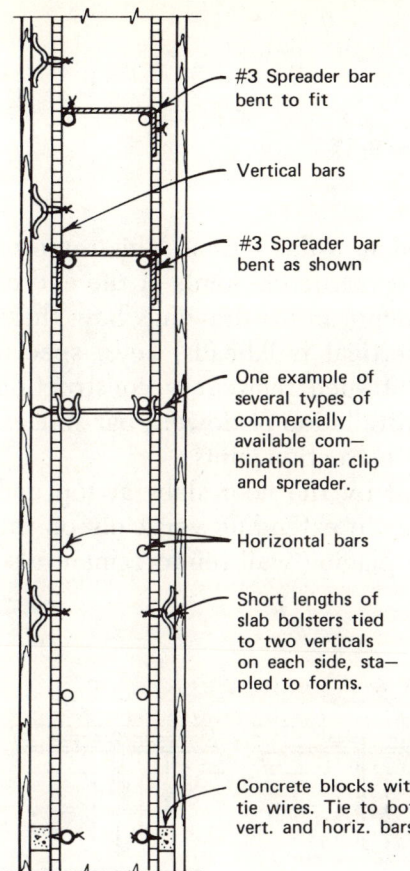

Figure 9–15 Composite details of wall mat spacers and spreaders.

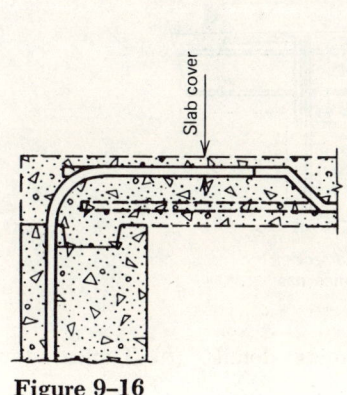

Figure 9-16

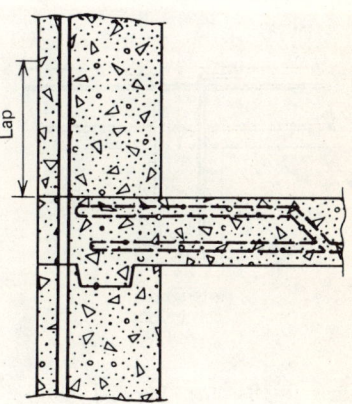

PLACEMENT OF STEEL 187

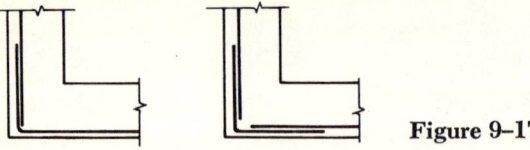

Figure 9-17

Only so much concrete can be poured in a day's time. This requires stopping at both horizontal and vertical construction joints at the end of a day or shift. The engineer usually specifies on his drawings how these joints are to be made. This includes vertical bulkheads, keys, special dowels, bar laps, and other provisions at such temporary construction joints. Follow the placing plans for the installation of dowels, bar splices, and other details to accommodate these construction joints.

Very often, basement walls are braced by the floor slabs at top and bottom and have the principal reinforcement extending vertically on the inside face. The previous description for placing wall reinforcement also applies to this kind of wall.

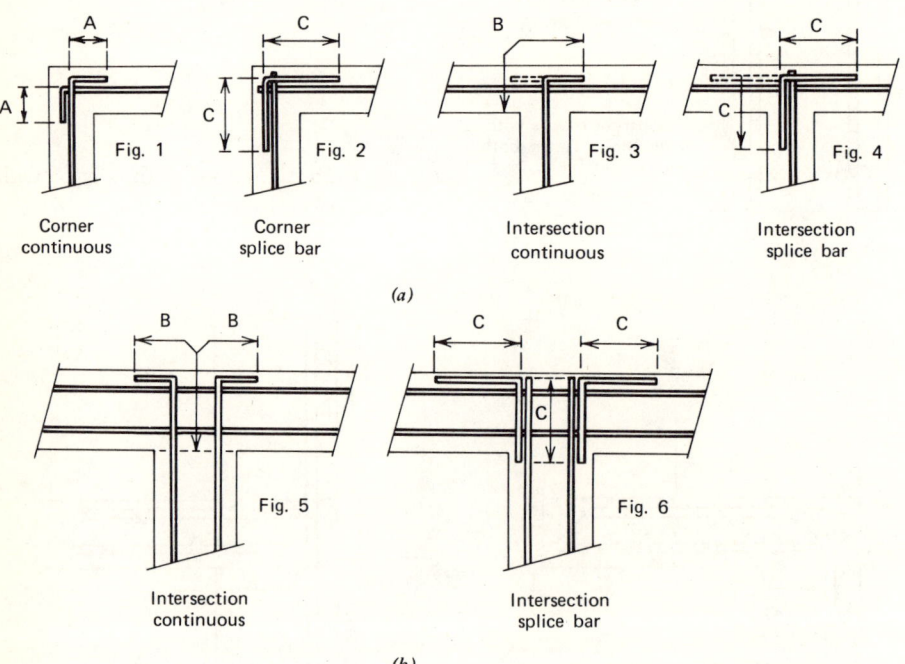

Figure 9-18 (a) Typical single curtain reinforcing details. (b) Typical double curtain reinforcing details.

Doors, windows, and other openings in walls require additional reinforcement. Additional bars are needed over, under, and alongside of door and window openings, normally extending well past the sides and jambs of the openings to prevent diagonal cracks. Often, diagonal bars are added and placed inside of, and wired to, the wall mat or mats already in place.

Cantilevered Retaining Walls. In general, the placing, supporting, and tying of bars in cantilever retaining walls and their footings is similar to ordinary building walls, so only the differences will be discussed here.

Bottom footing bars may extend all the way across the footing but usually begin at the toe of the footing and extend part way toward the heel, or may be bent upward at the point to extend as dowels at the back face of the wall.

Top footing bars are used, beginning at the heel of the footing and extending either all the way across the footing or at least a specified length beyond the back face of the wall.

Smaller longitudinal bars are tied to the main bars to complete the top and bottom mats.

Dowels usually consist of the bent-up ends of bars extending from the bottom at the toe of the footing. Where not bent up in this manner, separate dowels, as shown on placing drawings, must be provided. With either method, the dowels project above the footing just inside the back face of the wall. Dowels should be held in horizontal alignment until the footing is cast.

The stem bars are placed after the footing concrete has hardened and the forms have been built upon it for one face of the stem. The placing of the vertical and horizontal bars is quite similar to the method described for walls in general. The vertical bars are placed inside the back face of

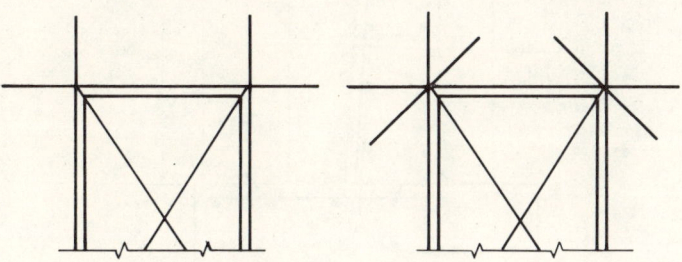

Figure 9–19   Reinforcement at wall openings.

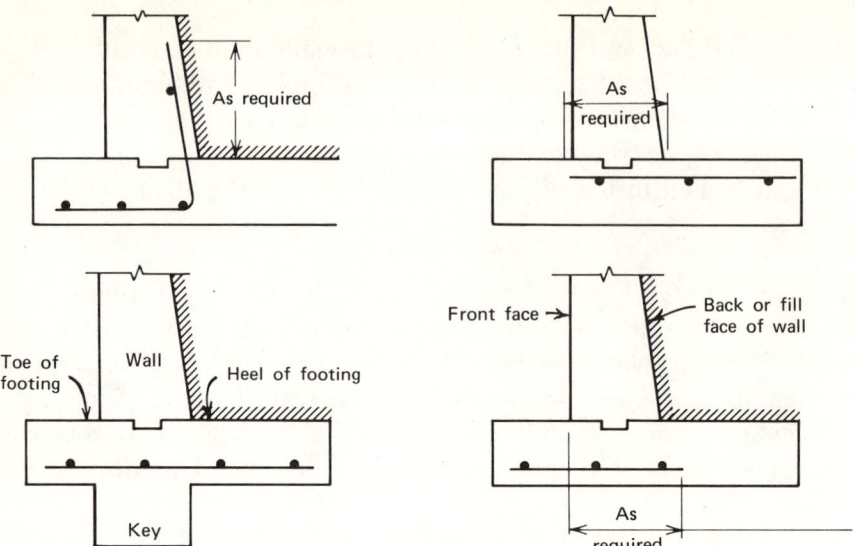

Figure 9-20 Cantilever wall footings.

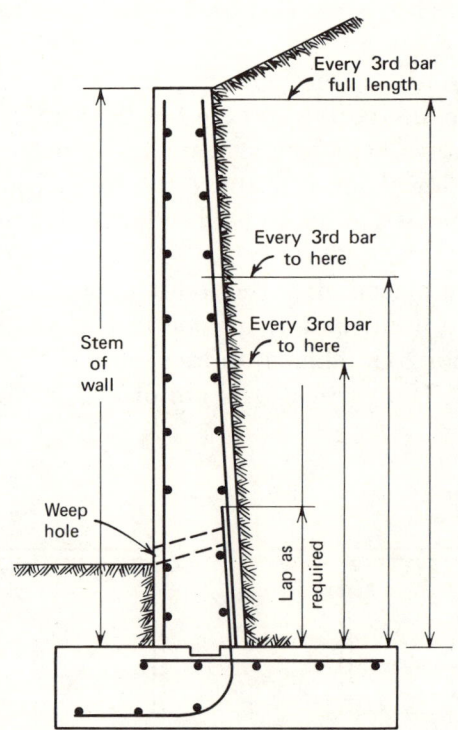

Figure 9-21 Cantilever wall.

190 REINFORCED CONCRETE, PRECASTING

the wall with the horizontal bars just inside of, and wired to, them. Usually, the dowels match the vertical bars in size and spacing, and the two are wired together.

Some retaining walls have a mat of bars without dowels in the outside or exposed face. The placement of these bars is the same as for ordinary walls.

Weep holes made by using pipe, drain tile, or other forms are often provided through the stem of the wall above the top of the footing to take care of drainage. These must not be disturbed or blocked when placing bars.

Columns

Column reinforcement includes vertical bars girdled by either a spiral or a series of ties. Where column bars start off at the same level, the complete column unit can be assembled at some convenient location. The column assembly can then be hoisted into position when needed. If the column unit is too heavy or the locations of splice points on the verticals are staggered, it is often better to assemble the column unit in place. A description of some of the accepted assembly practices follows.

Preassembly of Spiral Units. Spirals are shipped collapsed and must be opened into shape. To do this they should be "broken," that is, not only straightened up but bent over in the opposite direction so they will remain straight. Spiral spacers, which are usually shop-attached on opposite sides of the spiral, must be shifted so they are about equally spaced around the spiral. The spacer is usually a steel channel or angle punched with a series of "lips" to hold the spiral turns to proper pitch. (Fig. 9-22). Each fabricator selects the shape of spacer best suited to his purposes to provide the required stiffness.

In straightening up the spiral and adjusting the position of the spacers,

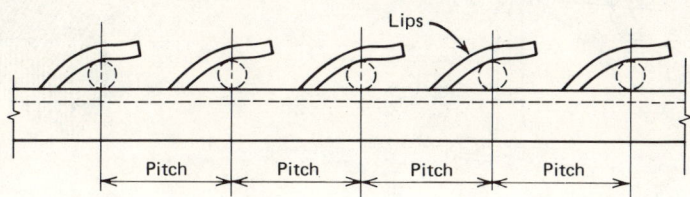

Figure 9–22   Typical spacer.

the spacer lips may have loosened, in which case they should be tapped with a hammer to tighten them around the spiral rod.

In preassembling, two of the vertical bars are placed inside of the spiral and across two supports. Since a spiral begins at a floor or footing level, the vertical bars at one end of the spiral must be located flush or projecting slightly beyond the spiral so that they extend at the opposite end as dowels. After the first two verticals are placed, the others are inserted in the spiral, and all are spaced equally around the circle, unless otherwise called for by the placing drawings. They are wired to the spiral to form a rigidly built cage that can then be lowered into place and wired to the dowels or verticals projecting from below.

Preassembly of Tied Units. Column ties with verticals are commonly assembled into cages by laying the vertical bars for one side of the column across two supports. Then, the proper number of ties are slipped over the group of bars. The remaining verticals are added, and the ties are spaced out as required by the placing drawings.

A sufficient number of intersections are wired together to make a

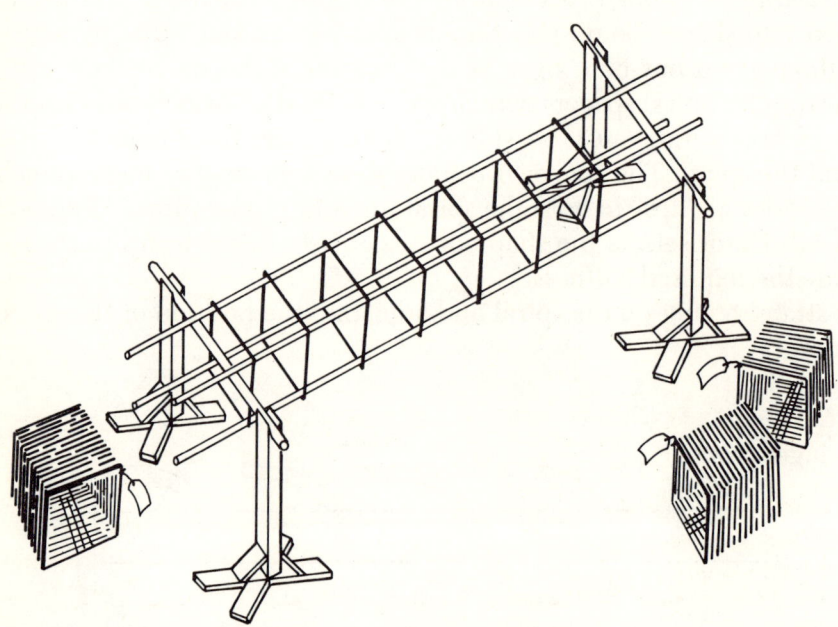

Figure 9-23

stiff, assembled cage that can be hoisted and placed as a unit. Sufficient stiffness can ordinarily be obtained in four-bar and six-bar columns by wiring every tie to every vertical at each intersection.

For columns with more than six bars, wiring at a minimum of three corners of each tie should be sufficient, using a staggered pattern on each successive tie, but making sure that each vertical is wired at least three times in its length. For large square or rectangular units, it may be necessary to install diagonal wire bracing, twisted and tensioned, to help keep the shape of the units while they are being handled and lowered into place.

Erecting Spiral and Tied Column Units Prior to Forms. Where form ties, inserts, or other obstructions cross the inside of the column form, it may not be practicable to lower a column unit into place in the forms. The unit is assembled as described above, then up-ended from the floor on which it is to stand, placed in correct position, and the vertical bars are wired to the projecting dowels. The form is afterwards placed around the column unit.

Assembled-in-Place Column Units. Preassembly of column units is usually preferred only for one-story-length vertical bars—all spliced at one point above the floor line. In some cases, splices must be staggered with splices of alternate vertical bars located at different levels above the floor. With staggered butt splices, large vertical bars are usually in two-story lengths. Usually, the lap splicing or butt splicing of all the verticals is completed, after which the ties are placed on the free-standing vertical bars.

Spacing of Column Ties. In preassembling tied column units, the individual ties or sets of ties are equally spaced, beginning about a half space above the floor, with the last ties at the top located below the lowest layer of bars in the deepest beam or joist framing into the column. This location must be determined from the drawings. If the vertical bars are off-set bent, the last tie is placed at the point of bend, and usually three additional closely spaced ties are added below this point. The placing drawings must be followed for the arrangement of ties required.

Height of Spirals. The height (or length) of a spiral is the distance out-to-out of coils, including the finishing turns, top and bottom, with the spacer length equal to that of the spiral, plus not more than one pitch dimension. The spiral extends from the floor or top of footing level to the level of the lowest horizontal bar in the slab, drop panel, or beam above. Where a column capital is used, the spiral can usually stop where the width of the concrete in the capital is twice that of the column. The ironworker must follow the placing drawings, of course, and it is mentioned here only as a matter of information to him. Sometimes, there may be variations in concrete level or in fabrication of the spiral, which could cause the top of the spiral to interfere with other bars. In that case the last turn or two of spiral can be united, removed from the spacer lips, pushed down, and then rewired to hold it in place.

Lap of Column Verticals. Column verticals should extend the distance called for on the placing drawings. Dowels formed by extensions of verticals from a lower level should usually come inside of the verticals being placed. With spirals, dowels may be lapped along the inside of the spiral, provided that, when lapped with the upper verticals, the minimum clearance required between such pairs of bars is maintained. The lapped bars should be wired together if not prohibited by the drawings.

Supports for Column Units. Tied or spiral column-bar units must be spaced from the forms to provide the required concrete cover. Wherever access permits, the bars are braced from the forms at three or four points as near to the top, bottom, and mid-height as may be practicable. Concrete blocks with embedded tie wires may be used and are wired to the ties or spiral. Blocks must be wired at intersections of the ties and vertical bars to prevent spinning when preassembled units are lowered into the

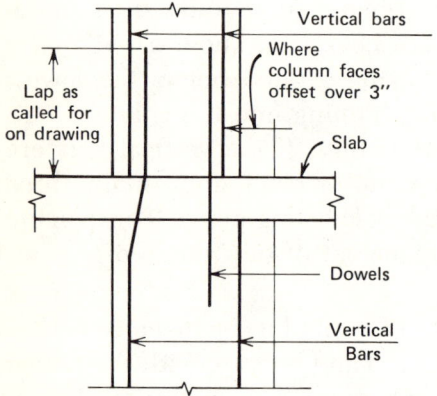

Figure 9-24

forms. Also, nails acting as spreaders are sometimes used as illustrated in Figure 9-25. Nails that are left exposed will rust and stain, and therefore should not be used where staining is undesirable.

PLACING BARS IN FLOORS AND ROOFS

Although threading bars in and around previously placed bars cannot always be avoided, it can be minimized by careful planning.

Since girders are usually deeper than the beams they support, the bottom lengthwise bars, truss bars, and stirrups are placed before any beam reinforcement is installed. Otherwise, the placement is similar to that for beams.

Beams may sometimes have all their reinforcement preassembled into a cage. Then this cage can be lowered into the beam form, making certain that all end clearances and bar extensions are correctly positioned.

Where beam reinforcement is preassembled or where lengthwise beam bars run through a column from each side and both directions, considerable difficulty can be avoided by positioning column verticals so they will clear the beam bars. This can be done by use of a template. This template must be installed before concrete is cast in the column. Some small adjustment of the column verticals may be required to provide for these clearances.

After column verticals are positioned and column concrete is cast to the bottom of the lowest beam or girder, the placing of beam and girder steel can be started. Beam bars are placed before slab or joist bars and after girder bars. When installing reinforcement, start at the bottom of

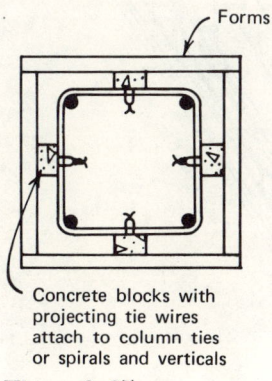

Concrete blocks with projecting tie wires attach to column ties or spirals and verticals

Figure 9–25

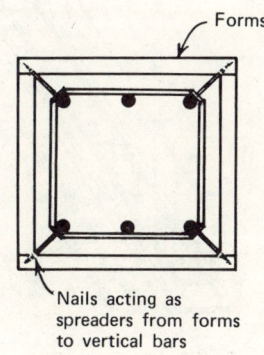

Nails acting as spreaders from forms to vertical bars

the lowest beam or girder and work up to the next lowest and so on. The steps are as follows (see Fig. 9-26).

1. Lower beam bolsters are properly located and spaced, resting on the bottom beam form.
2. Stirrups are placed with the closed end down, resting on the beam

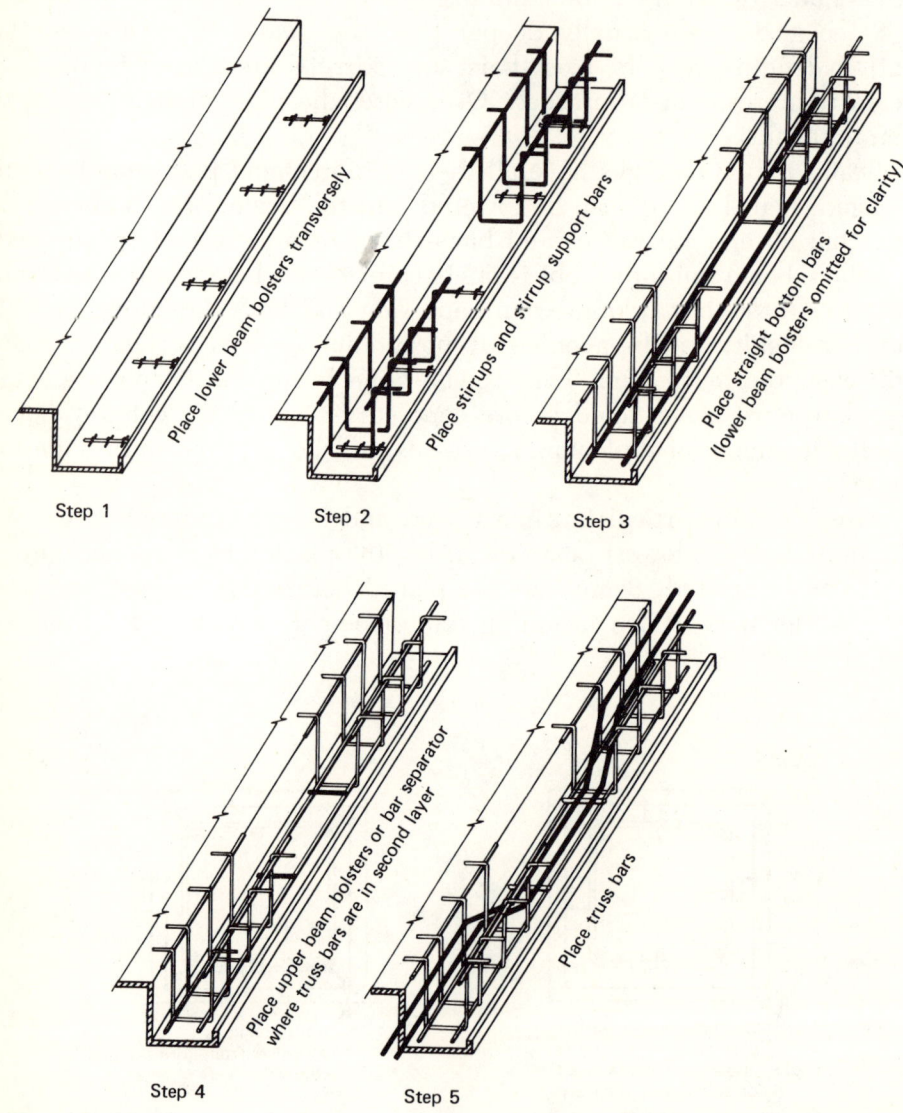

Figure 9-26

soffit and located opposite chalk marks made along the forms from spacings taken from the beam schedule. Be sure to start measuring from the right place, usually from the face of the column or wall. Special care should be used in locating stirrups and ties in spandrel beams. These beams may be subject to weathering, so be sure that the specified amount of concrete cover is provided over the steel at exterior surfaces to avoid the possibility of rusting. Stirrup support bars are placed under the hooks at the top of the stirrups.
3. Straight bottom bars are lowered into position inside the stirrups and rest upon the beam bolsters, with the ends of the bars extending the proper distance into each support. The stirrups are raised up and tied to the lengthwise bars at a sufficient number of intersections to produce a secure assembly. The beam bolsters are often wired to the bars to prevent displacement during positioning of bars and casting of concrete. When two layers of bars are required in the bottom of the beam, it will be shown in the beam schedule.
4. If the beam schedule or detail sections indicate that the bottom bars must be placed in two layers, upper beam bolsters or bar separators are laid in position and tied across the top of the bottom bars. These support the upper layer of bars.
5. Truss bars, if any, are lowered into place inside the stirrups upon the lower or upper beam bolsters, depending on whether there are one or two layers of bars. The bent-up ends are placed near the top of the beams to project the proper distance into the adjoining span. Any additional straight bottom bars required in the second layer are also placed as in step 3.

Bottom bars in the upper layer must be placed directly over those in the lower layer, not over the open spaces. When bars are in two layers, the straight bars are usually in the lower layer, and truss bars are in the upper layer. Sometimes, it is necessary to place some of the truss bars in each layer. This usually requires two depths of truss bars, the deeper ones being placed in the lower layer. The placing drawings will show any unusual or special placing requirements either in the beam schedules or by details on the drawings.

Closed ties are required in some beams instead of the open-type, U-shaped stirrups described. These ties may be specified in one piece, with hooks at a corner, or as cap ties in two U-shaped pieces that are lapped or hooked to form one tie.

Top bars in beams are placed after all other longitudinal bars are in place. Sometimes these bars are called for to be over the supports, and sometimes they are to extend the full length of the beam and into adjacent spans. They must be securely tied to stirrups and to column

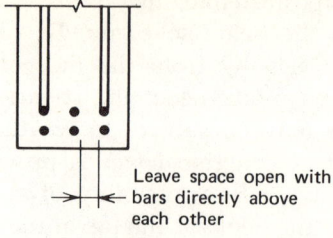

Leave space open with
bars directly above
each other
Figure 9–27

verticals. Some top bars are required for the sole purpose of providing support for the stirrups and ties. They are usually #3 (sometimes #4) bars and are designated in beam schedules as "Stirrup Support Bars." Ordinarily, they are just long enough to tie the first and last stirrup to each end of the beam, but, where stirrups occur throughout the beam, support bars are generally furnished full length between supports.

Joists

Joist bars are placed after all beam reinforcement is in place in the beams that support them.

Joist chairs are installed beginning 1 ft-0 in. from the edge of each support and at intermediate spacings as close to 5 ft-0 in. apart as possible.

The straight bottom bar is then placed on the joist chair, by threading

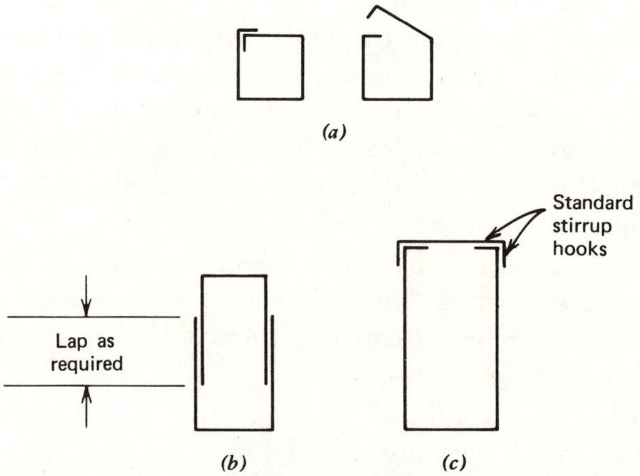

(a)
Standard
stirrup
hooks
Lap as
required
(b)
(c)
Figure 9–28 (*a*) Closed tie (placing difficult). (*b*) Two piece tie. (*c*) Closed cap tie or stirrup.

between beam stirrups and under top beam bars if necessary. The bottom bar must project into the supports at each end, as called for on the placing drawings and within proper tolerances.

The truss bar is placed on the same chairs, alongside the straight bar, with the bent-up ends crossing over all top bars in the beams and extending the specified distance into adjacent spans. Sometimes these bars must extend varying distances into each adjacent span.

If no truss bars are used, two straight bottom bars are placed as described. One or two top bars over the supports are then extended the correct distance into the adjacent spans and, where possible, wired to top bars or stirrup tie bars.

The extended ends of truss bars or straight top bars are held in position by support bars on individual chairs resting on the metal forms or on upper joist chairs, as shown on the placing drawings.

Some joists may have double or even triple the standard joist width and have either double or triple the amount of bars, or perhaps they may have a smaller number of larger-sized bars. Since joist chairs are made to carry two bars only, it may be necessary to place two or possibly three such chairs in line with each other across the width of the joist to handle the correct number of bars.

Distribution ribs, also called continuous header joists or bridging ribs, extend continuously the full length of the joist bay at right angles to the main joists and are the same depth as the joists. They are used for the longer joist spans to spread the load more evenly among the joists. There is no fixed rule as to the number of ribs used, but the engineer will generally use one at mid-span for spans between 18 and 24 ft, and two—at the third points—for longer spans. Bars in these ribs are shown on the joist floor plans and details, and they usually consist of one or perhaps two lines of straight bars in both top and bottom, lap spliced as necessary. The bottom bridging bars lie across the bottom joist bars, and the top bars are wired to the temperature steel in the joist slab.

Temperature reinforcement is placed after joist bars, sleeves, and con-

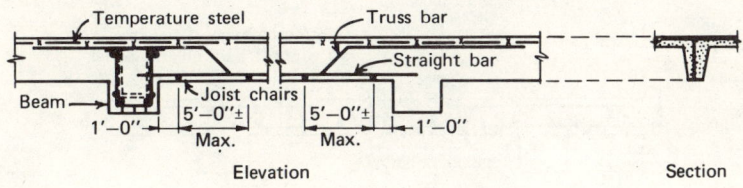

Figure 9-29 Typical bar placement in joists.

PLACEMENT OF STEEL

duits, for example, are installed and inspected. It may consist of either bars or welded wire fabric. If bars are used, they go on top of all other bars, with the spacer bars underneath them parallel to the joist bars. Temperature and spacer bars are usually #3.

Underfloor duct systems that carry electrical wiring are placed on shallow "pans" or special headers after beam reinforcement is placed but before joist bars are installed. It is necessary to work around these ducts, placing the bars over or under the ducts as required without disturbing them, since they are accurately set and leveled and must be kept in set position.

One-Way Slabs

Main reinforcement for these members consists of either alternate straight and truss bars or straight bottom bars and top bars. This reinforcement extends in one direction between beams. Temperature bars are placed at right angles to the main bars.

Slab bolsters are placed lengthwise to the slab panel with the lines spaced as close as possible to the 4 ft-0 in. maximum and with the two outside lines 6 in. from the edge of the beam. The first and last legs of each adjoining length of bolster are either locked or wired together.

Holes and openings in slabs are not desirable, but they cannot be avoided where ducts and piping are required. These openings are kept as small as possible, preferably so that the slab bars can be grouped with one half of the conflicting bar on each side of the opening at 2- or 3-in. centers running full length.

Where the opening is of such size that the space between bars from one side to the other is larger than three times the slab thickness (for example, larger than 18 in. for a 6-in. slab), the engineer may call for short header bars to be placed at right angles to the main slab bars at the other two slab edges. The slab bars then terminate at each side just as at an end support.

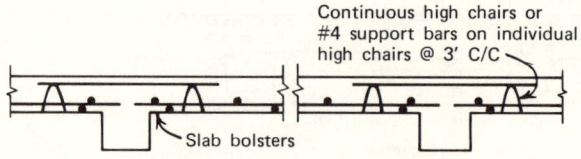

Figure 9–30 Straight top and bottom bars.

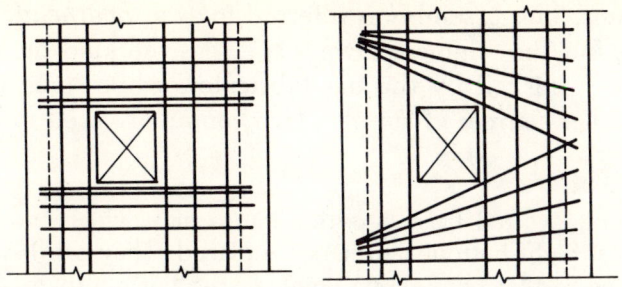

Figure 9-31 Alternate methods of placing rebars around an 8x12-in. opening.

For very small openings (6 to 12 in.), bars can be "fanned out" slightly to clear the opening, using the same number of slab bars as though no opening existed. Bar arrangement around openings is critical. The placing drawings should show the method to be used. Rebars should never be cut without the approval of the designer.

Two-Way Flat Slabs

Two-way flat slabs are generally built on flat, level deck forms with or without capitals or drop panels around the columns. The reinforcement is arranged in both directions in strips consisting of (1) straight and truss bars or (2) straight bottom and top bars. These strips are called column strips and middle strips. The width of each strip is one half of the span (center to center of columns). The column strips are more heavily reinforced, containing about 60% of the steel required in the bottom of the slab and more than 75% in the top of the slab. For this reason, column strips have either a greater number of bars at closer spacing, larger-sized bars or both, as compared with the middle strip. The column spacing in two-way slab construction is generally the same, or nearly so, in each direction, therefore the bar arrangement each way is usually the same.

To avoid building up too many layers of intersecting bars or the threading or weaving of bars, and to provide specified top and bottom cover, a proper order for placing of the bar supports and bars is necessary. Some field adjustment, particularly at the corners of intersecting strips or panels, may be necessary, but much less of this will result if a proper placing sequence is followed.

The placing order is always shown by notes or schedules on the placing drawings.

Without drop panels, it may be necessary to place shear reinforcement around columns. This assembly is referred to as a *shearhead*. Many kinds are available, but the engineer always specifies the kind he wants. Some kinds may be furnished by the bar fabricator, while some must be obtained from other sources of supply. Two commonly used types of shearheads are:

1. A group of inverted truss bars tied into a mat, which can be placed after the straight bottom slab bars are placed. They can be fabricated and placed so that bend-down points of the bar alternate at different distances out from the column. They may be preassembled on the job and set in place on bolster and high chairs, or tied to the bottom bars and afterward to the top bars of the slab.
2. A welded assembly of structural steel, usually with a pair of channels extending in both directions horizontally into the column strips and a pair of channels extending vertically into the columns above and below.

Sequence of Placing Bars and Bar Supports—with Straight Bars Only

1. Place continuous lines of slab bolsters in an E-W direction at 4 ft-0 in. maximum on center between columns. Begin spacing 1 ft-0 in. from center line of columns.
2. Lay N-S bottom bars in columns and middle strips.
3. Lay E-W bottom bars in column and middle strips.
4. Place three or more rows of #4 support bars (length 0.5L) at 4 ft-0 in. maximum on center in an E-W direction at each column head.
5. Lay N-S top bars in column strips.

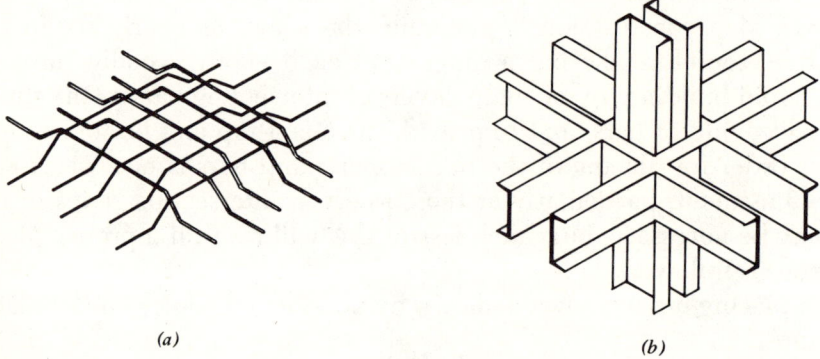

(a) (b)

Figure 9-32 (a) Tied, inverted truss bar assembly. (b) Steel shear head.

6. Lay E-W top bars in column strips.
7. Place three or more rows of #4 support bars (length 0.4L) between columns lengthwise in N-S and E-W column strips. Place two rows at all slab edges.
8. Lay N-S middle-strip bars within a width of 0.4L.
9. Lay E-W middle-strip top bars.

Sequence of Placing Bars and Bar Supports—with Straight and Truss Bars

1. Place continuous lines of slab bolsters in an E-W direction at 4 ft-0 in. maximum on center between columns. Begin spacing 0.1306L from center line of columns.
2. Place half-strip (0.25L) lengths of slab bolsters transversely in E-W column strips at 4 ft-0 in. maximum on center between columns. Begin spacing 0.130L from center line of columns.
3. Lay N-S bottom straight bars in column and middle strips.
4. Lay E-W bottom straight bars in column strip.
5. Place three or more rows of #4 support bars (length 0.4 L) at 4 ft-0 in. maximum on center in E-W directions at each column head.
6. Lay N-S column-strip truss bars and loose straight top bars. Adjust the spacing to place these bars within the bend-down point of E-W truss bars.
7. Lay E-W column-strip truss bars and loose straight top bars. Those bars resting upon N-S straight middle-strip bottom bars are to be tilted sideways to rest upon N-S top bars.
8. Place three or more rows of #4 support bars (length 0.4L) at 4 ft-0 in. maximum on center and high chairs at 3 ft-0 in. maximum on center in N-S and E-W column strips, parallel to the strips. Place two rows at all slab edges.
9. Lay N-S middle-strip truss bars within a width of 0.4L—adjust straight bars if necessary.
10. Lay E-W middle-strip straight and truss bars.

Placing drawings, bills of materials, and bundle tags should identify bar supports by nominal height, length, symbol of type of support, and class of protection. (Examples: $3\frac{1}{2} \times 10\text{-}0\text{-CHC-C}$ identifies a $3\frac{1}{2}$-in.-high, 10 ft-0 in.-long continuous high chair, plastic protected. $1\frac{1}{2} \times 5\text{-}0\text{-BB-A}$ identifies a $1\frac{1}{2}$-in.-high, 5-ft-0 in.-long beam bolster of bright basic wire.) Note that the last letter in the identification permits the ironworker to readily select the proper supports for placing.

Bar supports are intended to support the steel reinforcement and normal construction loads. They are not intended to support runways for concrete buggies or similar loads.

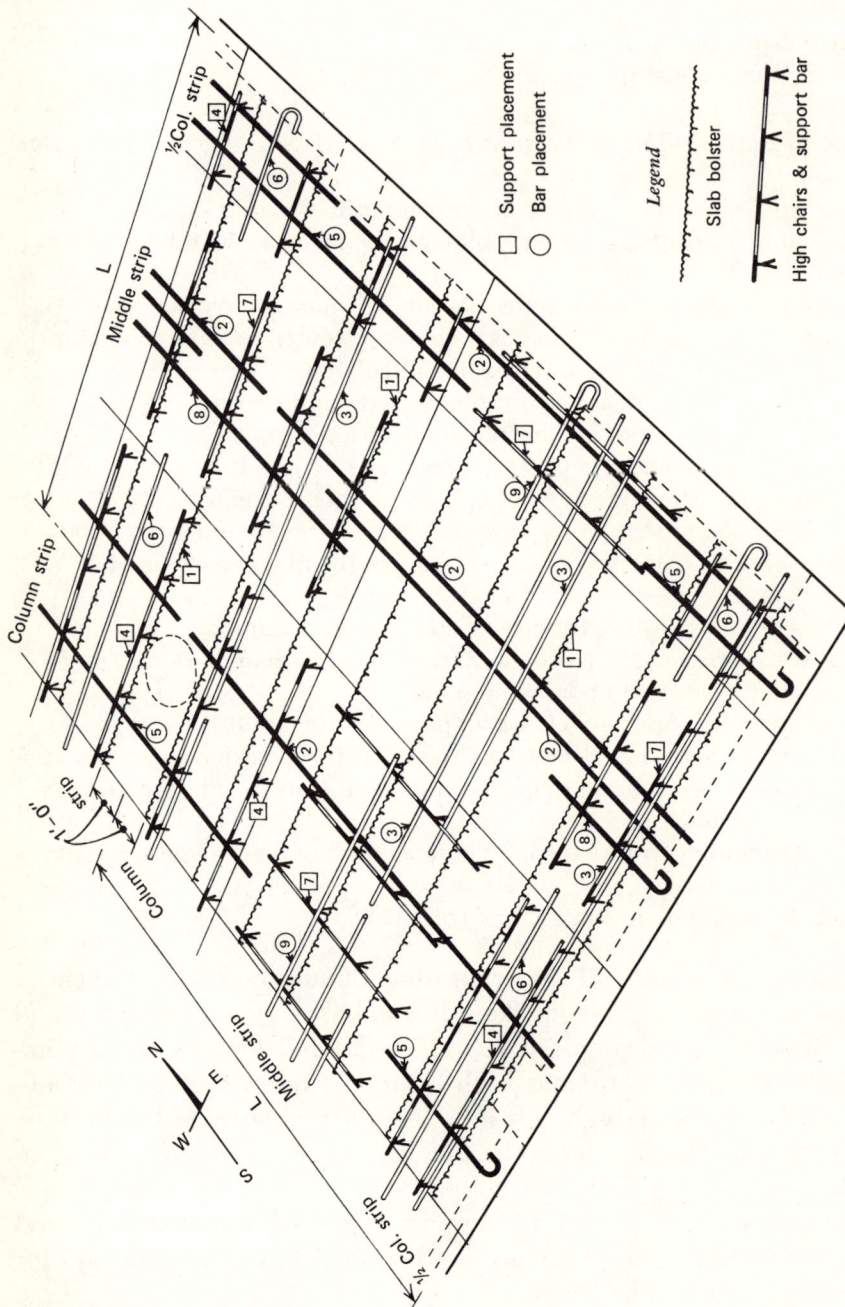

Figure 9-33 Two-way flat slab (with straight bars only, not all reinforcement shown).

204 REINFORCED CONCRETE, PRECASTING

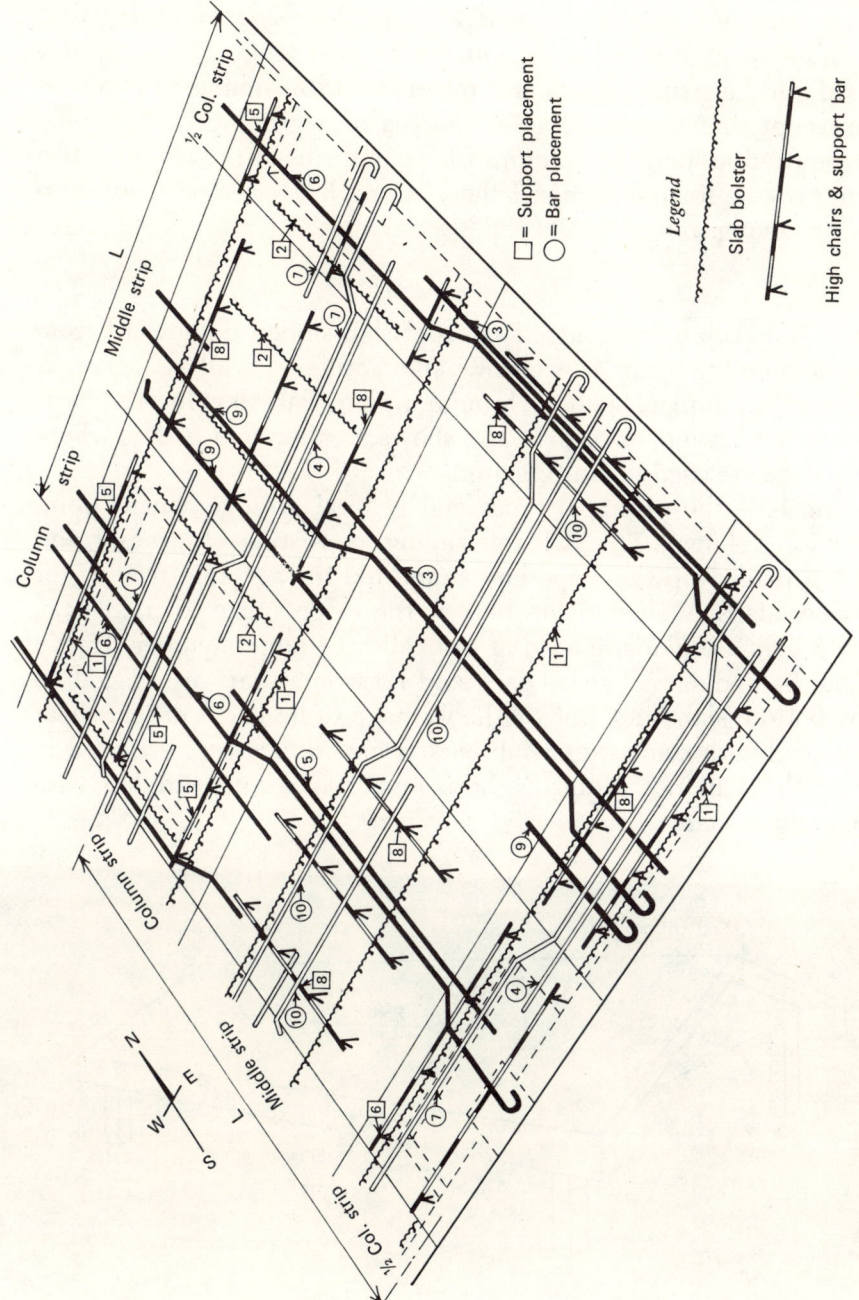

Figure 9-34 Two-way flat slab (with straight and truss bars, not all reinforcement shown).

PLACEMENT OF STEEL

Two-Way Waffle Slabs

The two-way waffle slab (so called because the dome forms produce a wafflelike appearance) is like the concrete joist floor system, except that the ribs run in two directions at right angles to each other. There are also column and middle strips, each a half panel in width, similar to two-way flat slabs except that a strip actually consists of a group of ribs. Dome forms are omitted around the columns for several ribs of the column strip in order to provide for the required shear strength. Here again, notes on the placing drawings must be followed.

Folded Plates

Figure 9-35 is a sketch of a single-span roof slab sloped downward from a ridge, with an edge beam at each side and a stiffener wall at each end.

Figure 9-36 is a typical section through a plate slab showing the location of the several layers of bars. It also shows a valley that occurs where these plate slabs are used in multiple units.

In Figure 9-37, the principal types and general arrangement of reinforcing bars are shown. The slab reinforcing, consisting of straight and truss bars, is placed up the slope and supported in the same way as for one-way floor slabs. V-shaped top bars at the ridge splice to truss bars. Hooked top bars are likewise used at the valley or edge member. Heavy longitudinal bars, diagonal shear bars, and horizontal bars are placed on top of the bottom slab bars, but the hooks have to lie flat. Vertical shear bars are placed with the bottom slab bars. Temperature bars are used in the part of the slab not requiring horizontal shear bars and lap with them where they occur.

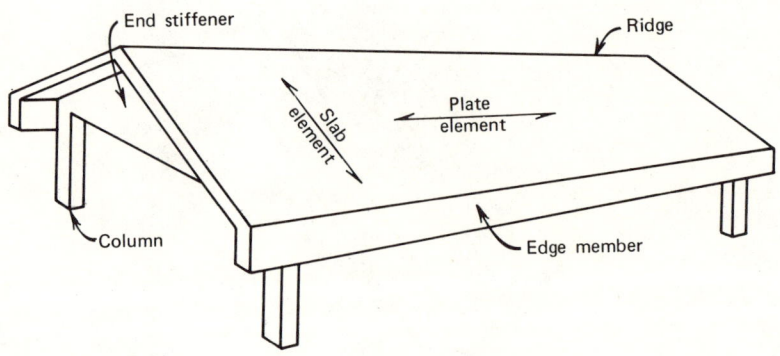

Figure 9–35

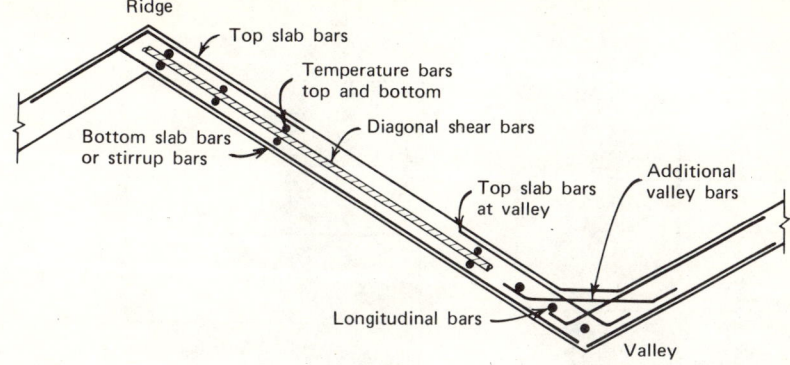

Figure 9-36

Arches, Barrel Shell Roofs, Domes, and Groined Vaults

These structures vary so much that it is virtually impossible to discuss a detailed reinforcing-bar placing procedure. The same general principles apply to these as to ordinary structures. The arrangement of bars can become quite complicated, however, and more than the usual amount of care must be used in following the placing drawings and details.

BAR PLACEMENT FOR OTHER STRUCTURES

Stairs

Bar placement for stair slabs is similar to other solid slabs. Sometimes the main bars extend crosswise onto a wall or other support on each side of the stairs. In that case, slab bolsters are placed lengthwise to the stair on lines approximately 4 ft on center, but with at least two lines in each stair from end to end. Cross bars are then placed on the bolsters and wired at

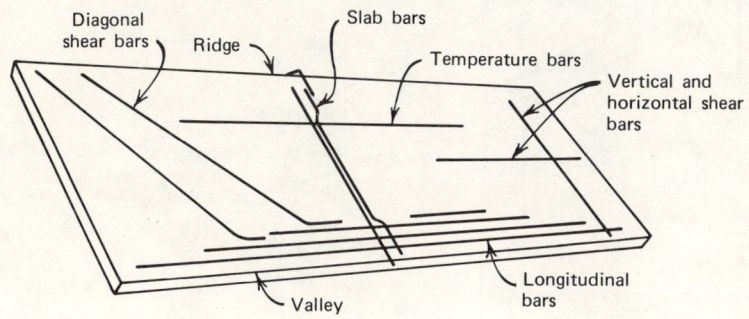

Figure 9-37 General bar arrangement.

PLACEMENT OF STEEL

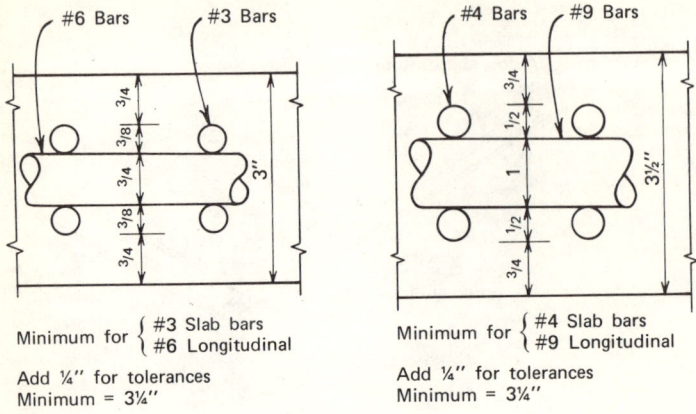

Figure 9-38 Example showing bar layers.

each intersection to prevent them from sliding. The longitudinal temperature bars are placed on top and wired to the cross bars to make a rigid mat. Wiring must be at a minimum of three points along the bar. As concrete is placed along this slope, men will be climbing up the mat, therefore it is important that it be fairly rigid.

When stair slabs span from end to end or between an upper and a lower landing, the main bars will run lengthwise and the temperature bars crosswise. Longitudinal bottom bars can be bent and extended into the bottom of the lower landing slab. At the upper landing slab, they should extend past the corner and be bent into the top. The bottom bars

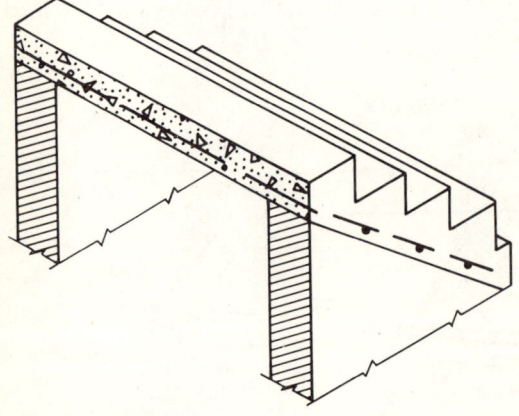

Figure 9-39

in the upper landing slab should extend past the corner and bend down into the top of the stair slab. Bars are always lapped as shown to avoid being pulled out of the concrete. Stair-landing bars may span in one direction, while the bars in the stair flight may run at right angles. Make sure that the heavier, closely spaced bars are placed exactly as called for. Check the drawings to make sure that any dowels or projecting bars for curbs, handrailing, or other features have been properly placed.

Bins, Tanks, and Grain Elevators

Circular Tanks—Slip-Form Construction. Most reinforced concrete tanks, silos, and elevators are constructed by a special slip-form machine. In this type of construction, bars and concrete are placed in one operation

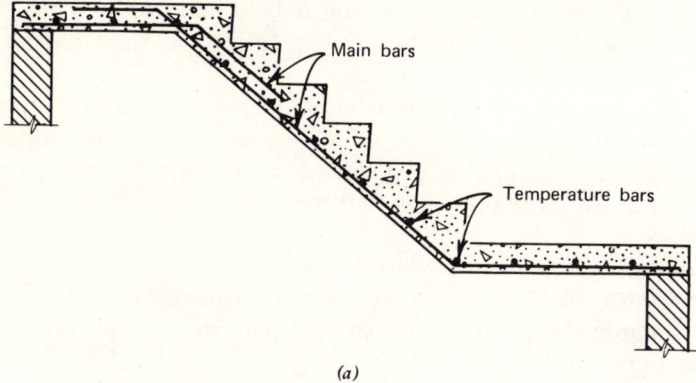

(a)

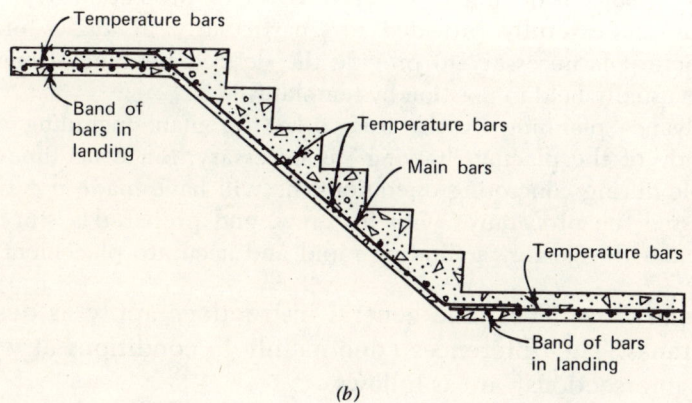

(b)

Figure 9-40 (a) Main bars one direction. (b) Main bars two directions.

PLACEMENT OF STEEL 209

while the forms are moving. Some important points in this type of operation are:

1. Horizontal ring bars must overlap each other by not less than the minimum specified amount.
2. Laps in adjacent lines of horizontal bars should be staggered.
3. Within practical limits, the spacing of horizontal rings should be as called for. There must be the specified number of rings within any given vertical section.
4. Anchoring and hooking of bars at openings and intersecting walls must be done exactly as shown on placing drawings.
5. Horizontal ring bars can be larger and more closely spaced at the bottom of the walls, varying to smaller and wider spacing at the top; or more commonly, the spacing is held constant and the size of bars is reduced. Care must be used to select the right bar size and to place it at the correct spacing.
6. No box, beam, or any obstruction should be allowed to prevent the proper number of lap of bars (though some reasonable adjustment in spacing is permissible). Watch for any special reinforcement around certain large openings that may be called for on the placing drawings.
7. Horizontal bars are often placed just outside of, and practically in contact with, the jack rods that are always centered in the wall. In cold climates, the temperature bars are placed as close to the outside face as permissible.
8. Due to the speed at which bars must be placed, only a minimum amount of tying of horizontals to verticals is done. This is to avoid disturbing the bars as the concrete is being placed.
9. The specified amount of concrete cover or protection over the bars must be carefully provided for, particularly in angles or corners where it is necessary to provide the desired bending strength. Bars are usually held in position by templates.
10. Advance planning by the ironworker foreman, including a careful study of the placing drawings, is necessary. Since no time is available during concreting operations, he will have made notations, discussed the procedures with his crew, and prepared a story pole or similar device for facilitating rapid and accurate placement of bars.

Rectangular Bins. The same general instructions apply as described for circular tanks. The differences (due mainly to conditions at wall corners and wall intersections) are as follows.

1. In the exterior walls, horizontal bars should be near the outside face, terminating near the end supporting walls with shorter bars near the

inside face at corners and diagonal corner bars spliced to the bars in the outside face.
2. In cross walls, horizontal bars are in two faces, and are anchored by hooks or bends around a vertical bar to provide reinforcement against outward pressure.
3. The distance from the surface of the concrete to the outside of the horizontal bars needs to be maintained carefully. Fig. 9–41 shows some of the steel needed for rectangular bins—much has been omitted for clarity. Also shown (Fig. 9–42) is a typical slip-form arrangement for positioning rebars.

Methods used in supporting steel vary considerably. In some cases, concrete is placed to the level where steel is required, the steel is laid on the wet concrete, and concreting is resumed. Another method uses concrete blocks to support single mats. Top mats in slabs up to 12 in. thick can be supported on wire chairs with sand plates or standees.

At interior construction joints, a keyed joint is used. The key, usually V- or trapezoidal-shaped, may be formed by using a metal or wood joint. If slip-dowels are used, the forms must have holes provided at the required spacing to allow the dowels to extend through. These dowels are usually #4 to #6, about 2 ft long, and spaced 12 in. apart. When the

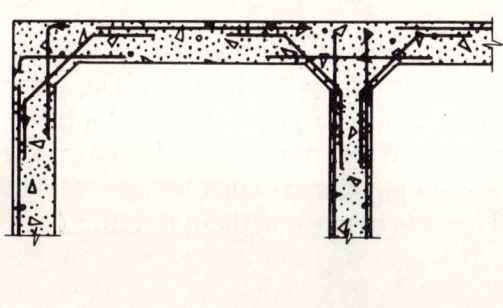

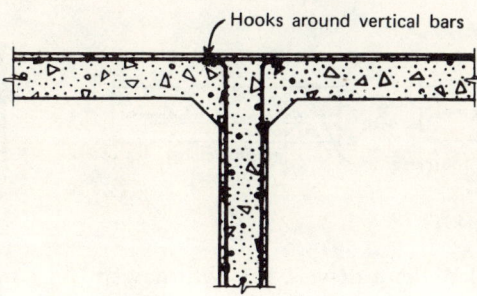

Figure 9–41   Anchorage at cross walls.

PLACEMENT OF STEEL   211

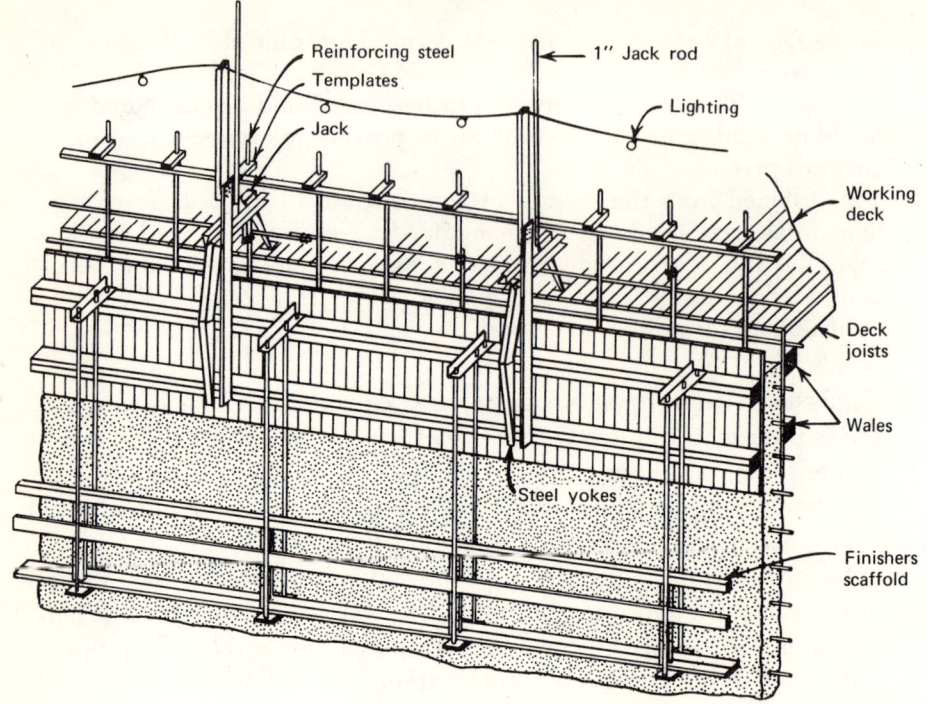

Figure 9-42

dowel extends through a form, it is supported by that joint in the center and by a dowel bar support at one of the ends. The dowel bar support is usually a formed stake that is driven a proper depth into the subgrade with a support hole near the top, or a wire stake with a spring-type clip at the top to hold the bar in position. It must be accurately aligned to make sure the dowels will be able to move without binding.

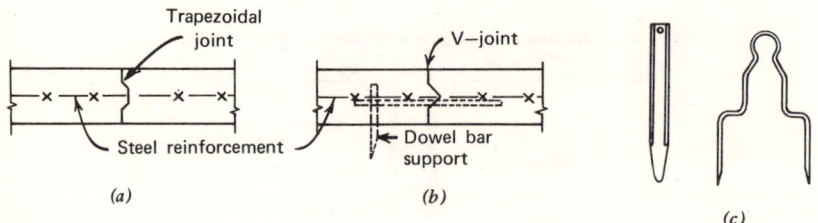

Figure 9-43 (*a*) Without dowels. (*b*) With dowels. (*c*) Examples of dowel bar supports.

Slabs on Ground

In building construction, slabs on ground are used for basement floors or for floors at grade where there is no basement. The subgrade must be properly prepared by compaction of the soil and provision made for proper drainage. All sewer pipes and underground duct work must be completed to the top of the slab elevation. Then steel placing and concreting can begin.

Premolded joint filler, $1/16$ to $1/4$ in. thick, is usually placed around the inside of all walls and columns. Control joints or interior construction joints are called for on column center lines, breaking up the floor into panels.

For residences and office buildings, the slab may be a minimum of 4 in. in thickness. Slabs used for trucking or other heavy commercial purposes, such as in industrial plants, may run from 5 in. to 10 or 12 in. in thickness. Bars are usually #3, #4, or #5 size, spaced 8 to 12 in. on centers each way. Heavy-purpose floors may require two mats, one near the bottom of the slab and the other near the top. Welded wire fabric, if used, is also correspondingly heavier. When only one mat is used, it is placed near the top.

Highway and Airport Pavement

Differences in the design of pavements among the states, counties, and cities involve thickness, joint types and spacing, and amount of reinforcing steel.

The thickness of a highway or airport pavement depends upon the subgrade support, strength of the concrete, the frequency and weight of wheel loads for the designed life of the pavement, and other factors.

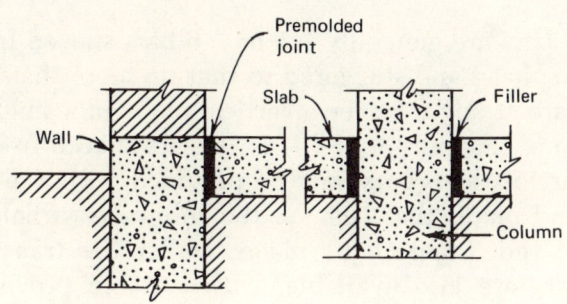

Figure 9-44

PLACEMENT OF STEEL

The width of pavements for airports is usually 150 ft for runways and taxiways. The reinforcement in continuously reinforced concrete pavements depends on the thickness of the pavement, the strength of concrete, temperature variations, and subgrade friction. In addition to the above factors, the reinforcement in jointed reinforced concrete pavements varies with the distance between transverse joints.

Pavements may be divided into three general types.

1. Continuously reinforced concrete pavement.
2. Reinforced concrete pavement with transverse joints.
3. Plain concrete pavement with weakened plane or sawed transverse joints.

Continuously Reinforced Concrete Pavement

Principles. In continuously reinforced concrete pavement (CRCP), properly placed reinforcement is vital to its performance. Each longitudinal bar must be made continuous by a proper lap splice.

Each longitudinal wire or bar must be able to carry its full yield-point load from end to end of the pavement with full strength at each line of splices. Careless placing, causing a line of splices too short or otherwise defective, can result in a local failure within the first week as the concrete hardens. The shrinkage movement of the concrete on both sides will accumulate at this point in a wide crack. Experienced superintendents, inspectors, and foremen learn to watch splicing carefully, but only the ironworker concerned can assure that each splice is properly made.

With CRCP, the continuous reinforcement eliminates the need of transverse joints, except at bridges. The reinforcement usually consists of deformed bars, but it may also be heavy plain or deformed welded wire fabric in flat sheets.

Bar-Reinforced

Longitudinal bars are generally #5 or #6 bars spaced from 4 to 8 in. apart with laps spliced and staggered so that no more than two-thirds of the bars terminate at any transverse vertical plane in a single traffic lane.

Transverse bars, generally #3, #4, or #5, are spaced from 12 to 48 in. apart. All or part of these bars may be provided with welded legs and, if necessary, sand plates that bear on the subgrade to hold the bars in proper position. Individual chairs placed under the transverse bars or continuous chairs are also available, with suitable provision made to support them on the subgrade.

In pavement that is reinforced with bars, it is not necessary to tie every intersection, but only a sufficient number to hold the bars in position, similar to the tying of any other mat. Longitudinal laps should be securely tied to insure the full length of lap specified.

Wire-Reinforced

Welded wire fabric is placed either on special chairs or on the surface of the concrete and is depressed into the concrete by means of special equipment designed for this purpose. A staggered pattern of end laps is desirable. End laps should be tied sufficiently to prevent dislocation while the concrete is being placed.

INSPECTION OF REINFORCING STEEL

Inspection is one of the most important steps in obtaining a thoroughly satisfactory job. The inspector represents the owner, architect, engineer, and governmental agency (city, county, state, or federal) in seeing that all details of the plans and specifications are properly carried out. In the performance of this function, the inspector must be prepared to interpret and explain the plans and specifications. If any details are lacking or confusing, he should consult with the engineer or architect for the correct interpretation or additional detail drawings. The inspector is obligated to get these questions cleared up quickly enough to allow the contractor to maintain his construction schedule.

The inspector's failure to detect or call attention to mistakes does not relieve the contractor or bar placer from the responsibility for following contract drawings and specifications.

The following lists contain points that should be checked by the inspector.

Check List Before Placing

1. Only approved placing drawings are being used and followed by ironworkers.
2. Reinforcing bars are of the grade specified.
3. Bar supports are of the type and finish specified.
4. All tests reports required by the specifications have been received.
5. Ironworkers are briefed on critical details that require special attention.

Check Items During Bar Placing

1. Number of pieces, size, length (if straight), shape (if bent), and grade of steel for all bars required in each structural member.
2. The proper amount of concrete cover between forms and steel for all members.
3. The type, finish, height, and location of all bar supports.
4. Proper positioning of bars in each member.
5. All lapped, welded, or mechanically connected splices.

Tolerances

Inspectors need to recognize that reasonable placing tolerances for spacing, positioning, and cover of reinforcing bars are necessary and should be permitted. Tolerances have been established to carry out the engineer's design requirements and, at the same time, provide leeway for efficient, economical bar placing.

Correction of Job Errors

The first basic decision to be made by the inspector when an error occurs in construction is how to correct it without delaying the job. In most cases, errors in reinforcing steel can be corrected by the addition of another bar or two. A telephone call to the structural engineer will usually result in an economical and immediate solution to the problem.

CHAPTER 10
PRECAST CONCRETE

Precasting means casting a concrete member at a place other than where it will be used and then moving it to the place where it will be installed. This is opposed to casting the unit in the position where it will be used (cast-in-place). For convenience, this discussion of precasting will be divided into two major topics: (1) on-site precasting and (2) precasting in plants.

ON-SITE PRECASTING

On-site precasting offers the following advantages over cast-in-place construction.

1. The cost of formwork can be reduced (with some techniques, this cost is almost eliminated).
2. Placement of concrete is easier, and finishing is simplified.
3. More accurate control of placing and finishing operations is possible.
4. Reinforcing steel is more accurately placed and more easily kept in position.
5. Precasting operations can continue, despite cold or rainy weather with only lightweight protection from the elements.
6. Precasting can be carried on at the same time that the foundation is being placed, with the double advantage of keeping the crews working

efficiently and having the building components ready to accept their full load as soon as they are in position.
7. Shoring, bracing, and scaffolding can be reduced or eliminated.

The disadvantages of this technique are:

1. The high initial cost of formwork. Because of the need for a number of reuses, forms are often made of heavy steel or concrete.
2. A large amount of space is required for both casting and storage.
3. Stripping the form from the unit may be a tricky job because of the size of the units and the need to prevent damage to either the expensive forms or the newly cast unit.
4. Precasting is a cost advantage only when the same size and shape is repeated many times in the building or when specialty items are in great demand.

Careful planning can eliminate many of these disadvantages. The initial cost of the form can be compensated for by the number of times it can be reused. If a form costs $1000 and can be reused 10 times without repair, the formwork cost per use is only $100.

Stacked Casting
The requirement for large areas of space can be circumvented in several ways. One of the most successful methods is casting one section directly on top of another. The first section is the form for the bottom of the second, and the only formwork that needs to be built is the edge of each slab and the bottom of the first section.

Because of the heavy loads involved with this technique, the supporting formwork must be substantial and the subgrade or supporting bed must be able to safely carry the total load. This is the same principle used in lift slabs, which will be discussed in detail later.

Form stripping can be made easier by (1) building the form so it will fold, (2) injecting water or air under pressure between the form and the concrete during removal, and (3) using a separating membrane (polyethylene) or compound (form oil).

Erection
Although it is possible to completely cure precast sections before they are erected, it is a common practice to shore and brace them in position when they are only 7 to 10 days old. This means that a considerable portion of drying shrinkage will occur after placement. Connections that would restrict this shrinking will cause cracking and should be avoided. But if

Figure 10–1 The floor for this building was cast first. It then served as the base of the forms for casting the tilt-up wall panels. Once they were braced in place, cast-in-place columns bound them together and filled the gaps between them.

connections can be made while still allowing the member some contraction, they are permissible.

One of the best and most commonly used connections is a dowel joint between wall panels in which one end of the dowel is prevented from bonding to allow the panels to shrink (Fig. 10-2). The space left by the shrinkage is filled with a joint sealant.

Special On-Site Forms

The variety of forms available to the contractor is almost unlimited. With careful planning, economical forms can be made of wood, concrete, plastic, steel, or any combination of these. Plastic-lined concrete forms have been used economically to produce many panels with a smooth surface and sharp details.

The procedure for making these fiberglass-lined forms involves forming a master pattern of the exact dimensions required for the finished product. The master pattern can be made of almost any material (wood, plaster, and even clay). The master pattern is covered with resin and

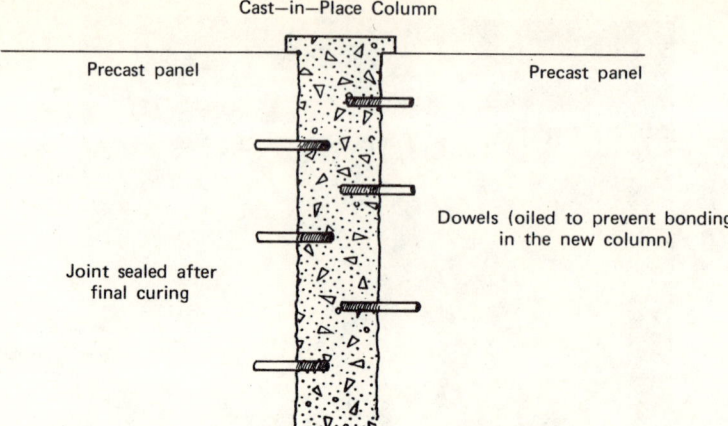

Figure 10–2

glass cloth built up in layers to the desired thickness. The fiberglass around the master pattern has anchors (usually roofing nails or something similar) embedded in it. When concrete is placed around the master pattern, the anchors become bonded to it and keep the fiberglass coating and the concrete together.

If more molds are needed, it is a simple process to cast one, cover the casting with fiberglass and place the concrete for another mold over it. This process can be repeated as many times as necessary. If, for example, you need 500 identical beams, it would be worthwhile to produce 20 or 30 molds in order to be able to cast the 500 beams as quickly as possible.

When used with care, fiberglass-lined forms have a long life. The fiberglass-lined forms for the Marina Towers in Chicago were used 60 times.

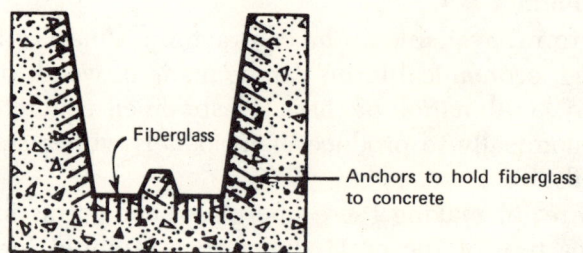

Figure 10–3 Fiberglass-lined concrete mold ready for reinforcing steel and concrete.

Preformed steel liners can be used in much the same way, but the liners are usually factory formed and are generally more costly than fiberglass.

All-steel forms, designed for many standard building units, are available at reasonable cost (usually for rent, lease, or purchase). The most common of these are the single- and double-T slab. These forms are equipped with a movable bulkhead that allows the contractor to cast the slabs in any length desired.

Tilt-Up Construction

This method of construction has several big cost-saving features. First, there is a great saving in formwork because only edge forms are needed since the previously placed slab becomes the bottom form and the top is troweled and finished in whatever texture is desired, just like any other slab.

Generally the procedure is a simple one, involving placing and finishing the base slab, then casting the wall panels on top of it in such a way that they can be tilted to their final position with jacks, cranes, or A-frames. The underside of the panels will have the same marks (in reverse) of the slab on which it is cast. This makes the finishing of the bottom slab doubly important.

A bond-breaker must be put on the bottom slab before a panel is cast on it. This can be a nonstaining liquid material that is brushed or sprayed on or a polyethylene film stretched so it is wrinkle-free.

The forming for small jobs can usually be done with four lengths of 2×6 used as edge forms. If the panel is being reinforced with light steel, the 2×6s can be notched or drilled to hold the rebars at the proper

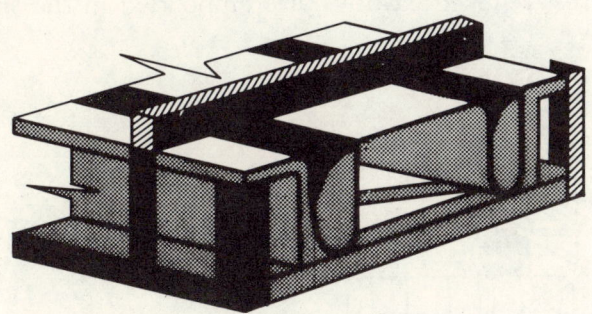

Figure 10–4 Double-T steel slab form with movable bulkhead to cast varying lengths in the same form.

PRECAST CONCRETE

height. With care in forming and stripping, the forms can be reused several times. Anchor bolts are embedded in the panel to attach the lifting (tilting) mechanism.

The anchor bolts for lifting can be placed in several ways. One technique is to wrap a bolt in paper or plastic.

This way the bolt can easily be knocked out of the panel and the hole filled with mortar after the panel is positioned and anchored. Another common method for small slabs is to embed the bolts in the top of the panel and use a section of angle iron to spread the lifting stress across the top. Whatever method of inserting anchors is used for attaching the lifting frame to the panel, the anchors should be located at the quarter points of the panel. That is, about $\frac{1}{4}$ of the distance from the top and $\frac{1}{4}$ from the sides. (Small panels can be lifted from the top.)

Equipment for tilting can vary from a simple pipe frame (Fig. 10-6) to heavy hoisting cranes. The pipe frame will handle panels up to 10 ft square and 4 in. thick. A winch, truck, or other vehicle is used to pull a cable attached to the top lifting bolts and passing over the high point of the frame. When the panel is vertical, the lifting frame holds it in place until it is firmly braced and ready for permanent attachment.

Panels are often held in permanent position by cast-in-place columns with a control joint on one edge, as shown in Fig. 10-7.

The control joint prevents cracking of the concrete that is caused by random movements of building elements from uneven settlement or by temperature differences between members.

Lift Slab

Lift slab is a spcialized construction technique in which all of the slabs to be used in a building are cast one on top of the other around the pre-erected building columns. Heavy steel collars, sometimes welded, but more often stress-relieved castings, are embedded in the slab when it is cast.

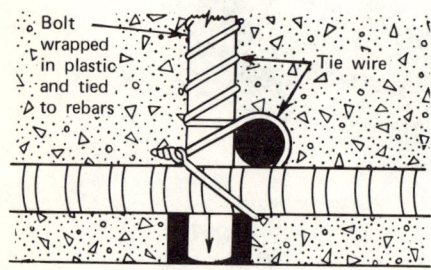

Figure 10-5

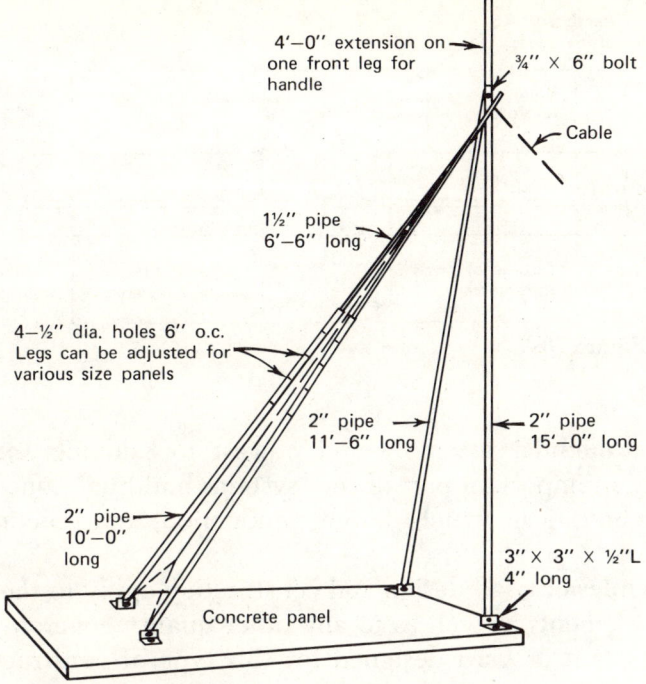

Figure 10-6

Special hydraulic jacks that are placed on the top of the columns lift the slabs to their final resting positions. The collars (with the slabs attached) are welded or keyed to the column for permanent attachment. As the slabs are jacked into position, they must be held to a horizontal tolerance of ½ in., which requires very careful control. Specialists who do this kind of work have developed systems of central control that monitor each jack and maintain the horizontal alignment to ¼ in.

Lifting rates vary from 5 to 15 ft per hour but are normally about 10 ft per hour. Sometimes, for the sake of column stability, slabs will be lifted and parked in a temporary position. After leveling and permanent attaching to the column, the space between the collar and the column is filled with a grout to prevent corrosion.

PLANT PRECASTING

Concrete has very few limits in size, shape, color, or texture and can be made strong enough for almost any job. For these reasons, concrete is well suited to precasting of building units. Any building unit can be

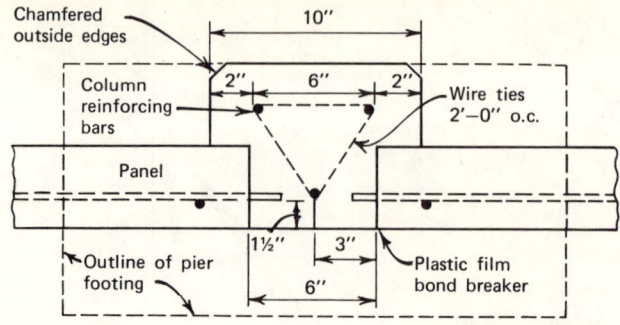

Figure 10-7

designed for modular assembly and precast to exacting specifications. Precasting is an important part of the "systems building" concept—a construction technique in which design, production, and erection are pre-engineered.

The principles of good design and construction apply to the precasting of building elements as well as to any other quality concrete work. Precasting in a plant or yard designed for this type of construction makes possible the economic use of a great variety of finishes and such special techniques as "sandwich" panels, steam curing, heavy external vibrating (Schokbeton process), and casting of large, intricately shaped panels.

Precast building units are designed for a specific job and are detailed in shop drawings for both manufacture and installation. The drawings will also detail the method of erection and any temporary bracing needed during erection. Shop drawings should also give all dimensions including tolerances and, if the units are to be prestressed, an allowance for elastic shortening will be included. The size, shape, and location of reinforcing steel and connections (as well as method of adjustment) will also be included. Drawings should show details on all inserts, reglets, and attachments, for example, and methods of anchoring and joint details. Precasting allows for much finer tolerances than are possible with on-site production. For this reason, every detail of the panel is stipulated in a drawing, and plus or minus tolerances are held to very close limits. A variation of ¼ in. in on-site production is no serious matter because, when the forms are rebuilt, that ¼ in. can be allowed for. In plant production, where many building elements are going to be cast in the same mold, a ¼-in. error will be compounded since each of the building elements will be off that much. This could add up to several inches of error in a tall building.

Forms

Forms for precast concrete are made of steel, wood, concrete, plastic, or a combination of these. Steel forms work best when the unit has a complicated shape, and many of them will be cast in the same mold. If steel forms are used, rivet heads and joints must be ground smooth because the surface of the steel will form the surface of the building unit. It is often necessary for steel forms to be capable of opening in order to remove the building unit. The forms must be checked occasionally for buckling or dimpling and must be coated with a nonstaining, light form oil before each usage. Steel forms are probably the best to use when units are to be steam cured.

Fiberglass can be molded into complicated shapes and reused many times. The material is strong, but it does require support to give it rigidity. Concrete surfaces cast against fiberglass can be glossy smooth or lightly textured.

Wood forms are easy to fabricate, but they are not as durable as steel or fiberglass. Wood forms must be treated with a sealant to protect the wood from the wet concrete and should be coated with a parting compound such as form oil before each reuse.

Concrete forms are durable but they are understandably heavy. They work well for many shapes and can be reused indefinitely if handled with care. Concrete forms require a sealant to prevent a bond between the form and the fresh concrete placed in it. The sealant can be wax, paraffin, form oil, or even a polyethylene film.

Figure 10-8 These precast concrete joists are just one of the many building components cast in a plant.

PRECAST CONCRETE

One of the greatest advantages of concrete forms is the simplicity of construction. A master pattern is made in exactly the size and shape that will be required. The shape can be intricate and complex, as long as there is sufficient draft to allow the cast unit to be removed easily. An edge form is built around the master pattern, a bond breaker put on the pattern, and concrete (with or without steel) is cast on top of it. This forms the master mold.

After the master mold has cured, it is lifted carefully from the master pattern, turned over, and placed where it will be used. Then a bond-breaker can be applied and the desired shapes cast.

If more molds are needed, it is a simple process to repeat the entire procedure with one of the members that has already been cast in the master mold acting as the pattern.

Fabrication

Mass production, assembly line techniques are common in precasting plants. Forming, casting, curing, handling, and storage are handled by crews of specialists who know their phase of the operation thoroughly. Because of the nature of precasting, much of the work is repetitious and errors can be spotted immediately.

Reinforcing steel, for example, can be cut, bent, bundled, tagged, and readied for placing. A special area is usually set aside for steel fabrications so that jigs can be set up on benches at a convenient working height and individual components assembled into complete reinforcing cages. The completed cages can be stockpiled for later installation in the forms. Frequently, inserts such as lifting eyes and connecting hardware can be attached to the steel cage, and electrical conduit can be tied in place before the cage is placed in the form.

While the reinforcing steel crew is making cages, another crew is preparing the forms for concreting. Previous castings are stripped out and moved to the storage area for additional curing. Forms are cleaned, oiled, and reassembled, and reinforcing steel is set in place. Cages are accurately positioned in the form, often with templates to insure accurate location, and tied so that they cannot move during the concrete placing.

Concreting follows conventional methods. The concrete used is generally of 2-in. slump or less. In most precasting plants, both internal and external vibrators are used. In many cases, high-early-strength cement (Type III) is used to permit a rapid turnover of the forms and casting beds.

Another technique used to increase turnover is steam curing. Perforated steam lines are run over the casting bed, and the lines and forms are covered with tarpaulins. The concrete is raised to a temperature of as much as 160°F for rapid curing.

With careful use of reinforcing steel, rich mixes, Type III cement, and steam curing, it is often possible to reuse molds daily. Some plants, particularly in Europe, have turned to using steam instead of mix water, producing concretes with temperatures of 140°F or more. These plants strip units from their molds as soon as 4 hours after casting. The units are cured with steam until they are stripped.

Many precasting plants continue to cure members after they are taken from the mold. This is usually done in a moist atmosphere, often under wet burlap.

After curing, units are sometimes stored for a period of time. Proper storage is necessary to prevent warping or distortion. Panels can be stored flat if they are fully supported on a flat surface. However, there is less danger of warping if they are stored on edge, provided the panel edge is fully supported and the panels are resting against a rigid, non-staining frame. Beams should be supported at their normal support points. Also, it is important that units be spaced so they can dry uniformly on all sides.

In handling and transporting precast units, great care is necessary to avoid marring or staining. Edges and corners are especially easy to break and should be protected with timber if it is necessary to lift the unit or tie it down to a truck. Ropes are less apt to damage surfaces than are cables or chains.

Finishing

Precast units are often made with exposed aggregates, special textural finishes, tinted concrete, or with surfaces that are acid etched, sandblasted or bushhammered.

Whatever color or texture is used in the building unit, uniformity of finish is very important. In order to assure uniformity of texture, form liners can be used. Rubber and plastic sheets are commonly used for this. For deeply textured panels, molded rubber form liners have worked very well.

Joints

Joints between precast panels are usually grooved and packed with a neoprene rubber sponge of low density or a similar material that is both durable and resilient. Thermo-setting plastics (plastics that harden at high temperatures) such as a polysulfide are preferred as joint sealants, although some thermoplastics (plastics that remain soft at high temperatures) have been found suitable. Oil base compounds eventually dry out and are not considered as suitable joint material between panels. Cement mortars are not flexible enough to form a good joint and for this reason are seldom used.

The outside ½ in. of the joint is usually filled with an elastic sealant. The sealant should be applied when the joint is clean and dry and only when temperatures are above freezing.

Connections

Connections are the hardware used for joining precast units to the building frame and to each other. Connection design is stipulated by the structural engineer because connections play a major role in the stiffness or flexibility of the building frame. Tragic failures of precast buildings have occurred because of lack of proper design or lack of care in the installation of the connecting hardware. Analysis for connections must give first importance to structural adequacy, second to architectural functions (the connection should be neat, clean, essentially invisible, and compact), and third to economy. Economy requires both ease of erection and ease of fabrication. For example, it's a good idea to design a connection so it can be fastened to temporary supports, thereby freeing the crane for other work while the permanent connection is being made. Permanent fastenings are usually welded, bolted, riveted—or a combination of these.

Sandwich Panels

In this type of construction the wall panel is cast in a flat position with a layer of structural concrete, a layer of lightweight insulating material, and a third layer of structural concrete. This way the panel has the double advantage of high strength and high insulating value.

The insulating material used must have the following desirable qualities in order to be effective: (1) the material must develop a good bond with the concrete; (2) it must be reasonably stiff; (3) it must have a reasonable compressive strength; (4) it should be resistant to absorption of moisture; (5) it should have no capillarity; and (6) it should be capable of little or no vapor transmission.

A material that meets all of these requirements can be used to produce thin, strong, good-insulating panels for building construction. Some of the materials meeting these requirements are foamed concrete, foamed glass, and some foamed or expanded plastics such as polystyrene and polyurethane. Densities of these insulating materials range from about 5 lb per cu ft to a little over 50 lb per cu ft.

The outside layers of the sandwich panels are connected by means of steel bars or by concrete spacers or ties. In general, properly located concrete ribs are more efficient than metal ties, but they can create cold spots on the walls.

Special Techniques
The *Schokbeton* process is a patented process where a so-called zero-slump concrete is placed in a rigid mold and the entire assembly, mold, and concrete is subjected to low-frequency vertical shock or impact that consolidates the concrete without segregating it. Speed of vibration is about 250 cycles per min.

Molds used in this process must be rigid and strong enough to withstand the rather extreme casting process. Mold dimensions are held to a tolerance of plus zero, and minus $3/32$ in. for 10 ft or over. The resulting concrete has high strength, high density, and low water absorption and produces castings with exact tolerances and sharp corners.

Reinforcing steel must be accurately placed and rigidly tied. The concrete used in this process has a compressive strength of not less than 6000 psi at 28 days.

The *extrusion* process makes use of the fact that compacted, vibrated "green" concrete will maintain its shape without a form to hold it. A low-slump dry mix is fed into a machine that compacts it, forces it into the desired shape, and squeezes it out much like a ribbon of toothpaste. The extrusion process is particularly well adapted to long-line, continuous manufacture of prestressed concrete.

Prestressing wires are stretched the full length of the casting bed, which may be as much as 700 to 800 ft long. The extruder, riding on metal rails at each side of the bed, moves along at about $3\frac{1}{2}$ ft per min., squeezing out a ribbon of concrete as it goes. Slab sizes are usually 48 in. wide and 6 or 8 in. thick and have hollow cores. As the slab emerges from the machine, it is covered with wet burlap or curing compound to keep it from drying out during the present time period. Curing is usually done by steaming so that the slabs are ready to be cut into the required length on the following day. Concrete saws are used, and the "planks" can be cut to any length desired.

Erection

Precast members are often subjected to severe loading during handling. For this reason, handling should be kept to a minimum. Units are lifted and carried from the casting beds to a storage area, from there to some form of transportation, and from the truck or train to the place where they will be erected in the building. In each of these moves, damage is a possibility, and great care must be exercised in lifting and handling. The structural engineer responsible for the design of the building will specify or approve the location and type of inserts for the lifting and handling procedures. Good rigging practices must be followed, and the members must be shored, braced, or connected adequately while the permanent connections are being made. A member capable of withstanding tremendous loads from a direction for which it was designed may be comparatively weak when loaded in other ways.

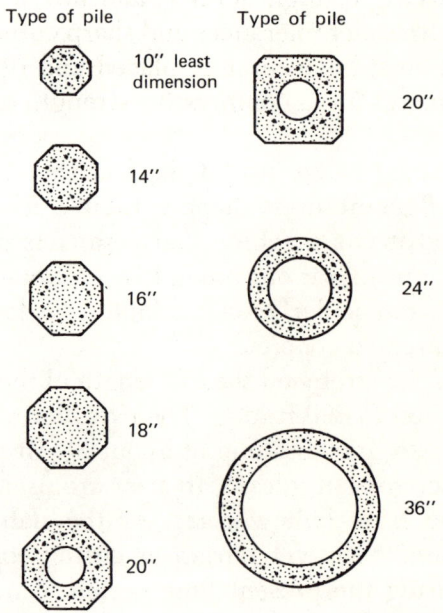

Figure 10–9   Some typical sections of pretensioned piles. Large piles frequently are posttensioned.

Precast Concrete Piles

Precast concrete piles are generally square or octagonal in cross section, although cylindrical spun piles, usually hollow, are also often used.

Piles are usually of a uniform section throughout their length. Octagonal and circular piles have the advantage of having the same flexural strength in all directions, so they can resist forces from any side. Lateral ties for octagonal piles can be a continuous spiral, and the pile itself can be made in wood or metal forms without edge chamfering. Square piles, however, are easy to form and have a greater surface area to volume ratio.

Piles are made in size to suit almost any condition, varying from a 6-in. square 30 ft long to a 54-in. square over 150 ft long. Reinforcing steel is usually preformed into cages and set into the form in one piece. Concrete should be at least 2 in.; and for piles exposed to sea water or cycles of freezing and thawing, 3 in.

The ends of the pile are called the head (the part struck by the driver) and the point. The head must be smooth and must have extra lateral reinforcement for distances of at least one diameter of the pile. The shape of the point (Fig. 10-10) depends on the soil and driving conditions. For

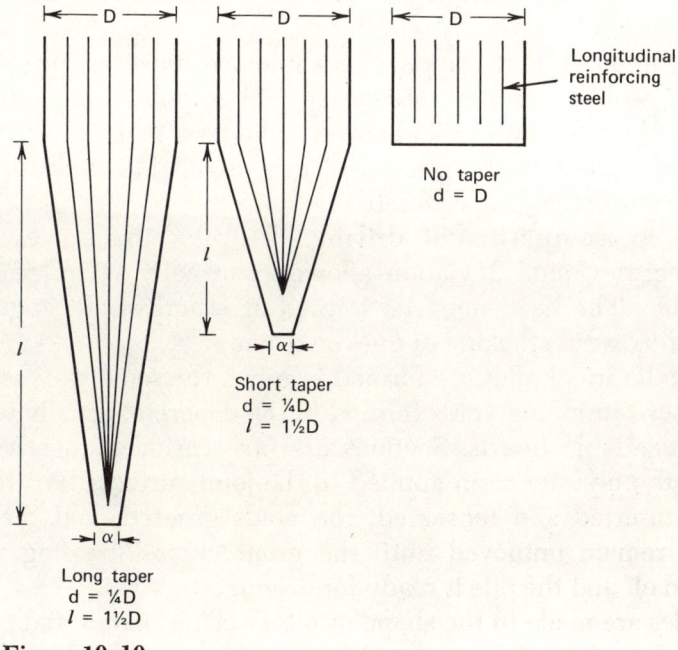

Figure 10–10

PRECAST CONCRETE 231

hard layers or packed sand and gravel, the point will have a long taper; for plastic soil, a short taper or perhaps no taper at all. Longitudinal steel will follow the taper, drawing together in the center of the pile. Extra lateral reinforcement is usually inserted in the tip. Prestressed piles are almost always square-ended.

Square piles are often formed in banks or tiers by covering the sides of the form with a heavy building paper or sheet plastic, which then becomes the form for the second group of piles after the first group has been cast and stripped.

Small concrete blocks are usually wired to the reinforcing cage to maintain accurate positioning of the steel from the sides of the form. The finished cage is usually 6 in. shorter than the finished pile to allow a 3-in. clearance at the head and point.

Mixes for concrete piles are usually rich (high cement content), containing 565 to 660 lb of cement per cu yd of concrete. The slump need not exceed 2 in. for vibrated concrete containing entrained air. When the water sheen disappears from the top surface of the fresh concrete, a wood-float finish is applied. Wood floating is usually sufficient for finishing procedures. Good curing practices must be followed. Concrete should have a compressive strength of not less than 4000 psi when the piles are driven. However, plans should be checked to determine what is actually specified for any job.

Handling imposes heavy stresses on a pile and must be done with care. Piles should be lifted only at designated pickup points to avoid bending stresses, shocks, and jolts. Field-cured cylinders, exposed to the same conditions of curing as the piles, will give a reliable indication of the compressive strength of the concrete.

In order to assure straight driving, the piles themselves must be straight. The maximum deviation allowed is usually ¼ in. per 35 ft of length of pile. The head must be formed at exactly right angles to the pile in order to avoid cracking or uneven driving.

Hollow cylindrical piles are formed in much the same way as concrete pipe with posttensioning voids formed in the concrete, usually with mandrels or expendable inserts. Sections are posttensioned together with a high-strength polyester resin applied to the joint surfaces. Posttensioning cables are inserted and tensioned, the voids grouted, and the sections allowed to remain unmoved until the grout cures. Stressing wires are then burned off and the pile is ready for driving.

Sheet piles are made in the shape of interlocking planks and in varying thicknesses from 6 to 12 in. and widths from 15 to 24 in. The interlocking is provided by a tongue-and-groove design. The point of the pile is beveled so that it will be forced against the previously driven pile, thereby assuring that it will remain in alignment during the driving operation.

Miscellaneous Precast Items

Concrete roofing tiles are made in a variety of machines that extrude, press, or vibrate the concrete. Color is attained either through the use of integral coloring material that tints all of the concrete or by adding a glaze applied to the surface with a pneumatic gun. These tiles are generally made 10½ by 15 in.

Railroad cross ties made of concrete have been installed in a number of locations in the United States and Canada. Ties most often used in this country today are prestressed, using high-quality, dry-mix concrete. Most are steam cured. One of the most recent and impressive uses of concrete railroad cross ties is in the San Francisco urban transportation system.

Other precast items include floor joists, manhole rings for construction, manholes for sewers and underground utility installations, septic tanks, parking bumpers, garden furniture, silo staves, fence posts and rails, and many others limited only by the ingenuity of the precasting manufacturer.

Most precast members are manufactured using rich mixes, low- or no-slump concrete, and some method of compacting to give a very dense product.

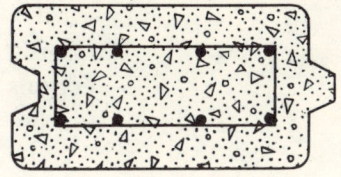

Figure 10–11 A tongue-and-groove interlocking sheet pile.

Figure 10-12 Los Angeles Dodgers' stadium is an all-concrete structure featuring precast units throughout.

CHAPTER 11
PRESTRESSED CONCRETE

In each of the following discussions, rectangular, nonreinforced, simply supported beams are used as examples. The principles are the same for any size or shape structure, but the actual design procedures are different (and somewhat more complicated) than the simple steps herein outlined. The selection of size, shape, number of cable strands, pattern of deflection, and manner of anchorage makes the design of prestressed members both a science and an art. In actual practice, safety factors are included that have been disregarded here for the sake of simplicity. This discussion is presented as a method of explaining what prestressing can accomplish and how it can be used to produce low-cost, high-quality building components.

The beams used as examples in this chapter would not be suitable for construction. These beams use no reinforcing steel and their size would have extremely limited application; also, the shear stresses in the beam in Figure 11-13 would be unacceptable.

About the turn of the century, engineers began using steel in concrete to construct building components capable of resisting bending and twisting forces. This technique produced a structure built of a low-cost material to carry compressive stresses combined with easily fabricated steel to carry tensile stresses. However, one of the disadvantages of the reinforced concrete method used at this time was that the concrete in the bottom half of members was not working at its full potential.

Prestressed concrete goes beyond reinforced concrete. First, all of the concrete on the tension side of the neutral axis is put under an initial compressive stress to eliminate any tension in the concrete from all future loads. The prestress is applied in such a manner that it creates a bending moment opposite to those that will be produced by applied loads. In the ideal design, this negative moment carries all of the dead load and creates a maximum allowable compressive stress in the concrete. As far as the stresses in the prestressed member itself are concerned, it is then a weightless structure with all its capacity available for live loads.

FORCES IN A PLAIN BEAM

The beam in Figure 11-1 is a concrete beam without steel. This is a solid, rectangular, simply supported beam, 9 in. high and 6 in. wide, that spans 15 ft.

In order to understand what prestressing does and how it works, it is necessary to understand some of the forces in a simply supported beam.

First, compute the dead load assuming normal weight concrete at 150 lb per cubic foot

Area of concrete $= A_c$

$A_c =$ width $\times$ depth $= 6$ in. $\times 9$ in. $= 54$ sq in.

Dead weight $= w$

$$w = \frac{A_c \times \text{weight of normal concrete per cubic foot}}{\text{Number of square inches in 1 sq ft}}$$

$$w = \frac{54 \times 150}{144} = 56.25 \text{ lb/ft}$$

The 150 lb per cubic foot for normal weight concrete is a safe assumption, so we can round off the weight of the beam to $w = 56$ lb/ft.

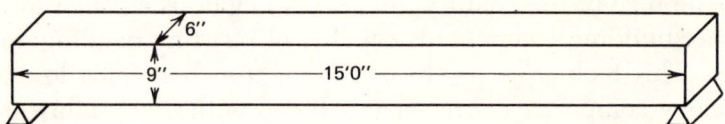

Figure 11-1 A 6 in. x 9 in. x 15 ft.-0 in. plain concrete beam, simply supported.

Bending Moment

Bending moment is the tendency of a structural member to rotate about a point or axis as a result of some force. The bending moment can be calculated separately for each of the many forces acting on a member. They are each taken into account by design engineers.

Since we are working with a simply supported beam, we are concerned only with the two major components—the bending moment as a result of the dead load (M_g) and (a little later) the bending moment as a result of live load (M_L).

Bending moment is expressed as a distance-weight such as inch-pounds or inch-tons or inch-kips. (Kip is a commonly used term for 1000 lb. Inch-kips is thousands of inch pounds. 20,000 in.-lb would often be shortened to 20 in.-kips.) The formula for determining the bending moment for a uniform load is

$$M = \frac{\text{Weight} \times \text{Length}^2}{8} \quad \text{or} \quad M = \frac{wL^2}{8}$$

The weight (w) used can be live load, dead load, or total load. In these examples, we will separate the stresses caused by dead and live loads so that we can alter the live loading without recalculating dead-load stresses.

A simply supported beam has a weight per unit of length (w) and a length (L), and each support carries ½ of the product of these, or $wL/2$. The bending moment is greatest at mid-span, so this is the point at which it should be calculated.

Forces are acting in two directions on this beam. The weight is pushing down and the supports are resisting this push—in effect, pushing up.

There is a point on any beam, under any load, at which the bending moment will be exactly the same if the load were concentrated as it would be if the load were uniformly distributed. For a uniformly loaded, simply supported beam, this point-of-equivalent-loading falls ¼ of the length of the beam away from the ends. These are the quarter points, or $L/4$.

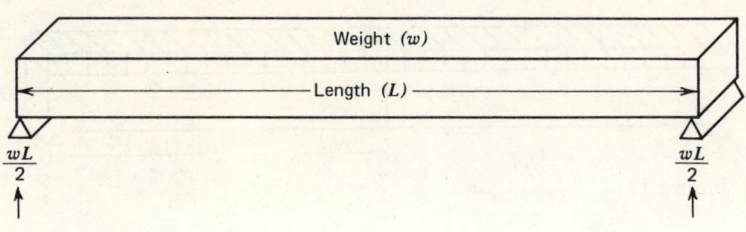

Figure 11–2

PRESTRESSED CONCRETE

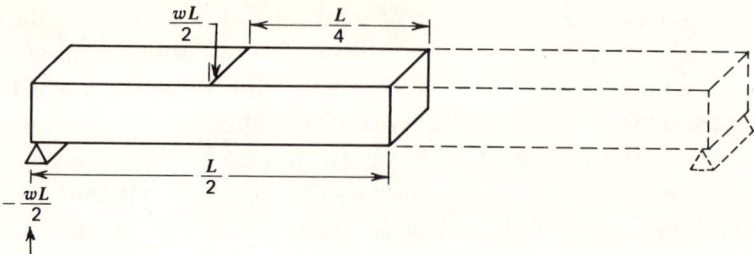

Figure 11-3

The loadings in Figures 11-4 and 11-5 would give the same bending moments, since there are 17 bricks at each quarter in the beam in Figure 11-5 and 34 spread over the whole beam in Figure 11-4.

The balance of forces can be shown mathematically if we give a positive sign to those forces causing compression and a negative sign to those causing tension.

Moment is equal to force times the distance through which the force acts. Since we have counteracting forces, the moment can be written as

$+\dfrac{wL}{2}$ (compressive force) $\times \dfrac{L}{2}$

(distance through which this force force acts) $- \dfrac{wL}{2}$

(tensile force) $\times \dfrac{L}{4}$ (distance through which this force acts—from one end to the ¼ point)

or $\left(\dfrac{wL}{2} \times \dfrac{L}{2}\right) - \left(\dfrac{wL}{2} \times \dfrac{L}{4}\right)$

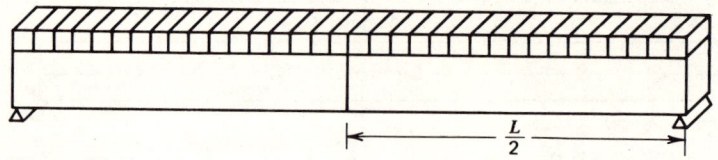

Figure 11-4

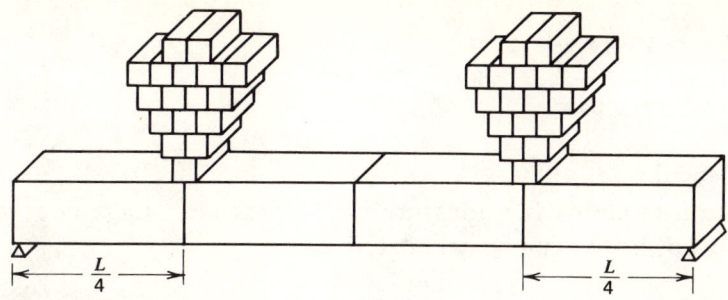

Figure 11-5

so:

$$M = \frac{wL^2}{4} - \frac{wL^2}{8}$$

$$M = \frac{2wL^2}{8} - \frac{wL^2}{8}$$

$$M = \frac{wL^2}{8}$$

The w can be dead load or live load or both. Let us take the dead load for the beam in Figure 11-1 and compute the bending moment.

Bending moment from dead load = M_g

$$M_g = \frac{w_g L^2}{8}$$

$w_g = 56$ lb/ft
$L = 15$ ft

$$M_g = \frac{56 \times (15)^2}{8} = 1575 \text{ ft lb}$$

To convert foot-pound to inch-pound,

$M_g = 1575$ ft-lb $\times 12$ in./ft
$M_g = 18,9000$ in.-lb

The foot-pound to inch-pound factor of 12 can be incorporated in the equation at the outset to eliminate some computation. This gives the formula:

$$M_g = w_g L^2 \times \frac{12}{8} = w_g L^2 \times \frac{3}{2}$$

$$M_g = 56 \times (15^2) \times \frac{3}{2}$$

$$M_g = 18{,}9000 \text{ in.-lb}$$

Section Modulus (Z)

The section modulus is a measure of stiffness and, for any rectangular, homogeneous beam, can be calculated as:

$$Z = \text{width} \times (\text{depth})^2 \times \frac{1}{6}, \text{ or, for the beam in Figure 11-1, } \frac{6 \times 9^2}{6} = 81$$

(The 1/6 is a constant for a rectangular beam.) Z is the same for top and bottom fibers or, $Z_t = Z_b$. The following shows how Z_t and Z_b are used to calculate stresses:

Section modulus in the top fiber = Z_t
Dead load bending moment = 18,900 in.-lb
Stress in the top fiber due to dead load = f_g^t

$$f_g^t = \frac{M_g}{Z_t}$$

$$f_g^t = \frac{18{,}900}{81} = +233 \text{ psi}$$

Section modulus in the bottom fiber = Z_b
Dead load bending moment = 18,900 in.-lb
Stress in the bottom fiber due to dead load = f_g^b

$$f_g^{fb} = \frac{M_g}{Z_b}$$

$$f_g^{fb} = \frac{18{,}900}{81} = -233 \text{ psi}$$

In concrete calculations, a compressive stress is indicated by a plus (+) sign and a tensile stress by a minus (−) sign.

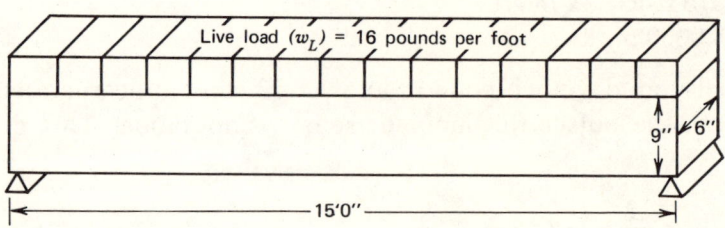

Figure 11-6

REINFORCED CONCRETE, PRECASTING

Since plain concrete has a tensile strength somewhere near 300 psi, it will support very little more weight than its own. If a 16-lb-per-linear-ft load is added, this beam will be close to collapse. We can find the live load stress for the extreme fibers this way:

Bending moment due to live load = M_L

$M_L = \dfrac{w_L L^2}{8}$ ft-lb $= \dfrac{3}{2} w_L L^2$ in.-lb

$M_L = 16 \times 15^2 \times \dfrac{3}{2}$ in.-lb

$M_L = 5400$ in.-lb

Stress in the top fiber due to live load = f_L^t

$f_L^t = \dfrac{M_L}{Z_t}$

$f_L^t = \dfrac{5400}{81}$ (Z_t is still 81)

$f_L^t = +67$ psi

Stress in the bottom fiber due to live load = f_L^b

$f_L^b = \dfrac{M_L}{Z_b}$

$f_L^b = \dfrac{5400}{81}$

$f_L^b = -67$ psi

The stresses due to dead load and live load are simply added to get the total stress in the extreme fibers. This beam will carry only 16 lb per ft.—not enough to be of any use.

$f_g^t + f_L^t = +233 + 67 = +300$ psi
$f_g^b + f_L^b = -233 - 67 = -300$ psi

A stress diagram of this beam would appear as in Figure 11-7.

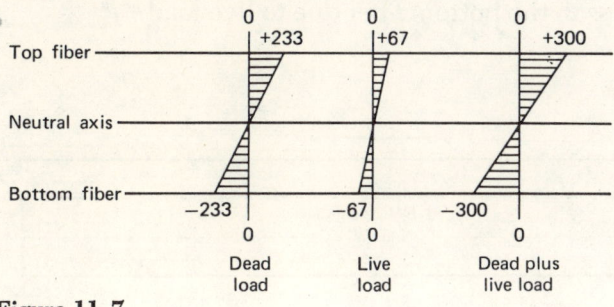

Figure 11-7

FORCES IN PRESTRESSED BEAMS

A beam that is subject to bending moments has a compressive force in the top and a tensile force in the bottom. Since these forces are opposite and graded from maximum positive at the top to maximum negative at the bottom, there is a point or plane at which these forces are balanced. This is called the neutral axis. In an unreinforced rectangular concrete beam, the neutral axis is $d/2$ or midway between the extreme fibers.

If we add 54,000 lb of prestressing force to this beam at the center of gravity of the concrete section, it will add 1000 psi of compressive force over the entire cross section of the beam. With a live load of 184 lb per ft, this will give a stress diagram similar to Fig. 11-10.

Stress in the top fiber due to prestressing = f_F^t
Stress in the bottom fiber due to prestressing = f_F^b

$f_F^t = \dfrac{F}{A_c}$ $\qquad\qquad f_F^b = \dfrac{F}{A_c}$

$f_F^t = \dfrac{54{,}000 \text{ lb}}{54 \text{ sq in.}}$ $\qquad f_F^b = \dfrac{54{,}000 \text{ lb}}{54 \text{ sq in.}}$

$f_F^t = +1000$ psi $\qquad\qquad f_F^b = +1000$ psi

Bending moment due to live load = M_L

$M_L = w_L L^2 \times \dfrac{3}{2}$ in.-lb

$M_L = 184 \times 15^2 \times \dfrac{3}{2}$ in.-lb

$M_L = 62{,}100$ in.-lb

Compressive stress in the top fiber due to live load = f_L^t

$f_L^t = \dfrac{M_L}{Z_t}$

$f_L^t = \dfrac{62{,}100}{81} = +767$ psi

Tensile stress in the bottom fiber due to live load = f_L^b

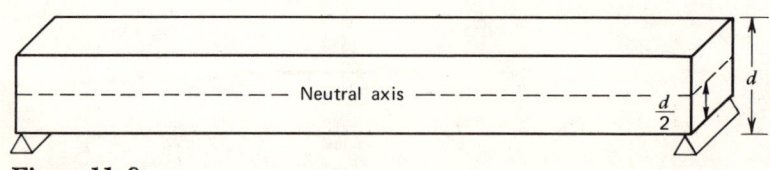

Figure 11-8

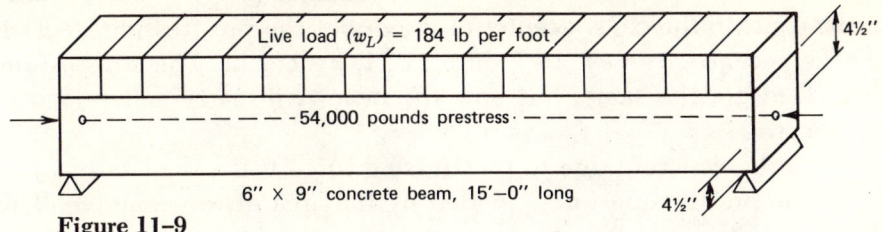

Figure 11-9

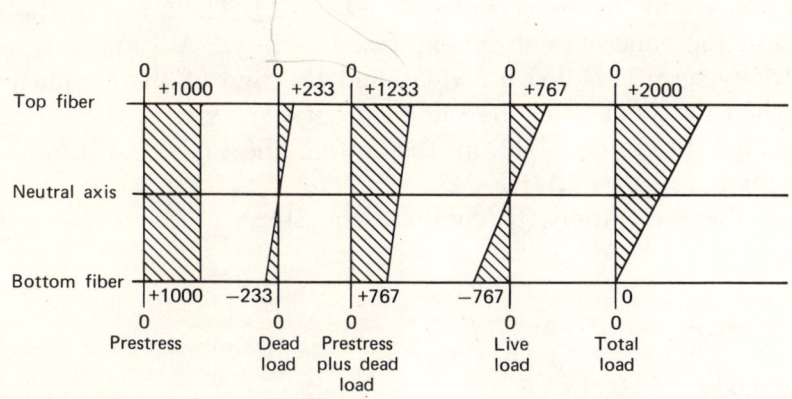

Figure 11-10

$$f_L^b = \frac{M_L}{Z_b}$$

$$f_L^b = \frac{62,100}{81} = -767 \text{ psi}$$

The prestress force is plus (compressive) for both top and bottom fibers. Summation of stresses gives

$$f_{(g+L+F)}^t = +233+767+1000 = 2000 \text{ psi}$$
$$f_{(g+L+F)}^b = -233-767+1000 = 0$$

Under the same loads as our other beams, this one could carry 184 lb per ft without even putting the concrete in tension, since it would have a net stress of +2000 psi in the top fiber and zero stress in the bottom fiber.

Eccentric Prestressing

Beyond this, it is possible to take the same beam, apply the same prestressing force, and make no changes except to move the prestressing

PRESTRESSED CONCRETE 243

force 1½ in. *below* the center of gravity of concrete (c.g.c.). The distance from the c.g.c. to the c.g.s. (center of gravity of the prestressing steel) is called the eccentricity (e). In Figure 11-11, $e = 1½$ in. The stresses due to dead load are the same, but now the beam will carry a live load of 424 lb per ft.

Stress in the concrete due to prestressing force was calculated as $f_F = F/A_c$, or the prestressing force divided by the area of the concrete. This is also true with eccentric prestressing, but a modifying term is introduced to account for the eccentricity ($\pm Fe/Z$). The stress in the top or bottom of the concrete can be expressed as $f_F = F/A_c \pm Fe/Z_{(t \text{ or } b)}$. This eccentricity term is called a *couple,* and the sign of the couple will be ($+$) when the steel is on the same side of the c.g.c. as the fiber being investigated. In Figure 11-11, for the bottom fiber: $f_F^b = F/A_c + Fe/Z_b$; and for the top fiber: $f_F^t = F/A_c - Fe/Z_t$.

Using these equations to compute the stress in the extreme fibers, we get

$$f_F^t = \frac{F}{A_c} - \frac{Fe}{Z_t}$$

$$f_F^t = \frac{54{,}000}{54} - \frac{54{,}000 \times 1.5}{81}$$

$$f_F^t = 1000 - 1000 = 0, \text{ and}$$

$$f_F^b = \frac{F}{A_c} + \frac{Fe}{Z_b}$$

$$f_F^b = \frac{54{,}000}{54} + \frac{54{,}000 \times 1.5}{81}$$

$$f_F^b = 1000 + 1000 = 2000 \text{ psi}$$

When this beam was prestressed at the neutral axis (centroidal prestress), there was a compressive stress of 1000 psi in the bottom fiber,

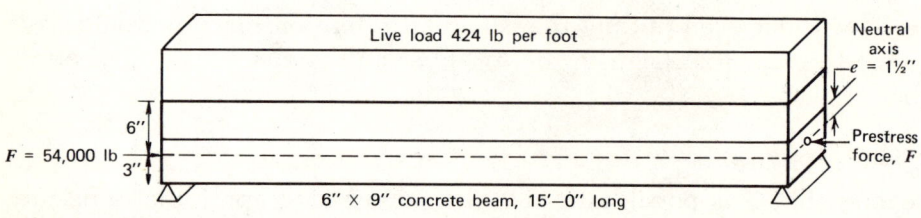

Figure 11-11

and it was able to carry a load of 56+184=240 lb per ft. Since we have twice the compressive stress in the bottom fiber, this beam should be able to carry twice the load, or 480 lb per ft, or a live load of 480−56=424 lb per ft. We will now check the stresses using a live load (w_L) of 424 lb per ft.

Moment due to live load = M_L

$$M_L = w_L L^2 \times \frac{3}{2} \text{ in.-lb}$$

$$M_L = 424(15)^2 \times \frac{3}{2} \text{ in.-lb}$$

$M_L = 143,100$ in.-lb

Compressive stress in top fiber due to live load = f_L^t

$$f_L^t = \frac{M_L}{Z_t}$$

$$f_L^t = \frac{143,100}{81} = +1767 \text{ psi}$$

Tensile stress in the bottom fiber due to live load = f_L^b

$$f_L^b = \frac{M_L}{Z_b}$$

$$f_L^b = \frac{143,100}{81} = -1767 \text{ psi}$$

Summing stresses gives:

In the top fiber: $f_L^t + f_g^t + f_F^t = 1767 + 233 + 0 = +2000$ psi
In the bottom fiber: $f_L^t + f_g^t + f_F^t = -1767 - 233 + 2000 = 0$

A diagram of the stresses appears in Figure 11-12.

This beam has the same cross section and total prestressing force as the beam in Figure 11-9, yet it is carrying more than twice the live load with the same net unit stress. This increase in carrying capacity is a result of moving the prestressing force down into the area of the beam that would be in tension.

Dropping the Prestressing Tendon
Since the steel in the beam in Figure 11-11 is the same distance from the c.g.c. for the full length of the beam, the stresses due to F are the same at all points on the beam. Because there is no bending moment at the supports due to applied loads, the net stresses at the supports are due to F. The e chosen in the Figure 11-11 beam (1.5 in.) was selected to give maximum prestress in the bottom fiber without putting the top fiber

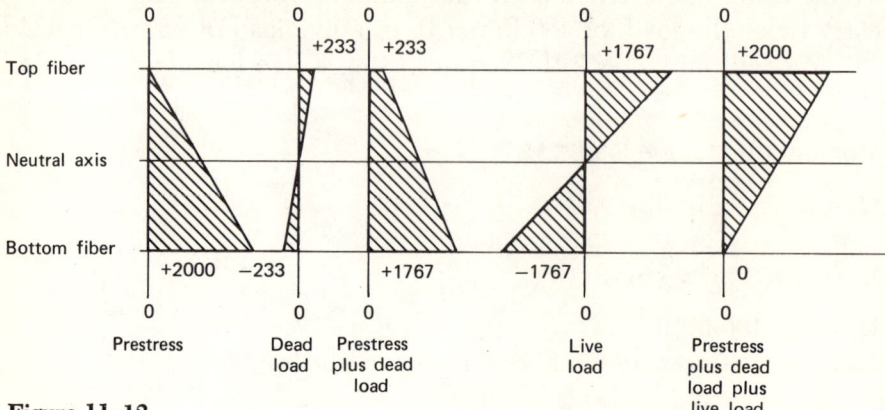

Figure 11-12

into tension. If e were to be increased, the f_F^t would become a tensile stress.

A small amount of tensile stress in the top fiber, at mid-span, is acceptable since the dead-load bending moment at that point will counteract it. There would be no dead-load bending moment (M_g) over the supports, however, and these areas would be in tension. This is generally not acceptable since the concrete would then have tension cracks as a result of prestressing. A greater e would carry a larger load and be even more efficient if some method could be found to reduce the tension near the supports.

The benefits of increased eccentricity can be realized if the prestressing tendon is "draped" as shown in Figure 11-13.

It was found that $f_F^t = 0$ when $e = 1.5$ in., or the stress in the top fiber due to the force imposed by 54,000 lb of prestressing 1.5 in. below the neutral axis is equal to zero. The eccentricity can be increased until the top-fiber tensile stress at mid-span, due to prestressing, is equal to the

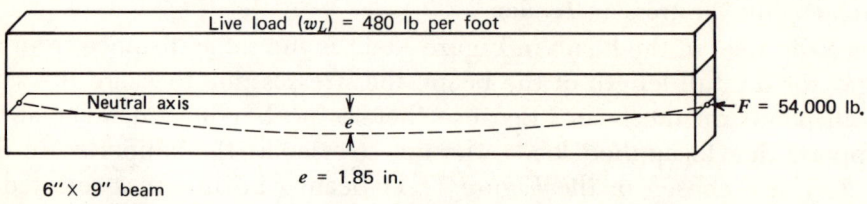

Figure 11-13

246 REINFORCED CONCRETE, PRECASTING

dead-load stress at mid-span, or until $f_F^t = f_g^t$. We know from previous calculations that $f_g^t = +233$ psi. The additional e necessary to give a tensile stress equal to this can be calculated using the formula: $f_g^t = Fe/Z_t$:

$$+233 = \frac{54,000e}{81}$$
$$233 = 667e$$
$$e = \frac{233}{667}$$
$$e = 0.35 \text{ in.}$$

This means e at center span can be $1.50 + 0.35 = 1.85$ in. below the c.g.c. With this eccentricity the beam should support a live load of $424 + 56 = 480$ lb per foot without tensile stress. In the beam in Figure 11-9, prestressing created a compressive stress of 1000 psi in the bottom fiber, and the beam was able to carry a total load of $56 + 184 = 240$ lb per ft. The beam in Figure 11-11 had a compressive stress of 2000 psi in the bottom fiber and was able to support $2 \times 240 = 480$ lb per ft or a live load of $480 - 56 = 424$ lb per ft. The beam in Figure 11-13 has a $+2233$ psi stress in the bottom fiber and should be able to carry a total load of $480 \times 2233/2000 = 536$ lb per ft or a live load of $536 - 56 = 480$ lb per ft. It was calculated this way:

Stress in the top fiber (due to prestress) at mid-span $= f_F^t$

$$f_F^t = \frac{F}{A_c} - \frac{Fe}{Z_t}$$
$$f_F^t = \frac{54,000}{54} - \frac{54,000 \times 1.85}{81}$$
$$f_F^t = 1000 - 1233 = -233 \text{ psi}$$

Stress in the bottom fiber (due to prestress) at mid-span $= f_F^b$

$$f_F^b = \frac{F}{A_c} + \frac{Fe}{Z_b}$$
$$f_F^b = \frac{54,000}{54} + \frac{54,000 \times 1.85}{81}$$
$$f_F^b = 1000 + 1233 = 2233 \text{ psi}$$

These same formulas can be used to find the f_F for either fiber at any point on the beam simply by substituting the known value of e. At the supports the value of e is zero because the strand is anchored at the c.g.c.

Stress in the top fiber at the supports (due to prestress) would be the same as in the Figure 11-9 beam—$f_F^t = f_F^b = F/A_c$—since the value of the couple ($\pm Fe/Z$) is zero because $e = 0$. At the supports, the f_F value is

+1000 psi, and will vary from this to −233 at the top and −2233 at the bottom as you approach the center of the span.

The bending moment due to a live load (M_L) of 480 lb per ft is 162,000 in.-lb, as shown below:

$$M_L = w_L L^2 \times \frac{3}{2} \text{in.-lb}$$

$$M_L = 480(15)^2 \times \frac{3}{2}$$

$$M_L = 162{,}000 \text{ in.-lb}$$

Stress in the top fiber due to live load = f_L^t

$$f_L^t = \frac{M_L}{Z_t}$$

$$f_L^t = \frac{162{,}000}{81} = +2000 \text{ psi}$$

Stress in the bottom fiber due to live load = f_L^b

$$f_L^b = \frac{M_L}{Z_t}$$

$$f_L^b = \frac{162{,}000}{81} = -2000 \text{ psi}$$

A stress diagram for the beam in Figure 11-13 is shown in Figure 11-14.

The stresses under dead load only (mid-span) are summarized
$$f_{(g+F)}^t = +233 - 233 = 0$$

Fig. 11-14 Y-160......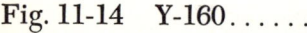

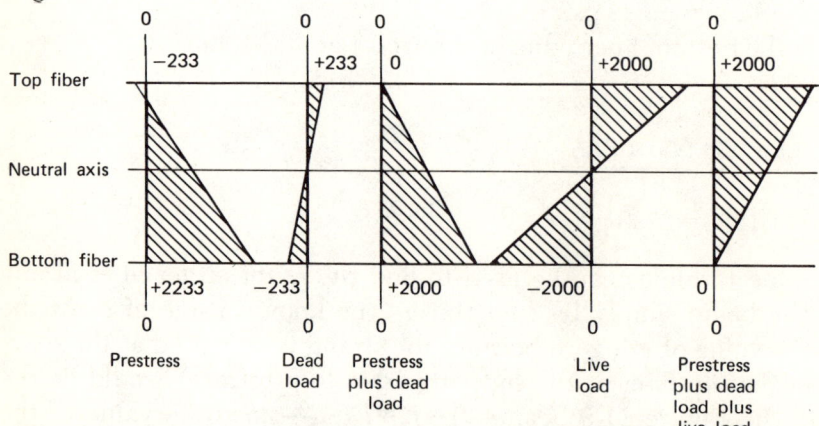

Figure 11-14

$f^b_{(g+F)} \quad -233+2233 = +2000$ psi

Summations of stress under all loads (mid-span) are

$f^t_{(g+F+L)} = +233-233+2000 = +2000$ psi
$f^b_{(g+F+L)} = -233+2233-2000 = 0$

By draping the tendon in a curve, the live load capacity is increased from 424 lb per ft to 480 lb per ft, while still using the same sized beam and the same prestressing force.

PRESTRESSING

Prestressing is basically a way of adding stress to a material so that this added stress offsets the stresses caused when the member is loaded. For example, a nuclear reactor shielding wall may be globe shaped for greatest efficiency and have prestressing steel wrapped around the outside trying to squeeze the concrete in toward the middle. As working pressures are built up inside this "ball" of concrete, trying to push the concrete away from the center, the two forces are counteracting.

In any prestressing operation the procedures of good concreting practices must be followed. These include using proper materials and correct mixing, placing, curing, and handling. Care must also be used in the placing of posttensioned strands or wires. In the case of pretensioning, the strands must be released at the proper time in the proper sequence.

Prestressed units sometimes require conventional reinforcing steel to control certain types of cracking. A small amount of extra stirrup (shear) reinforcement near the end of a pretensioned beam is effective in reducing the occurrence of cracking caused by prestressing.

Basic Methods

There are two basic methods of stressing concrete: pretensioning and posttensioning. Pretensioning is most often done in mass production techniques in a central casting yard. In pretensioning, the stressing strands are pulled tight in long beds before the concrete is placed in the forms. After the concrete has hardened, the load on the strands is released, transferring the stress to the concrete. Since the strand was stretched before the concrete was placed around it, the diameter was somewhat reduced. After the concrete has hardened and the tension is released from the strand, it will try to spring back to its original shape, which includes increasing the diameter. The hardened concrete prevents the strand from returning to its original length and the stress is transferred to the concrete.

Forms for this type of construction are usually continuous, separated by bulkheads to produce members of required length. In some yards, concrete is placed continuously in a long form and the hardened section is sawed to appropriate length.

Pretensioning beds may be as much as 600 ft long, but the most economical length seems to be between 300 and 400 ft. The length of the prestressing bed is dependent primarily on economic considerations and the availability of space.

In posttensioning, relatively small tunnels or ducts are formed by placing conduits in the concrete without prestressing strands in them. The concrete is allowed to harden without tension from the prestressing strands. When the concrete reaches specified strength, usually in 3 to 7 days, the steel is stressed and anchored to the ends of the concrete member. The annular spaces between the strands and the ducts are either grouted or left clear. The end anchorages spread the compression over the face of the concrete member, which is being pulled together by the prestressing steel.

There are advantages and disadvantages to each of these methods of stressing. Pretensioning usually presents greater economy when many

Figure 11-15 Posttensioning of precast, prestressed concrete bridge girder. Hydraulic jacks are used to stretch cables and induce precompressive stresses in the concrete.

identical members are going to be formed. However, large units may create a transportation problem and require forming and posttensioning on the job. Pretensioning beds require abutments that are able to withstand perhaps as much as several million pounds of tension and are understandably expensive. Posttensioning can sometimes increase the effective span of cast-in-place slabs or beams and, at the same time, offer deflection control, which can aid in developing continuity in precast construction.

Materials

Currently, the only material that meets all of the requirements for prestressing strand is stress-relieved, high-tensile-strength wire or strand (or, less frequently, high-strength rods) made of steel. (New materials are being developed that offer great promise but are not yet at the point of development to be able to replace steel in the process of prestressing.) Whatever material is used, it must have the following characteristics:

> *High tensile strength.* It should be capable of resisting tension stresses of as much as 250,000 psi.
>
> *High yield strength.* The material must be capable of being stretched under heavy loads and returning to its original shape when the load is removed. The elongation (called strain) is proportional to the applied load (called stress) up to a point called the proportional elastic limit. With steel, if the load is released before the proportional elastic limit is reached, the wire or strand will snap back to its normal length. If the load is carried beyond the elastic limit, the wire is permanently stretched to a degree depending on the load. This is called permanent set. In prestressing concrete, steel is generally elongated to just below the elastic limit, usually about 175,000 psi.

The modulus of elasticity (E) is a measure of the elastic ability of a material. It is equal to the unit stress divided by the unit deformation of strain. The modulus of elasticity of steel for prestressing is in the neighborhood of 28 million psi. By contrast, aluminum has a modulus of elasticity in the approximate range of 10 to 10½ million psi. For pure copper, it is 15 to 16 million psi or roughly half that of steel. Concrete used in prestressing has an E on the order of 4 million psi.

Permanent Tension. If rubber is stretched even a small amount for a long period of time, it will take a set and begin to lose its ability to return to its original shape. This cannot occur with a prestressing material for obvious reasons.

The following types of steel are the ones most commonly used.

1. *Small-diameter strand* (½ in. or less). These are made up of six wires wrapped helically around a center wire. The center wire is slightly larger than the six outer wires to make sure that all six bond to the center wire and not to each other. The small diameter seven strand is used mostly for pretensioning. (ASTM designation A416.)
2. *Cold-drawn single wire.* These are uncoated and stress relieved, or hot-dip galvanized, used in groups of two or more eccentrically parallel wires for posttensioning. (ASTM designation A421.)
3. *Cold-stretch alloy steel bars.* These are used mostly for posttensioning.
4. *Large-diameter strand.* These consist of 37 or more wires, used for posttensioning.

Small-diameter Strand. Placing a large number of single wires can be expensive. To reduce this cost, cables are stranded, almost always as seven wires—six twisted around a center wire. These strands are available from ¼ to ½ in. in diameter with an ultimate breaking strength of about 250,000 psi. Some manufacturers also provide a special 270,000-psi strand, approximately 17% stronger than the standard ASTM 416 strand, permitting reduction in material and placement cost.

There is a tendency for the helical wires to untwist as they are loaded, causing the prestressing strands to exhibit greater elongation than plain wire. The overall effect of the strand is an ultimate strength a little less than the seven wires stressed separately.

Prestressing Wire. Wire for prestressing is made by cold-drawing high carbon steel bars. The bars are drawn through a series of dies to reduce the diameter and increase the tensile strength. The strength of the wire is increased at each drawing and, the smaller the diameter, the higher its ultimate strength. As the wire comes from the final die, it is very stiff and

Table 11–A
PROPERTIES OF SMALL-DIAMETER SEVEN-WIRE PRESTRESSING STRAND

Diameter (in.)	Approximate Weight (lb/ft)	Steel Area (sq in.)	Minimum Breaking Strength	
			lb	psi
¼	0.122	0.036	9,000	250,000
⁵⁄₁₆	0.198	0.058	14,500	250,000
⅜	0.274	0.080	20,000	250,000
⁷⁄₁₆	0.373	0.109	27,250	250,000
½	0.494	0.144	36,000	250,000

has very little stretch before it breaks. It is difficult to handle and will take a set if rolled into a coil. The cold-drawing process also produces some locked-in stresses that are not desirable. To relieve these drawing stresses and improve the physical properties of the wire, it is heat treated after the final drawing. The wire is heated to about 600°F for about 30 seconds. This mild annealing process produces a slight reduction in ultimate strength but more than doubles the stretchability of the metal before it will break. As an added benefit, the heat treating will remove any oils left on the wire during the drawing process and leave a clean, oil-free surface for bonding to the concrete.

The heat treating is done either by passing the wire through a bath of molten lead or an oven of the proper temperature and length. Some wire is used for unbonded posttensioning, which means it may be subject to corrosion because it is not bonded to the concrete. These wires are often furnished with a coating of zinc (galvanized). In passing the wire through a bath of molten zinc, the stress-relieving heat treatment and the corrosion-protecting zinc coating are accomplished at the same time.

Since the cold-drawing process tends to increase the strength of the wire with each additional drawing, wire strength will vary with diameter. The smaller-diameter wires, which have been drawn more often, will show a higher ultimate tensile strength than the large-diameter wires. Table 11-B shows some of the variation in strength between smaller wires and larger wires.

High-Strength Alloy Bars. For posttensioning, smooth round bars made of high-tensile-strength alloy steel are available in diameters from ⅜ to 1½ in., and in lengths up to 80 ft. The bars are manufactured from high-alloy, hot-rolled steel rounds that are heat-treated and then cold-

Table 11-B
PROPERTIES OF SELECTED SIZES OF STRESS-RELIEVED WIRE

Diameter (in.)	Approximate Weight (lb/ft)	Steel Area (sq in.)	Minimum Breaking Strength	
			lb	psi
0.1055 (12 gage)	0.0297	0.00874	2,440	279,000
0.135 (10 gage)	0.0486	0.01431	3,830	268,000
0.162 (8 gage)	0.0700	0.02061	5,340	259,000
0.192 (6 gage)	0.0983	0.02895	7,260	251,000
0.2253 (4 gage)	0.1354	0.03987	9,770	245,000
0.250 (¼ in.)	0.1667	0.04909	11,780	240,000
0.2625 (2 gage)	0.1838	0.05412	12,880	238,000

stretched by loading the bar to 90% of its ultimate tensile strength. The cold stretching increases the yield strength of the bar and also acts as a proof load to assure that defective bars are eliminated before they are used. These rigid bars are much easier to place in the posttensioning ducts but are not as strong as prestressing wire or cable. Commercial high-tensile prestressing bars are produced in two grades: "regular" and "special." The regular grade has a minimum ultimate tensile strength of 145,000 psi, while the more expensive special grade has a minimum ultimate tensile strength of 160,000 psi.

Large-Diameter Strands. Strands larger than ½ in. in diameter are used for posttensioning and are arbitrarily classed as large-diameter strands. They are capable of applying extremely high prestress forces up to a third of one million pounds per tendon. These strands are constructed exactly the same way as cables used in suspension bridges and, for that reason, are often referred to as "bridge strands." They are available in sizes from ⅝ in. up to 2 in. in diameter. These strands use 7 to 91 wires and have breaking strengths from 54,000 to over 500,000 lb.

Large-diameter strands are furnished bright (uncoated), stress-relieved and non-stress-relieved, or galvanized. Because of the greater number of wires, large-diameter strands tend to stretch more than small-diameter strands. They are generally furnished as strand assemblies with anchorage sockets attached at each end to form a specific length assembly.

Camber

Prestressed concrete has a natural tendency to bend so that the midpoint is higher than the ends. This curvature is called camber and results from the application of the prestressing load below the center of gravity of the concrete. Both pretensioned and posttensioned members will exhibit camber.

It is difficult to know exactly how much a member will camber since camber depends largely on the modulus of elasticity of the concrete and increases with time because of creep. The main quality control consideration is the difference in camber between adjacent members in the structure. For example, box beams in a bridge deck, to be covered with a cast-in-place slab, could tolerate a difference of 1 in. between beams, while adjacent tees in a building could tolerate only ½ in.

Camber can be measured while a member is still on the stressing bed, just after detensioning of pretensioned steel or stressing of posttensioning steel, using the soffit plate of the bed itself as a reference.

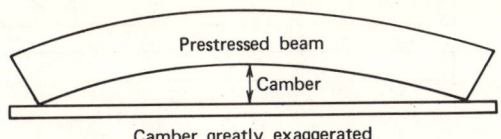

Figure 11-16

Cracking

One of the principle reasons for prestressing is to reduce cracking. However, some cracking does occur during the manufacturing process which, if slight enough, may have no effect on the structural usefulness of the member. Some minor cracking will result from the shrinkage of the concrete. Because low-slump concrete is normally used, shrinkage cracking is minor and not important. Short, discontinuous cracks, usually less than 3 in. long, have been seen near the bottom of I-beam girders. These are shallow shrinkage cracks and are not serious.

Cracking may occur as a result of failure to follow proper curing and detensioning procedures. Vertical cracks can result from shrinkage and cooling before the member is detensioned. If steam curing is used, hold-downs and strand anchorages should be released immediately after the curing is discontinued while the member is still warm.

Cracking in pretensioned members can be controlled by proper use of conventional reinforcement. This will not prevent cracking, but it will minimize the size and number of cracks.

Pretensioning

In pretensioning, steel tendons are placed in forms and stretched by means of jacks. The concrete is placed around the stretched steel and, after it reaches a certain strength, the tendons are cut loose from the anchorages. This transfers the load to the concrete as a compressive force that is maintained by the wedging action between the concrete and the steel (Figure 11-17).

Casting Beds. Casting, or stressing, beds may range in length from 20 to 650 ft. The reason for the long beds is to permit many forms to be set up in a line when the cross section and tendon pattern in the units is the same. One typical plant has four casting beds ranging from 20 to 450 ft in length. In this plant, each bed consists of a concrete soffit plate supported on I-beam rails that are set in the concrete vase. The raised soffit

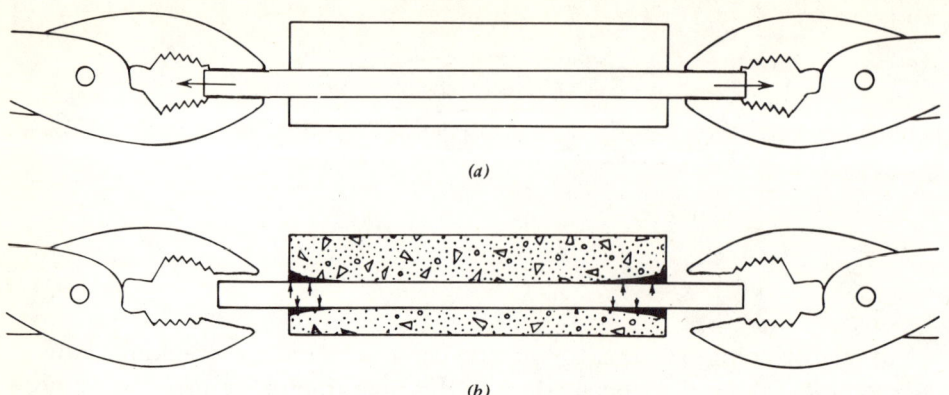

Figure 11-17 (a) Prestressing wire stretched and ready for concrete. (b) Stretching released after concrete has hardened causes a wedging action near the ends which maintains the tension in the steel.

plate provides access to the underside of the form for easy installation and release of hold-downs for draped or deflected strands.

Most precasting beds are made of concrete because the top of the bed acts as the bottom of the form (soffit plate). Concrete can be troweled to a smooth surface to attain members with smooth bottoms (or soffits).

The horizontal loads imposed on the end anchorages by the fully stressed strands in a bridge girder may approach 2 million pounds and may require extremely heavy construction to resist these loads. End anchorage may be of reinforced concrete or steel, extending down to transmit the force to the ground or to the concrete of the casting bed. The prestressing strands pass through openings in the end anchorages to the hydraulic stressing equipment. In some cases, heavy steel forms may be used to resist the stressing tension.

Many prestressed units are designed with deflected or draped strands. Pretensioned, deflected strands require hold-downs in the forms to hold the strands in the proper pattern. There are two general methods for doing this. Figure 11-18 is a hold-down through which the strands are threaded and then tensioned. Figure 11-19 is a hold-down that pulls already tensioned strands or cables into the proper deflected pattern.

Members in a long pretensioning, precasting bed are separated by means of bulkheads installed in the forms prior to the placing of concrete. There are a number of different types available and, for limited use, plywood can work well, but for constant reuse, steel is the best. Usually, bulkheads are anchored to the strands to prevent movement

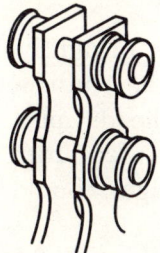

Roller unit   **Figure 11–18**

during the concrete placing. When draped strands are used, the bulkheads will also serve as a "hold-up" for the strands.

Forming. Usually, only steel or steel and concrete in combination are recommended as formwork for pretensioned, prestressed building members. Other materials can be used for the formwork, but steel and concrete have been found to be the most economical. Forms should be rigidly braced and anchored, and capable of withstanding the vibration required

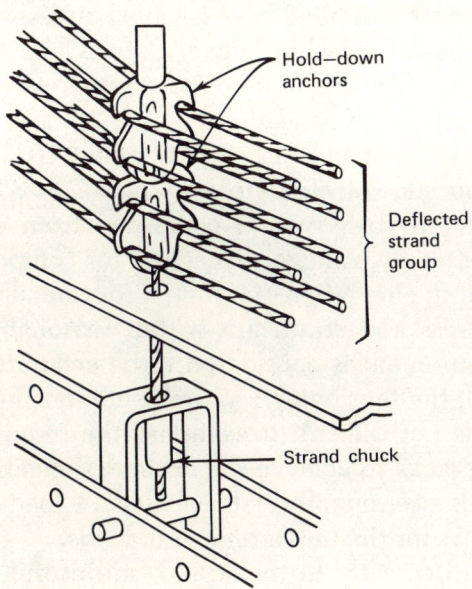

Figure 11–19

PRESTRESSED CONCRETE 257

to place fairly stiff concrete without bulging or deflecting. Forms should be easy to clean and should be cleaned after each casting.

Joints in the forms must be smooth and tight to prevent unsightly offsets, sand streaks, or other blemishes on the surface of the constructed member. Joints between form sections, soffits, and bulkheads must be tight and, if necessary, gasketed. Certain types of chamfer strips of rubber or plastic can also double as gaskets. All 90° edges should be chamfered.

Most forms for pretensioning work are sufficiently rigid so that form ties are not necessary. If form ties are used, they must be the breakaway kind that will leave no metal exposed on the surface of the concrete.

Form oil or parting compound is applied before each usage. Special care is necessary when applying form oil in order to avoid getting any of it on the prestressing strands. Any form oil that does get on the strands *must be removed immediately* with a suitable solvent.

In many large structural members, void-formers are used to displace concrete, thereby reducing the overall weight with negligible loss in load-carrying capacity. Voids may be cylindrical, square, or rectangular. Accurate positioning of the void-formers and a method of securely holding them in place are necessary.

Tensioning. The more pull that is put on a piece of steel, the more it will stretch. The pull or the applied load is called stress, and the stretching is called strain. The relationship between stress and strain is known as modulus of elasticity. The modulus of elasticity tells you how much the steel will stretch under a given load.

In prestressing work the steel is stressed to about 70% of its ultimate strength. Since the modulus of elasticity can vary 7 or 8% between lots of steel, possible error in stress values computed from strain measurements can be introduced. By using jack pressure for computing stress and measuring the amount of elongation (strain), a continuous check is maintained. As long as stress and strain are within reasonably close agreement, the stress measurement is considered to be accurate. Knowing the properties of the steel, the tensioning requirement, and the dimensions of the stressing bed, it is not difficult to compute the required pressure to apply to the stressing jacks to achieve the desired elongation. As a check on stressing, load cells occasionally will be used. A load cell makes use of electrical strain gages for the measurement of loads.

Strain gages applied directly to the strands are useful in determining the uniformity of tension along a deflected strand. This procedure is most often used when checking a new tensioning system, but it is not practical as a routine check.

A great deal of energy is stored in a pretensioned strand, and the strand can cause great damage if failure occurs. For obvious reasons, special safety precautions must be followed. Clearing the area of non-essential personnel during the stressing operation is a sensible precaution.

Tensioning Method. In pretensioning, the strands are anchored at one end of the bed and attached to long-ram-travel hydraulic jacks at the other end. Strands may be tensioned together or singly.

In multiple-strand tensioning, jacks pull an anchor that moves a template to which the strands are secured. When the proper tension is reached, as indicated by jack pressure and elongation, the strands are held at the proper elongation by adjustable end anchors, and the jacks are moved to the next bed.

Single-strand tensioning permits the use of smaller, more portable jacks and is used more often now than multiple-strand tensioning. A center-hole jack may be used in which the strand to be tensioned is fed through the center hole of the jack at the outer end. After stressing, another anchor on the strand holds it against the stressing bed anchorage, and the jack is released and moved. In either case, elongation must be measured to a tolerance of ⅛ in. or 1% of theoretical elongation, whichever is smaller.

Temperature Effect. Normally, the small amount of difference in temperature between steel and concrete is not enough to cause a change in the stressing of the tendons. If the steel is cold and the concrete placed around it is warm, the steel will expand and reduce the tension. If the steel is warm and is covered with cold concrete, the steel will contract and increase the tension. Total length change in the steel resulting from temperature change can be expressed by the following equation.

$\triangle t = etl$, in which
$\triangle t$ = length change of steel due to temperature change, in.
   e = linear coefficient of expansion of steel = 0.0000065 in. per in. per °F
   t = temperature difference, °F
   l = length of strand subject to temperature change, in.

Unless there is a large temperature difference the effect can usually be disregarded.

Discrepancies. Small discrepancies can be expected between calculated and measured elongations. These discrepancies will be caused by slippage in the gripping devices or anchors at the ends of strands, movement of anchoring abutments, elongation of anchor bolts, and the tendency of strands to untwist. All of these discrepancies can be calculated.

Computing Prestress. The size, number and distribution of tendons, and the prestress to be applied are computed by the designer. The information must then be translated into jack pressure and inches of elongation for use in the prestressing plant.

The following calculation is based on tensioning a group of strands simultaneously. If single strand stressing is being used, the equations are still valid using $N = 1$ and $J = 1$.

Let:
A = area of one strand, sq in.
a = net area of jack ram, sq in.
d = unit deformation of strain, in. per in.
E_s = modulus of elasticity of strand, psi
$\triangle$ = total change in length l, in. (elongation)
J = number of jacks
L = net length of stressing bed, ft
l = length, in.
M = measured movement of jack-end anchors, in.
N = number of strands in one group
P = jack pressure, psi
S = unit stress applied to strand or unit stress required, psi
T = total slippage, in.
W = total load on N strands, lb

The unit stress, S, is determined on the basis of the physical properties of the steel being used, and this will be the basis for the field computations. The modulus of elasticity, E_s, is determined from laboratory tests and is the basis for determining the change in length of the strands in the bed.

To obtain the jack pressure reading, P, which is required to produce the unit stress in the strand, we first use the relationship $W = ANS$; then,

$$P = \frac{W}{Ja} = \frac{ANS}{Ja}$$

In using E_s to determine the total elongation of the strands, it is necessary to consider the amount of slippage of the strands in the anchors—

both at the dead end of the bed and at the jack end. As the load is applied, there will be a slight movement in the anchors at each end as they grip the strand. This is normally about ¼ in. To find what it is, mark each strand under no load, just where it emerges from the anchor. After the bed has been stressed, measure the distance that the marks have moved away from the ends of the anchors. The sum of the slippages at each end is the total slippage, T.

From the relationships, $d = \dfrac{S}{E_s}$ and $d = \dfrac{\Delta}{l}$, we get

$$\Delta = \dfrac{Sl}{E_s} = \dfrac{12SL}{E_s}$$

Δ being the total required elongation in a bed of length L. Movement of the ram is given by the relationship,

$$M = \dfrac{12SL}{E_s} + T$$

The slippage, T, should be checked for each operation. Once the pressure and elongation have been set up, there will be no change unless a change is made in the other constants.

Example. A certain plant has a stressing bed 400 ft long, using one jack with a net ram area of 32 sq in. It is planned to stress a string of beams using 16 strands of 0.035-sq-in. area. The steel has an elastic modulus of 28 million psi, and the strands will be stressed to 175,000 psi. Slippage in the anchors is ¼ in. at the jack end and ¼ in. at the dead end. Compute the required jack pressure and movement of jack-end anchors.

From this, we get:

$A = 0.035$ $\qquad N = 16$
$a = 32$ $\qquad J = 1$
$E_s = 28,000,000$ $\qquad S = 175,000$
$L = 400$ $\qquad T = ½$ in.

$$P = \dfrac{ANS}{Ja} = \dfrac{0.035 \times 16 \times 175,000}{1 \times 32} = 3062 \text{ psi}$$

$$M = \dfrac{12SL}{E_s} + T = \dfrac{12 \times 175,000 \times 400}{28,000,000} + ½ = 30.5 \text{ in.}$$

In practice, it is necessary to apply some load to the strands in order to take up the slack and give a reliable starting point for measuring elongation. This is done by applying a load of about 100 psi to the jack in order to take the droop out of the strands. The stress in the strands (S) can be computed as follows.

Using values from the above example, for a 100 psi initial jack load:

$$s = \frac{PJa}{AN} = \frac{100 \times 1 \times 32}{0.035 \times 16} = 5714 \text{ psi}$$

$$\frac{s}{S} = \frac{5714}{175,000} = 0.325 = \text{portion of total stress at 100 psi}$$

Elongation at 100 psi $= (M-T) \times 0.0325 = 30 \times 0.0325 = 0.98$ in.

Deflection Strands. The above computations apply to straight pretensioning strands. Many prestressed units have deflected or draped strands. Elongation and load computations are the same for the draped strands as they are for the straight strands, except that the distance L may vary a little, and allowances must be made for friction.

Because of the length of the bed and the amount of friction from the hardware involved in prestressing draped strands, it is seldom possible to apply the full prestressing load by jacking at one end only. Usually, the full load is applied to one end, then the jack is moved to the other end of the casting bed, and the full load is applied again. The sum of the two elongations should equal the computed total elongation within about 5%.

Concreting. After the stressing tendons have been stretched, the reinforcement placed, and the forms oiled and secured, the concrete is placed in the usual way, following good construction and control practices. The concrete is placed in lifts, usually not over 16 in. deep, and it is thoroughly consolidated by both internal and external vibration; cold joints are avoided.

Test cylinders are cast at the same time, using the same concrete. The time for releasing the tension on the prestressing tendons is determined by the test specimens. Test cylinders are cured in the same way as the member.

Curing. Because of the requirement for high compressive strength in a short period of time, accelerated curing is usually practiced in pretensioning yards. These methods include using steam, hot water, or hot oil. Good accelerated curing can be effected with a combination of raised temperatures and excess moisture. If the heat is supplied by wet steam, sufficient moisture is usually present and requires only impervious tarpaulins to keep the moisture from being lost. If the heat is supplied by hot water or hot oil, moisture must be supplied by soaker hoses, fog spray, wet burlap, or other means. In general, the best curing is obtained with wet steam under tarpaulins.

The start of high-temperature curing should be delayed 3 or 4 hours after the last of the concrete is placed. Temperatures should be raised gradually and should not exceed 150°F. After the curing (as determined by field-cured cylinders) the concrete is allowed to cool to air temperature. When air temperature is below 50°F, the concrete should not be allowed to cool faster than 5°F per hour.

Detensioning. When the concrete reaches the strength specified for transfer of stress, strands are released or detensioned. This can be done by gradually releasing the jacks when multiple-strand tensioning has been applied, or by cutting the strands individually in a predetermined pattern developed for each type of unit and stressing bed.

The relationship between curing, detensioning, stripping of forms, and release of hold-downs is important in the control of cracking. Forms should be loosened as early as possible in the curing cycles and, in any case, while the concrete is warm and before detensioning. However, the concrete should not be allowed to cool appreciably before detensioning, or cracking may result.

Multiple-Strand Release. While the practice of multiple-strand tensioning and release is falling into disuse, there are some casting yards that feel that the advantages of stressing all of the tendons at the same time is worth the additional cost. In multiple-strand detensioning, jacks are used to take the entire stressing load and then are gradually released. When the stressing tension is released from the hydraulic jack, the members will move. The amount of movement will depend on the length of the tendon between members and between the last member and the dead anchorage end. When the strands are released, the concrete members must be free to move in order to prevent damage.

Single-Strands Release. A sequence of cutting the strands from the anchorages, which keeps stresses nearly symmetrical about the axis of the members, is most often used. The strands are heated with a low oxygen flame until the metal softens, loses its strength, and parts gradually. Best results are obtained if the heating can be done at both ends of the strand simultaneously.

Detensioning Deflected Strands. If the member weighs twice as much as the hold-down force, the hold-down devices are released first; then the strands can be detensioned either with multiple- or single-strand release.

For light members weighing less than twice the hold-down force, the draped tendons must be released first at each end; then the hold-downs and, finally, the straight strands, are released. Light members can be detensioned in the same way as heavy members if weight or restraint is applied directly over the hold-down points.

Handling and Transportation. Types of equipment used for handling precast members include cranes, lift trucks, and straddle carriers of various types.

Girders, beams, and T-sections are usually transported to a job by trucks. Long girders can be hauled with tractor and bogie combinations. With especially long beams the bogie may be equipped with a steering device similar to that used on hook-and-ladder fire trucks.

It is important that prestressed concrete members be picked up, stored, and transported in the position in which they will be used. Disregard of these precautions can result in a damaged member and even complete failure.

Because of the large forces intentionally induced in prestressed concrete, the top center portion of a beam may well be in tension. The dead weight of the beam is used by the design engineer to offset these tensile stresses, but it will be effective only if the beam is handled and supported at the ends. Lifting a prestressed concrete beam by a sling around the midpoint may very well result in a beam breaking into two pieces. The pickup points are usually indicated on the construction drawings.

Precast members can be stacked for storage if they are supported at their bearing points. For simply supported, pretensioned beams, this support point is no more than 1 ft from the end of the beam.

Posttensioning
The second of the two methods of prestressing is posttensioning. In posttensioning, openings or ducts are placed in the concrete through which the stressing wire is pulled. After the concrete has hardened, the steel is stretched to its designed elongation and anchored. In some cases the duct is grouted.

The effects of posttensioning and pretensioning are much the same, although posttensioning is generally more costly. Posttensioning is most often used in cases where the members are too large to be transported from a prestressing site. For a variety of reasons, both pretensioning and posttensioning are sometimes used on the same member. When both techniques are used, as much of the stress as possible is applied with pretensioning because it is less costly. This is usually done with the straight strands being pretensioned and the draped strands posttensioned.

The same care in production is necessary for posttensioned members. It is essential that the concrete be properly mixed, placed, compacted, and cured in order to insure satisfactory performance. The record of the prestressing industry is an impressive one. Although there have been a few failures, these have been traced, almost without exception, to improperly designed connections between members and not to the failure of prestressed members themselves.

The holes or ducts for the wire are formed in the concrete in any of several ways. For large prestressed members, flexible metal tubing is most often used. The tubing is placed in the form, sealed so that cement paste cannot leak into it, and then tied generally in a parabolic drape or curve. Flexible tubing, which can be removed from the concrete after it has hardened, has also been used successfully to form ducts for posttensioning steel. Rubber tubing, which is inflated or made rigid by inserting steel bars, is placed in the forms in the same manner as the metal tubing. After the concrete has hardened, the rubber tubes are deflated or the steel coils removed, and the tubes are then pulled from the concrete.

Individual wires are usually used in posttensioning rather than the strand commonly used in pretensioning. Wire is forced through the duct and anchored at one end of the member. The other end is stressed with hydraulic jacks and anchored, and in most cases the duct is grouted to protect the steel from corrosion and to act as a bonding agent.

One type of posttensioning uses wire accurately measured and assembled into groups in flexible tubing. The ends of the wires are deformed into buttons that bear against a special bearing plate. The entire assembly is secured in the form, and concrete is placed around it and allowed to cure. After the concrete has developed the necessary strength, a special jack pulls the perforated plate, thus elongating all of the wires. The plate is then anchored or shimmed, and the tube may be grouted. This method keeps the anchored end of the prestressing strand from being visible since the perforated plate is hidden under a concrete cover.

Sometimes, several members are posttensioned together to act as a single unit. A notable use of this technique in recent times was for the pilings for the Lake Pontchartrain Bridge. Cylinders for the pilings were cast in a centrifugal forming machine with 12 holes equally spaced around the perimeter of the cylinder for the posttensioning tendons. Six of these 16-ft sections were posttensioned to form one 96-ft pile.

A small preload is usually applied to take the slack out of the tendons in the same way as pretensioning. The design stress is then applied with the proper allowances for friction. The tendons are anchored at the ends and elongation is accurately measured. It is important that elongation and jack pressure be measured. This discrepancy must be within 5%.

After tensioning, the ducts are usually grouted to protect the tendons from corrosion and to provide a bond between tendon and concrete. Grouting should be done as soon as possible after stressing—in any case, no more than 48 hours later. Ducts are provided with entrance ports and openings for the discharge of air and grout.

Ducts to be grouted are flushed with water and blown out with air to make sure they are clean before grouting. Grout may be pumped in at one or more points, depending on the length and configuration of the duct. Grout-pumping machines are equipped with bypass valves to maintain a constant flow of grout through the pump, even when the grout is not being introduced into the duct. Grout is pumped until it flows steadily from the discharge port; then the discharge port is closed and grout pressure is gradually increased up to as much as 125 psi for 10 or 15 seconds. The entrance is then closed and the grout is allowed to cure. Materials for grout (cement, sand, and water) must, of course, be accurately measured. After a thorough mixing, the grout is passed through a screen before entering the pump. Sometimes, a small amount of aluminum powder (about 1 teaspoonful per sack of cement) is added to expand the grout.

Circumferential Prestressing. Recently developed techniques enable cylindrical objects such as large concrete pipe or concrete tanks to be posttensioned. Tanks up to several million gallons capacity have been successfully prestressed with the wrapping process. A conventional cylindrical tank is constructed on a concrete slab. The joint between slab and walls is usually flexible, with leakage prevented by a rubber or plastic waterstop in a joint. A machine, suspended from the top of the tank, travels around the wall wrapping the high-tensile-strength wire at the proper degree of prestress. The machine is operated so that the wires are spaced properly from bottom to top. After the prestressing wires are installed, the entire tank is shotcreted for corrosion protection. This is a highly specialized process and should be attempted only by contractors specializing in this type of work.

Chemical Prestressing

Portland cements can be formulated so that they expand as the early hydration process occurs. These are called expansive, shrinkage-compensating, or self-stressing cements.

The process of placing enough steel in expansive cement concretes to produce high stressing is so far only in the experimental stages. Under favorable circumstances, when the concrete is properly restrained with adequate reinforcement, a prestressing load can be induced by chemical means similar to those induced by mechanical prestressing. Although still experimental, the process shows promise for the future.

ant
SECTION FOUR
CONCRETE MASONRY

CHAPTER 12
MANUFACTURING, MATERIALS, AND MATERIALS HANDLING

Concrete masonry units are designed and made for use in all types of masonry construction and in sizes and shapes to fit different construction needs. Generally, building units are designated as hollow load-bearing concrete block, solid load-bearing concrete block, hollow non-load-bearing concrete block, concrete building tile, or concrete brick.

Concrete blocks are made in a variety of machines that consolidate a relatively dry concrete mix into the desired size and shape. The manufacture of units involves six basic steps, each of which will be discussed in this chapter.

1. Selection of proper raw materials.
2. Accurate batching of these materials.
3. Thorough mixing.
4. Proper compaction and consolidation in a molding machine.
5. Curing.
6. Storage and handling.

MATERIALS

All concrete masonry contains portland cement, aggregates, and water. In addition, coloring pigments, air-entraining compounds and curing accelerators or retarders are sometimes used for special circumstances.

Cement can be standard Type I or II. Where high early strength is necessary or to reduce curing time, Type III cement may be used. Air-entraining cements, portland pozzolan cements, and special block cements are also used in block making. White cement is often used for decorative block when light color is important or when the block will be tinted.

Aggregate should be (1) clean, and free of particles or substances that will interfere with strength or cause surface imperfections, (2) sufficiently strong, (3) durable, to resist freezing, and (4) be uniformly graded from fine to coarse to produce workable and uniform mixes.

The Besser Company, a manufacturer of block-making machinery, has compiled the following chart (Table 12-A) showing many of the aggregates used in the production of block and the physical characteristics that they can be expected to impart to units made with them. Also included is a brief description of where the aggregates originate and how they are manufactured.

As indicated in Table 12-A, normal or heavy aggregates include sand, gravel, crushed stone, and blast furnace slag. Sand should be graded so

Table 12–A
AGGREGATES COMMONLY USED IN THE MANUFACTURE OF CONCRETE MASONRY UNITS AND THE GENERAL CHARACTERISTICS OF BLOCK MADE WITH THESE AGGREGATES

Aggregates	Source and Manufacturer	Characteristics of Block Made with Aggregates
Dense Aggregates 1. Sand and gravel	Natural deposits. May be used in natural state, washed and screened; or crushed, washed, and screened	High compressive strength, low absorption, dense smooth texture, high durability, and good sound impedance properties. Weight: (8 × 8 × 16-in. unit) 40–50 lb
2. Crushed stone	Natural deposits of trap rock, granite, colomite, limestone or sandstone. Querried, crushed, and screened to size	High compressive strength, low absorption, dense smooth texture, high durability, and good sound impedance properties. Weight: (8 × 8 × 16-in. unit) 40–50 lb
3. Air-cooled slag	By-product of blast furnace operations. Molten slag from furnace run into pits or into banks and allowed to air-cool and harden. Excavated, crushed, and screened to size	High compressive strength, low absorption, high durability, good fire resistance and insulating value, good sound impedance properties. Weight: (8 × 8 × 16-in. unit) 35–43 lb

Table 12-A (continued)

Aggregates	Source and Manufacturer	Characteristics of Block Made with Aggregates
Lightweight Aggregates 1. Cinders	By-product of high temperature combustion of coal or coke. May be treated, crushed, and screened or used in natural state	Good strength, high insulating and fire resistance properties, medium to coarse texture, good nailability, usually dark color, good durability, and good acoustical properties. Weight: (8 × 8 × 16-in. unit) 25–35 lb
2. Expanded slag (Trade names: Waylite, Celocrete, Superock, Foamed Slag, and others)	By-product of blast furnace operations. Molten slag, expanded to produce a cellular product through contact with a limited quantity of steam or water and, in some cases, agitation. Crushed and screened.	High strength, high insulating and fire resistance properties, medium to coarse texture, good nailability, usually dark color, and good acoustical properties. Weight: (8 × 8 × 16-in. unit) 25–33 lb
3. Granulated slag	By-product of blast furnace operations. Molten slag chilled suddenly by immersion in water. Crushed and screened	High strength, high insulating and fire resistance properties, medium texture, high durability, good nailability, dark color, and good sound impedance. Weight: (8 × 8 × 16-in. unit) 30–38 lb
4. Expanded shales and clays (Trade names: Haydite, Beslite, Rocklite, Lelite, Aglite, and others).	Raw materials from natural deposits heated to incipient fusion and expanded by rotary kiln or sintering method to to produce a cellular product. Cooled, crushed, and screened	High strength, high insulating and fire resistance properties, medium to coarse texture, high durability, good nailability, and good acoustical properties. Weight: (8 × 8 × 16-in. unit) 18–35 lb
5. Pumice	Natural deposits of light to gray-colored cellular, glassy lava of volcanic origin. Used in natural state or crushed and screened	Good strength, excellent insulating and fire resistance properties, excellent nailability and sawability, good durability, medium texture, and good acoustical properties; may be dark colored. Weight: (8 × 8 × 16-in. unit) 20–30 lb
6. Scoria	Natural deposits of reddish to black cellular, glassy lava of volcanic origin. Used in	Good strength, excellent insulating and fire resistance properties, excellent nailability and sawability

Table 12-A (continued)

Aggregates	Source and Manufacturer	Characteristics of Block Made with Aggregates
	natural state or crushed and screened	good durability, medium texture, and good acoustical properties; may be dark colored. Weight: (8 × 8 × 16-in. unit) 20–30 lb
7. Vermiculite	Micaceous material found in natural deposits expanded by being subjected to high temperatures	Excellent insulating and fire resistance properties, good acoustical properties, and excellent nailability and sawability. Weight: (8 × 8 × 16-in. unit) 15–30 lb

NOTE: Other lightweight aggregates are used, such as perlite and diatomite. However, the above list covers all of the principal aggregates now in use.

that 100% passes the No. 4 sieve; 15 to 25% should pass the No. 50 sieve, with 5% passing the No. 100 sieve. The fineness modulus of the sand should be between 2.50 and 3.00, and a small amount of silt (less than 5%) is usually not objectionable. All coarse aggregate for concrete block should pass the 3/8 in. screen, no more than 5 percent should pass the No. 8 screen, and 25 to 40% should pass the No. 4 screen.

Grading for lightweight aggregate, either natural or manufactured, should be about the same as for normal weight aggregate. Many lightweight aggregates are angular and harsh, and some (especially the very lightweight ones) are apt to be fragile. The desirable properties for aggregates apply to lightweight as well as to normal. All aggregates should be free of sulfur compounds, excess lime, iron particles, pyrites, and any other materials that might cause staining or popping. Frequently, special "block mixes" are used, which contain all the necessary coarse and fine material in one gradation.

Segregation must be controlled for all aggregates, since this would result in a variation in the grading. Grading variations can cause variations in workability of the mix, texture, and strength of the finished block as well as the yield number of blocks per yard of aggregate.

Water for concrete block has the same requirement for purity that is necessary for any concrete. Water fit to drink is generally suitable for use in concretes. This means it should be free from injurious amounts of oil, acid, alkali, organic matter, or chemical waste. Concrete block manufacturers use a "dry mix."

Admixtures play an important role in block production. Admixtures are anything added to the mix other than aggregates, cement, and water. These can be materials to speed or retard curing, to reduce shrinkage after curing, to color or tint the block, or to make the block stronger after a high-pressure, high-temperature curing. Other admixtures are included to aid in air entraining or to improve the workability of the mix.

Whenever an admixture is considered, it should be carefully evaluated using the materials and methods that will be used in the production of block. The effect of the admixture on all of the manufacturing operations and properties should be carefully noted. The amount to use and the conditions under which it should be used will depend on mix materials, equipment, weather, and other factors. Claims for performance of admixtures have a tendency to go somewhat beyond actual performance. The manufacturer must be the judge as to which, if any, admixtures best meet the conditions of his operation. Trial batches should be made and must be batch-tested prior to any commitment to an admixture.

Pozzolanic and siliceous materials such as fly ash and silica flour are used in block manufacture. Silica flour is a binder component that reacts with lime under autoclaving curing conditions to make stronger concrete.

Handling of Materials

For economic reasons, most plants use bulk cement delivered by rail cars or trucks that are equipped with pneumatic pumps to convey the cement through a pipe to a silo or bin. Some plants receive cement in a ground-level hopper from which screw conveyors or bucket elevators carry it to the silo. Cement in bags is rarely used except in very small plants.

Maintaining a steady flow of cement into the batching and mixing machinery is essential to proper proportioning. Cement is a superfine powder that exhibits many characteristics of a liquid, but it can lump. Many bins have low-pressure air jets in the bottom to fluff the cement and keep it free flowing. Pressure for these types of aerators should not exceed 5 psi, and the air must be as dry as possible.

Problems of moving and handling aggregates are somewhat different from those encountered in the handling of cement powder. Cars or trucks discharge aggregates, usually in ground-level hoppers from which inclined belt conveyors or vertical bucket lift elevators lift the aggregate to overhead storage bins. Occasionally, a "clam shell" will be used to unload cars and put the aggregate into a stockpile or move the aggregate from stockpiles to bins, but this system is seldom used today because of the high cost of labor involved.

Movement of aggregate is aided considerably by the proper installation of external vibrators on hoppers and bins. Vibrators are vastly superior to a hand-powered sledge in moving the materials in the bin and are obviously less detrimental to the bin itself. The cost of installing and operating a vibrator on an aggregate bin is a great deal less than the cost of a laborer swinging a sledge hammer and possibly damaging the bin.

MIXES

The combined aggregate (fine plus coarse aggregate) should be proportioned so that the fineness modulus is between 3.5 and 4.0. A usual mix will be 55 to 70% sand with 45 to 30% gravel. These proportions might be changed to 40 to 75% sand and 60 to 25% crushed rock. The strength of the block for any given cement content is higher when the aggregate grading is more coarse. However, sufficient fine material is necessary to produce a workable mix that will give regular and uniform surfaces. The proportions of aggregate and cement must be determined by trial mixes at the plant. Workability, cracking, surface texture, green strength, density, absorption, strength and color of the cured block, economy, and the general appearance should be considered. As much coarse aggregate should be used as the mix will tolerate while still maintaining workability. This may be as little as 25% of the total aggregate, but it is generally between 40 and 50%. Trial mixes should consist of varying portions of coarse to fine aggregate combined with different amounts of cement. A rich mix will probably be about 1 part of cement to 10, 12, or 16 parts of aggregate.

An aggregate that will be used for making block should be graded from 0 to 3/8 in. and, for control grading, should be divided into fine and coarse material. The fine material usually ranges from dust to 1/4-in. size, and the coarse material ranges from 1/4 to 3/8 in. However, for a given fine and a given coarse material, the proportion of each must be determined.

The group of curves in Figure 12-1 illustrate a simple method that can be used to determine the amount of fine and coarse material necessary to obtain a block of high strength and low absorption with a particular aggregate. These curves are based upon an extensive test program in which the effect of gradation had to be determined. The test began with a mix that had only relatively fine sand with a fineness modulus of about 2.0, and ended with the coarsest possible gradation with a fineness modulus of about 4.5. With these different gradations the same cement content per block was used. The proportion of fine material was decreased and

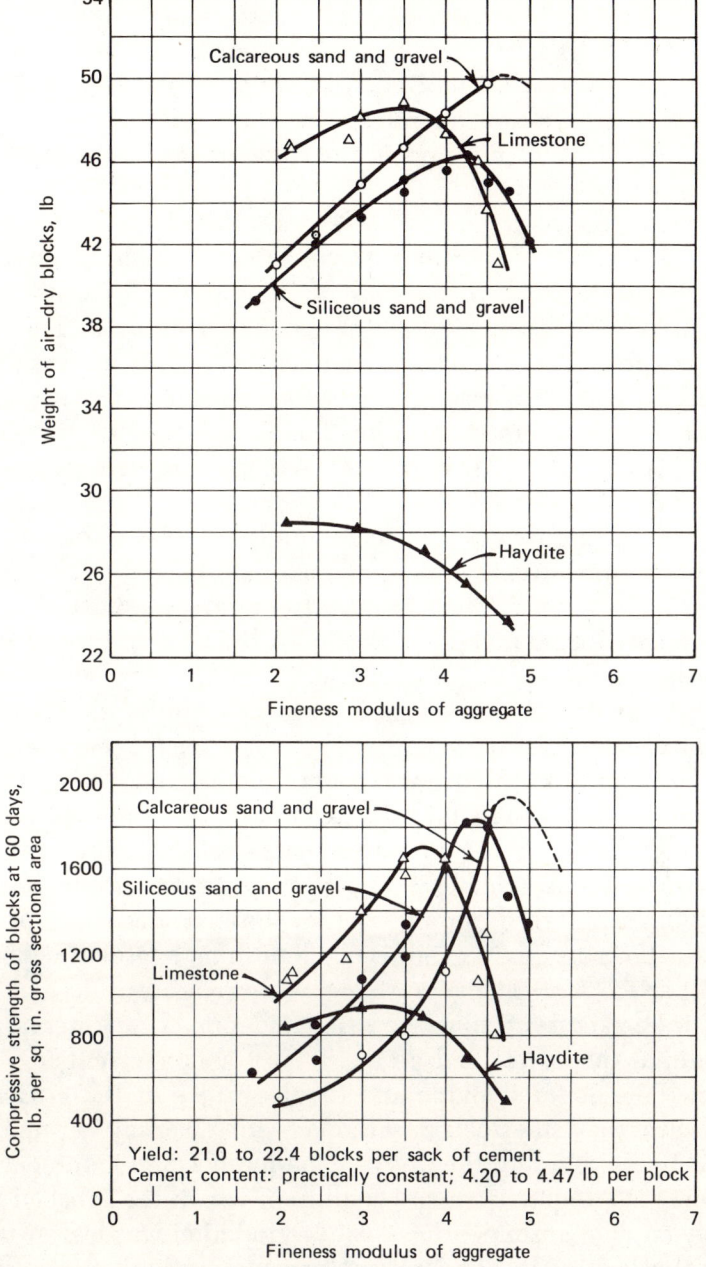

Figure 12-1 Relation of fineness modulus of aggregate to block weight and compressive strength.

the amount of coarse material was increased as much as possible while making sure that the block would hold together after it was withdrawn from the mold. The blocks were all cured in the same manner for a period of 60 days—they were kept moist at 70°F in a fog room for 5 days, then exposed to the air in the laboratory for the remaining 55 days. The strength of the block at 60 days is indicated in Fig. 12-1 in the lower group of curves.

In that graph, there are curves for several different types of aggregate. The curve of calcareous sand and gravel shows that the strength of the block made with a fineness modulus of 2.0, which is all sand with no coarse particles, was about 500 psi of gross bearing area. This strength increased progressively as more and more coarse aggregate was used until a strength of about 1800 psi of gross area was attained with a fineness modulus of 4.5. By merely adjusting the proportions of fine to coarse aggregate in the block, the strength of the block was increased over three and one-half times.

Tests were repeated with a siliceous sand and gravel, beginning with fine sand and progressively adding coarse material until cohesiveness of the block was threatened. The curve for this aggregate indicated that the strength increased from 600 psi to a peak of 1800 psi and then decreased. In this particular case, strength decreased because there was more coarse material in the block than the block could hold and still retain a dense concrete mixture. The concrete became somewhat honeycombed and therefore weakened, since concrete containing air voids cannot have the same strength as concrete that is solid and dense.

Next is the curve for the fine and coarse crushed limestone. Again, starting with the finest material, strength increased from 1000 psi to around 1700 psi before dropping off sharply to a low value of 800 psi. In this case, the crushed fine and coarse limestone aggregate consisted of elongated particles of a slivery shape with rough surfaces and sharp edges. As in the case of siliceous sand and gravel, an increase in the coarse material finally formed air voids and a honeycombed structure. Because of the elongated shape and rough surface of the crushed limestone particles, they interlocked with each other permitting more coarse material to be used before the loss of cohesion became imminent. With a manufactured aggregate such as Haydite, in which the crushed particles are roughly cubicle in shape, the effect of gradation was less pronounced. It is clear that aggregates of different types, sizes, and shapes affect the strength of concrete masonry units.

The curves in the upper diagram illustrate the important effect of variations in grading on the weight of the block. The curves in Figure 12-1 also show the relationship between weight or density of the concrete and the compressive strength. For example, the curve for calcareous sand and gravel shows that the weight of the block increased progressively with an increase in coarse material from an initial 41 lb to a maximum of about 50 lb per block. The corresponding strength curve for calcareous sand and gravel in the lower diagram shows that strength increased sharply at the same time that the weight increased.

In the case of the limestone aggregate, the weight curve indicates that the weight increased from about 47 lb to a maximum of about 49 lb and then dropped off rapidly to 41 lb. The corresponding strength curve for limestone aggregate shows that the maximum strength was at or near the point of maximum weight. With siliceous sand and gravel (and also with Haydite), there is again a very close relationship between weight and strength. Thus, the heavier the block becomes due to gradation, the more dense it becomes and the greater will be its strength.

To obtain optimum strength with a given aggregate, fine and coarse aggregate are combined in trial batches until the maximum weight in the freshly molded unit is obtained, keeping the cement content the same. Usually, it is a good plan to start with about 80% of fine material and 20% of coarse material by weight and to increase the amount of coarse material until maximum weight is obtained. Once the strongest combination of fine and coarse material has been established, the amount of coarse material can be increased or decreased to obtain the particular texture desired.

Another way of getting at the optimum proportion of fine and coarse aggregate is to first make dry, rodded weight determinations of the fine and coarse aggregate. By combining these aggregates in different proportions, the proportion that gives the maximum dry rodded weight of the mixture can be obtained. This is a good procedure to follow before starting to make block with cement and compacting it in the particular block machine being used.

Having determined how to get the optimum proportion of fine and coarse aggregate, the next step is to use the proper cement content. Figure 12-2 shows the relation of cement content to compressive strength when using either calcareous or siliceous sand and gravel or using Haydite. In both cases we have a linear relationship. If the cement content is increased from 4 lb per block to 8 lb per block, the strength will be increased about 1600 psi to about 3200 psi for calcareous or siliceous sand and gravel and, in the case of Haydite, from about 1000 psi to about 2000 psi. In general, doubling the cement content approximately doubles the compressive strength if grading remains the same.

The proper amount of water can be judged in several ways. One way is to select a handful of fresh concrete from the mixer and slide a trowel back and forth over its surface several times. If a water sheen develops on the surface with this troweling action, then the water content is about right for block manufacture. Another way is to watch the mix as it is being turned over by the paddles of the mixer. If there is a definite reflection of light from a water sheen on the concrete just as it leaves the paddle, this indicates that moisture is present on the surface and the amount of moisture in the mix is just about adequate. A third method is to observe the surface of the concrete immediately after the block is being withdrawn from the block mold. If a water sheen (sometimes called a water web) develops on the surface, this also indicates that there is enough water in the concrete. If a water sheen is not indicated by any of these three methods, there is a good chance that the degree of compaction, strength, watertightness, low absorption, and other characteristics that help to build up the structure of the hardened concrete block have not been obtained.

When using lightweight aggregate, it has been found that the aggre-

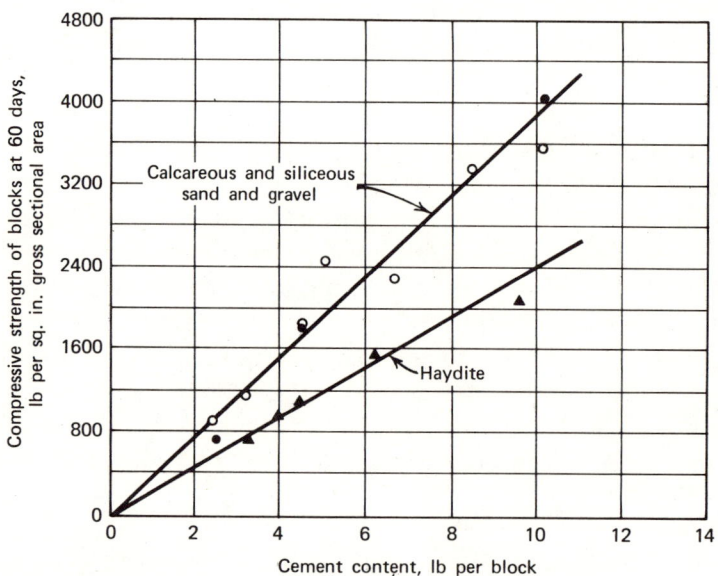

Figure 12-2

280 CONCRETE MASONRY

gate should first be mixed with about one-half the total amount of mixing water. The water will tend to fill the voids of the aggregate, and when the cement is added, it then will coat only the surface of the aggregate particles and not be lost in the voids. If the cement and aggregate were mixed dry, some of the cement would fill in the voids and not have enough water for hydration.

Zero-slump concrete is used so that the masonry will maintain its dimension as soon as it is removed from the compacting machine.

One of the best clues to the final quality of the block is the behavior of the mixture in the machine, the appearance of the block as it comes from the machine, and the ease of handling of the green block. Once a mix has been decided upon, every effort should be made to keep the proportions the same and all of the equipment in good operating condition. When using lightweight aggregate, it may be necessary to reduce the amount of fine material in order to obtain the required lightness, texture, insulating value, and acoustical properties. This is usually accomplished at some sacrifice in strength and increase in drying shrinkage.

BATCHING AND MIXING

Batching is done in various ways. It can be done in batchers by weight, on belts by volume, or by determining the load on the mixer motor. In most modern plants, batching is done in automatic or semiautomatic batching machines. A weigh batcher is used that measures the weight of each ingredient (except water). The batcher then moves over the mixer and discharges its load while water is measured into the mixer. Admixtures, if any, are generally introduced with the water. Mixers are pan, turbine, horizontal-shaft pug mill, or rotary. A great variety of mixers are available, and detailed information about the latest models on the market can be found in the current *Concrete Industries Yearbook.*

Control of the mix water is best accomplished with some type of moisture meter. One of these has probes in the mixer that sense the moisture content of the batch by electrical resistance. A dial gauge shows the moisture content and enables the operator to adjust the amount of water. The meter can be made to actuate automatic controls that regulate the amounts of water and sand in the batch. Probes should be cleaned regularly and the equipment kept in adjustment to be sure that they are operating properly.

Adequate mixing is essential. In order to obtain adequate mixing, the mixer should not be overloaded, the time for mixing should be sufficiently long, and the proper consistency of the batch should be maintained through control of the water. The mixer is overloaded if the batch covers the mixer shaft. The length of mixing time depends on the machine, batch proportions, and materials and must be determined by trial. With lightweight aggregates, mixing time may run about 7½ min., including 1 min. for prewetting the aggregate before adding the cement. Water control for lightweight aggregate mixes can best be done by the use of automatic machinery.

Adequate mixing will do four things for block: (1) minimize web cracks under the core bars, (2) assure uniform texture between block batches and prevent texture variations, (3) contribute consistent strength, and (4) assure uniform color. After all ingredients are in the mixer, they should be mixed for a minimum of 5 min. Prewetting or premixing of lightweight aggregates may, of course, add to the amount of mixing time.

Allowing 2½ min for charging, discharging, and adding water, a 7½ min cycle with a 50-cu-ft mixer will produce approximately 400 cu ft of loose concrete per hour. This is enough material for 1000 8-in. blocks.

MAKING THE BLOCK

After mixing, the concrete is discharged into the block machine hopper from which controlled quantities of concrete are fed into the molds. Vibration and pressure consolidate the concrete in the molds, which rest on a steel or wooden pallet during the molding process. As soon as the concrete is compacted and struck off, the mold box lifts, and the pallet holding the green block emerges from the machine while the new pallet is being pushed in from the other side. The pallet containing the green block then travels to the curing area.

Output of block machines is based on the number of 8 × 8 × 16-in. blocks (or equivalent amount of concrete) produced by the machine per cycle. Most block-making machines operate on about 6, 7, or 8 cycles per min.

The pallet on which the green block is molded and handled (usually made of steel but occasionally of wood) is a rectangle, most often 18½ × 26 in. The 26-in. length permits working space to mold three 8-in. blocks; four 6-in. blocks; six 4-in. blocks; or 28, 32, or 36 bricks, depending on the mold design. The machine can be used to produce any of the common modular units by changing the mold box.

Many steps in block molding are automated, even on some of the small machines. Push-button controls permit the one-man operation of machines with capacities up to 1000 8-in. blocks per hour. Block machines are ruggedly built, since they are subjected to moisture, vibration, and abrasion. Proper maintenance is essential to keep them operating efficiently—thorough cleaning and lubrication at every shift are important. To assist in this, compounds are available that can be sprayed or brushed on the molding surfaces to prevent adhesion of concrete. If pallets are used, they must be kept clean. Those that show excessive wear should be discarded. Block should be checked for density and dimensions at regular intervals to make sure that the proper quality is being maintained.

The process of moving the block off the machine is called offbearing and, in many plants, is either semiautomatic or fully automatic. Offbearing can be done manually, but the cost is usually considered prohibitive. The loaded pallets are placed in steel curing racks that are transferred to the curing kiln or autoclave. Racks are transported by lift trucks, by small rail cars, in trains—each rack having wheels and coupling to the preceding rack, or by hand if the curing racks are on casters.

CURING

Curing can be accomplished in two ways: (1) with saturated steam at atmospheric pressure or (2) with high-temperature, high-pressure steam in a pressure vessel called an autoclave. These curing methods are used in concrete block plants because they reduce storage space for finished units and allow quicker delivery.

The presence of adequate water for hydration is always important. More rapid hydration of the cement is accomplished by increasing the temperature. Systematic control is necessary to raise the temperature slowly enough so that the block does not develop critical internal stresses through differential thermal expansion.

Atmospheric Pressure Curing

A normal schedule for atmospheric steam curing is pretty much as follows.

1. *Preset, also called delay or holding time.* This is a period between molding and steaming. Hydration of the cement continues during this time and stabilizes the block. Block are stored in the steaming chambers at normal temperatures and prevented from drying by the application of a light fog if necessary. During cold weather, a small amount of heat might be necessary to maintain a temperature above 60°F.

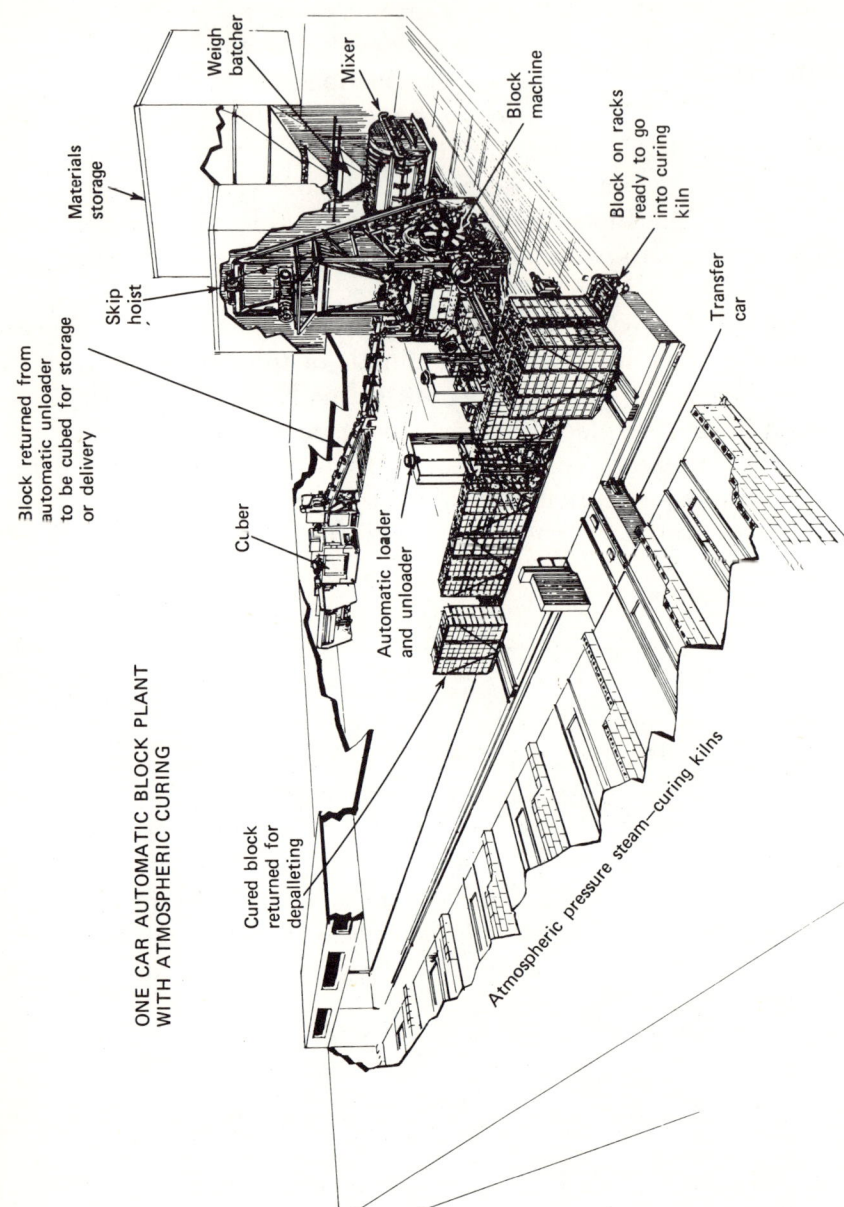

Figure 12-3 One car automatic block plant with atmospheric curing. (Courtesy of Besser Co.)

284 **CONCRETE MASONRY**

2. *Period of rising temperature.* During this time, the blocks are enclosed in the curing chamber, more commonly called a kiln. Steam heat is introduced to raise the temperature to about 120 to 180°F. The temperature rise is controlled at a rate of 20 to 60°F per hour depending on the size and the type of aggregate used for the block.
3. *Maximum temperature period.* The temperature in the kiln is held constant for whatever amount of time is necessary for the block to develop adequate strength. The time required will vary up to 18 hours.
4. *Soaking period.* After the maximum temperature is reached, steam is shut off, the kiln is kept closed, and the block "soaks" in the kiln as the temperature slowly drops. However, if the maximum temperature is maintained sufficiently long, soaking is not necessary and is uneconomical.
5. *Cooling period.* The block is permitted to cool to atmospheric temperature at a controlled rate—again dependent on the size, shape, and character of the block. Rapid cooling results in dryer block.

In one typical plant, after a holding of 2 to 4 hours, the temperature in the curing chamber is raised at a rate of 40°F per hour to a maximum of 175°F. The temperature rise is discontinued at this point, and a soaking period of about 12 hours (or longer if tests show an advantage in a longer period) follows, during which the temperature is lowered at a rate of not more than 1°F per min. Block that have thin walls or small dimensions can be heated and cooled at a much faster rate. Recording temperatures or automatic temperature control equipment is necessary for product uniformity in the curing chamber.

Many block plants moist cure at atmospheric pressures by means of wet steam that maintains a temperature of about 135°F. Steam from the boiler is released through perforated pipe laid along the floor between the racks. If the temperature exceeds 135°F, the steam is shut off or reduced to prevent the units from becoming too hot, causing dehydration and insufficient curing. Block will always show a film of moisture while curing with low-pressure atmospheric steam. Less than two days of atmospheric steam is customary if tests show that adequate strength has been obtained. When the kiln is opened, there should be a period of several hours for the concrete to adjust to the temperature of the outside air in order to prevent cracking (most likely to occur in cold weather).

Under high-temperature conditions both heat and moisture may be lost to the outside atmosphere if the kilns are not well sealed or well insulated. More moisture is required to maintain 100% humidity at higher temperatures than at low temperatures. At 70°F, for example, about one and one-half pounds of water is required to keep each 100 lb of air at 100% humidity. However, 15 lb of water are required at 140°F, and 66 lb are required at 180°F.

Under steam curing conditions, the heat of hydration will raise the concrete temperature in the same way that it does under normal curing. Therefore, there is a time when the concrete temperature reaches the temperature of the kiln, and it would then rise above it if control measures were not taken. After the concrete reaches the ambient temperature, the amount of heat supplied should be reduced gradually in order to minimize the temperature difference between the concrete and the atmosphere in the kiln, and to permit a gradual cooling of the concrete.

Heat of hydration may cause the concrete temperature to exceed the temperature of the kiln, and moisture can be lost from the concrete if water is not supplied. Wet steam is used, or if heat is provided by hot water or hot oil, additional moisture is added from fog sprays.

High-Pressure Steam Curing (Autoclaving)
High-pressure steam curing can be used for any concrete product, but it is most often applied to masonry units.

The increased temperatures are sometimes necessary in order to cause a chemical reaction. The high temperatures required for this type of moist curing can be reached only by raising the pressure in the curing chamber. Autoclaving includes a period of temperatures of about 360°F in saturated steam. The cycle can be varied, but a typical sequence starts with a preset or holding period at room temperature for 2 to 4 hours. The autoclave is then loaded and sealed, and the temperature and pressure is raised over a period of about 3 hours by means of saturated steam to about 360°F and 140 to 150 psi. These temperatures are held for 5 or 6 hours. Pressure is quickly released during the final 20 to 30 min., which rapidly dries the block. Variations in timing and temperatures are used to suit the materials available and appearance desired. The best curing cycle is chosen to give the desired lightness of color and optimum production timing.

The autoclave generally used for this kind of work is a long tanklike steel cylinder with a diameter large enough to accept a rack of block. The rapid pressure release used with this type of curing removes enough of the moisture from the block to bring them to a relatively stable air-dry condition by the time the curing cycle is complete.

EQUIPMENT

The machinery used to produce, cure, and handle concrete masonry products has become more and more sophisticated during the past few years and, in some plants, has become completely automated. Machines in use include (1) level controls for dry bulk materials, (2) bin level indicators, (3) elevator buckets and conveyors to load large overhead storage bins, (4) vibrators to insure an even flow of materials from storage bins, (5) skip hoists, (6) moisture monitors to control mix water amounts, (7) systems for automatic batching, (8) mixers, (9) block machines to mold the concrete masonry units, (10) machines to clean off the concrete build up on pallets, (11) block delivery conveyors, (12) off-bearing hoists to remove units from the molding machine, (13) racks to hold the block during production, and (14) de-burring brushes to remove burrs of concrete from newly formed block. There are also (15) crushers to smash aggregate to size and (16) self-dumping hoppers to deliver materials. Some plants use (17) automatic loading and unloading machines to put block on racks before curing and to remove them from the racks after curing, (18) automatic cubing machines, and (19) turnovers to turn the block from a vertical to a horizontal position. All of these machines (and more) are available to the block producer.

Several years ago, the automatic block machine was considered a labor-saving miracle. Today, a great deal of machinery is automated to the point that it only needs to be turned on at the start of the production day and will perform its function with a minimum of help from workmen.

Continuous-feed mixers were once quite popular, but with the increase in the variety of aggregates used, they have frequently been replaced by batch mixers. The continuous-feed mixer does offer several advantages to the producer, however. These are:

1. Lower initial cost.
2. Lower operating cost.
3. Less manpower requirement.
4. Smaller space requirement.

In continuous mixers the ingredients are placed in a mixing trough that has a long, horizontal screw. The mixer works continuously, discharging the mixed concrete at the far end of the machine.

Batch mixers are used with great success in many block plants. Automatic controls can eliminate the need for a man to weigh and measure raw materials. Even the amount of mix water can be accurately controlled automatically. This type of equipment is costly to install, but the saving in labor and time more than repays the initial investment.

No matter what type of mixer is being used, care should be taken to avoid the build up of hardened concrete on the interior. Clean blades aid in the proper mixing of the concrete. When the blades become worn, they should be replaced.

If there is a delay between the mixing operation and the molding of the concrete, the batch can be used as long as it retains workability. If, however, workability is lost and cannot be restored by further mixing, the batch should be discarded.

There are many types and sizes of block molding machines. Some are operated by hand, some operate semiautomatically, and some are completely automatic. Generally, one block machine can turn out between 800 and 1100 blocks per hour. Most machines compact the dry-mix concrete by vibration and pressure. Most block machines can easily be adapted to the production of several sizes of block by the insertion of various molds. It is not necessary to have a machine for each block size being manufactured.

The size and capacity of the racks used to hold the block during curing are dependent on the size and shape of the kiln or autoclave and on the production capacity of the molding machine. Block are stacked on the racks with air space between them to allow for complete circulation of moist air during curing. The selection of the method of curing concrete masonry units of an important one, and many considerations must be taken into account.

Although the initial investment in the construction of a series or bank of kilns is high, they give service for many years. The planning of the kiln is an important step in determining the effectiveness of the curing operation and in the assurance of a smooth production flow within the plant. It is best to obtain the services of a specialist before construction is started. The following factors should be considered.

1. Plants' physical layout.
2. Traffic flow between kiln and block machines.
3. Space available for kiln.
4. Most economical and serviceable construction materials (usually block).
5. Size of racks (if kiln is to be built to accommodate existing racks).
6. Number of units per rack and number of racks per kiln required.

7. Production rate of the block machine.
8. Capacity of lift truck or other mechanical device to transfer units from molding machine to kiln.
9. Capacity of boiler.
10. Flow and distribution of steam for curing.

The kilns should not be larger than necessary since the unused space is unproductive. The daily production requirements should be determined and the kilns designed to handle them. There is a favorable growth forecast for concrete masonry, therefore, additional kiln space should be provided for increased production during the next few years. This can often be done by building several smaller kilns instead of one large one.

Kilns should be insulated to avoid heat loss. Well-sealed kilns are easier to control, and control of humidity and heat level is very important to good curing. Kilns with doors at both ends facilitate the loading and unloading operation.

Special provisions should be made for the exhausting of steam. It should not be allowed to escape into the general work area. It is usually simple to install an escape duct so that the steam can be exhausted into the outside air.

When steam is introduced too forcibly into the kiln, it will hit against walls and condense. On the other hand, if the velocity is too low, there will be insufficient steam distribution within the kiln. The best velocity depends on the kiln dimensions. If the available velocity of the steam is limited, the kiln design should also be limited to allow for proper steam velocity. The solution to the problems of steam distribution requires careful attention to the system and needs for the specific kiln being used.

Water droplets should not be ejected from the steam nozzle. These droplets would gather on the block and cause surface imperfections. Care should be taken to have steam that is saturated (100% relative humidity).

The initial investment is higher for autoclaves, but the improvement of product quality and the increased daily production can offset the cost. There is also a significant saving in cement.

Autoclaves are from 50 to 60 ft long with a diameter of from 6 to 11 ft. An autoclave is nothing more than a cylinder sealed at both ends. Steam pressures can be as high as 250 psi and temperatures as high as 485°F. A pressure of 150 psi and temperature at 365°F is more usual, however. The higher pressure ranges demand autoclaves that are much more expensive to build and operate. Research shows that pressures of around

150 psi are adequate to assure speedy and complete hydration. Practical use is 125 psi at 350°F.

Although the pressure within the autoclave must be watched closely, the temperature is the more critical factor. Gauges should be installed within the autoclave to keep close check on the temperature rise. It is also advisable to attach several thermocouples to the blocks themselves at various places in the autoclave, at least during trial runs.

The size and location of the autoclave is determined by block machine capacity; traffic to the autoclave; the length of loading, curing, and unloading times; steam requirements for normal and peak load; space limitations; and the future plans of the producer.

The size and capacity of the steam boiler selected depend on the load requirements of the kiln and the number of kilns fed by the boiler. The capacity should be great enough to insure adequate maximum temperatures within the required time schedule. Because boilers work best at less than maximum capacity, it is prudent to install a boiler larger than needed.

PRECARBONATION

It has been found that concrete exposed to carbon dioxide will shrink. Tests by the Portland Cement Association (PCA) and the National Concrete Masonry Association (NCMA) have shown that precarbonation of concrete masonry units can reduce further shrinkage by as much as 50% under subsequent exposures to cycles of wetting and drying or to carbon dioxide.

The practice in the industry has been to introduce waste flue gas in the kiln after the blocks have been cured and are in the drying period. During this period, block must be exposed to conditions that dry the block to a favorable moisture level and maintain that level during treatment and, of course, provide a sufficient carbon dioxide concentration to affect rapid carbonation.

Carbonation is beneficial in the manufacture of concrete block if it is applied after the curing period. Carbonation is seriously detrimental to *uncured* concrete.

Conclusions based on PCA/NCMA tests were: (1) the relative humidity in the kiln should be between 15 and 35%; (2) the CO_2 content should be as high as possible, but in any case not less than 1.5%; (3) the temperature should be between 150 and 212°F; (4) a 24-hour treatment is generally required, but under optimum conditions shorter periods may suffice; and (5) potential shrinkage reduction of more than 30% may be obtained under favorable conditions.

CUBING

Cubing is the process of assembling blocks into convenient groups for easier handling. After completion of curing, racks of block are transported to the cubing station where the cubes are assembled either by hand or by automatic machinery. The size of the sube will depend on local customary usage and the size of the block that make up the cube. It usually requires four laborers to cube the production of one block machine, but a semiautomatic cuber can do the job with two men. Fully automatic machinery may require periodic supervision. After cubing, the blocks are moved to a storage area.

Fork lift trucks are extensively used for moving racks and cubes of blocks. By cubing the block on pallets or cubing them so that the tines of the fork lift slide through the cores of the bottom tier the entire cube can be lifted onto the stockpile or the truck that will be used to transport the block to the building site.

PROPERTIES OF CONCRETE MASONRY

While concrete block is similar to structural concrete in many respects, there are important differences. In general, the quality of the product depends on the qualities of raw materials, mixing, and workmanship.

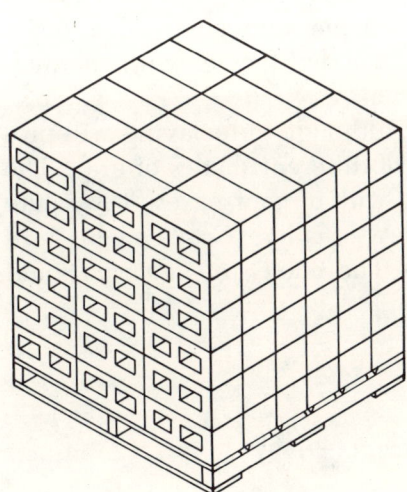

Figure 12-4   After curing, the block is cubed for ease in handling and transporting. Many plants cube the block on pallets, some use handling equipment that doesn't require a pallet. The pallet shown here contains 108 standard-sized 8x8x16-in. blocks.

However, block contains less cement (2½ to 4 bags per cu yd) and less water (2 to 4 gal per sack) than most cast-in-place concrete. The aggregate is rarely larger than ⅜ in. and is frequently lightweight. Curing for block is almost always at elevated temperatures.

Compressive strength is determined on the entire block and computed on the gross area of the block, that is, on the overall dimensions including core space. Compressive strength based on net area (excluding core space) is about 1.8 times the gross area value. Compressive strength is influenced principally by the type and amount of cement per unit, the type and grading of aggregate, degree of compaction, age of the specimen, curing procedures, and moisture content at the time of test.

Masonry, as with any concrete, undergoes changes in volume due to changes in temperature and moisture content. One of these changes is the original drying shrinkage of the green concrete. If block are laid up in a wall before they have taken all of their initial drying shrinkage, tensile stresses will be developed that are almost certain to crack the wall. By proper curing and drying in the yard, the moisture content of the block can be reduced to an equilibrium condition with the surrounding atmosphere, and cracking tendencies can be minimized. The selection of materials and curing methods, such as autoclaving, will produce block with greatly reduced potential shrinkage.

Texture is the disposition of particles and voids on the surface of the unit. The texture desired can be obtained by adjusting (1) the gradation of the aggregate, (2) the amount of mixing water, and (3) the degree of compaction in the mold. In addition, special face molds and rubber or plastic mold liners are available for special finishes. The units also can be ground or treated with chemicals to expose the aggregates. The desire for color and texture has led to the use of bonded facings with thermal-setting resin binders and glazed surfaces.

Color depends mainly on the materials, although autoclaving will produce lighter shades of block. Cements are of different shades of gray. The use of white cement with white aggregate will, of course, result in white block. Pigments make possible a wide variety of colors. Gray cement with pigments can be used for darker shades. The choice of aggregate and color is especially critical for split block.

CHAPTER 13
SHAPES, SIZES, AND CONSTRUCTION DESIGN

Concrete masonry has become one of the most dramatic wall materials available. In addition to the standard rectangular units, many new shapes, sizes, and textures have appeared on the market. Equally important, architects and builders are now using conventional units to create striking patterns and designs for beautiful buildings.

Four recent developments in concrete masonry units are molded-face block, screen block, ground-face block and scored block. Molded-face block has a geometrical design that protrudes $3/8$ to $1/2$ in. beyond the face of the block. Some of the molded-face block have companion units. Used together, these units allow an unlimited number of designs.

Screen block are usually 4 in. thick. They are used in outside walls to attain privacy, or screen out the hot sun or unattractive areas. Screen block are also used within buildings to form dividers and low walls.

Ground-face block are made by grinding the face of block to expose the aggregate, providing interesting effects. Block that measures 8×16 in. may be scored to simulate 8×8-in. block for architectural effects.

Stonelike split concrete units, pierced block, and irregular-slump units are some of the other interesting units available. Many are patented and produced on franchises granted by the patent holders. Although all units are not available everywhere, local producers usually have a wide range of attractive materials from which to choose.

BLOCK WEIGHTS

Normal-weight masonry units are made of portland cement and aggregates. A hollow 8×8×16-in. normal-weight unit weighs between 40 and 50 lb. Lightweight units are made of portland cement and natural or manufactured lightweight aggregates such as expanded shales, slate, and slag. Their weight is usually between 25 and 35 lb.

ASTM SPECIFICATIONS

Masonry units are made to comply with specifications of either the American Society for Testing and Materials (ASTM) or the federal government. ASTM specifications for Grade A, hollow, load-bearing units require that the compressive strengths of 5 units average 1000 psi of gross area. To meet this requirement, an 8×8×16-in. unit must withstand 128,000 lb or 64 tons.

The specifications also cover the permissible water absorption of the units. To comply with the ASTM specifications, units may not absorb water at a rate higher than 15 lb per cut ft of dry concrete.

A third ASTM requirement is that the average moisture content of 5 units shall not exceed 40% of the total absorption. This is specified to prevent the use of wet units.

If their moisture content is high, units will shrink excessively, which could cause the wall to crack. Only moisture-stable, well-cured units should be used in masonry walls.

Concrete masonry units should be covered from the time they are delivered to the job until they are laid in the wall. Polyethylene and other plastic film materials are reasonable in cost and make excellent protective covers.

SIZE AND SHAPE

Concrete masonry units are classified as solid or hollow units. A solid unit is one in which the core area is 25% or less of the gross cross-sectional area.

Unit sizes are actually ⅜ in. shorter than their normal dimension to accommodate the mortar joint. The nominal 8×8×16-in. unit is actually

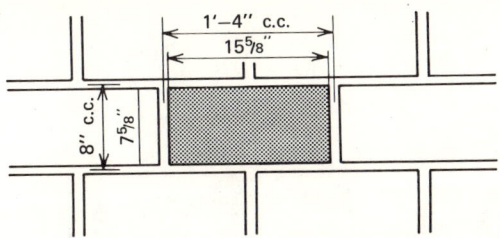

Figure 13–1 An 8x8x16-in. block is actually ⅜ in. smaller in all dimensions to accommodate the mortar joint.

7⅝ × 7⅝ × 15⅝ in. With a ⅜-in. mortar joint, the laid-in-the-wall length of the unit is 16 in. and its height is 8 in.

Typical shapes and sizes of concrete masonry units are shown in Figure 13-2. Both heavy and lightweight units can be obtained in these shapes and sizes. Many others are available or can be made to order. The units described here are the ones most commonly used by masons today.

Stretcher Block
This is the most commonly used block in backup for composite walls, farm buildings, garages, light commercial buildings, and for all types of buildings where stucco is to be used as the exterior surfacing. Either two-core or three-core block is available, and each is equally satisfactory for building purposes.

Corner Block
This block is used with the stretcher block for corners and for simple window and door openings.

Jamb Block
This block is used with stretcher and corner block around window openings. The recess in the block allows room for the various casing members such as those used in a double-hung window.

Full Cut Header Block
Header block is the same as stretcher block except that a shelf has been provided to facilitate bonding with a masonry surfacing.

SHAPES, SIZES, AND CONSTRUCTION DESIGN

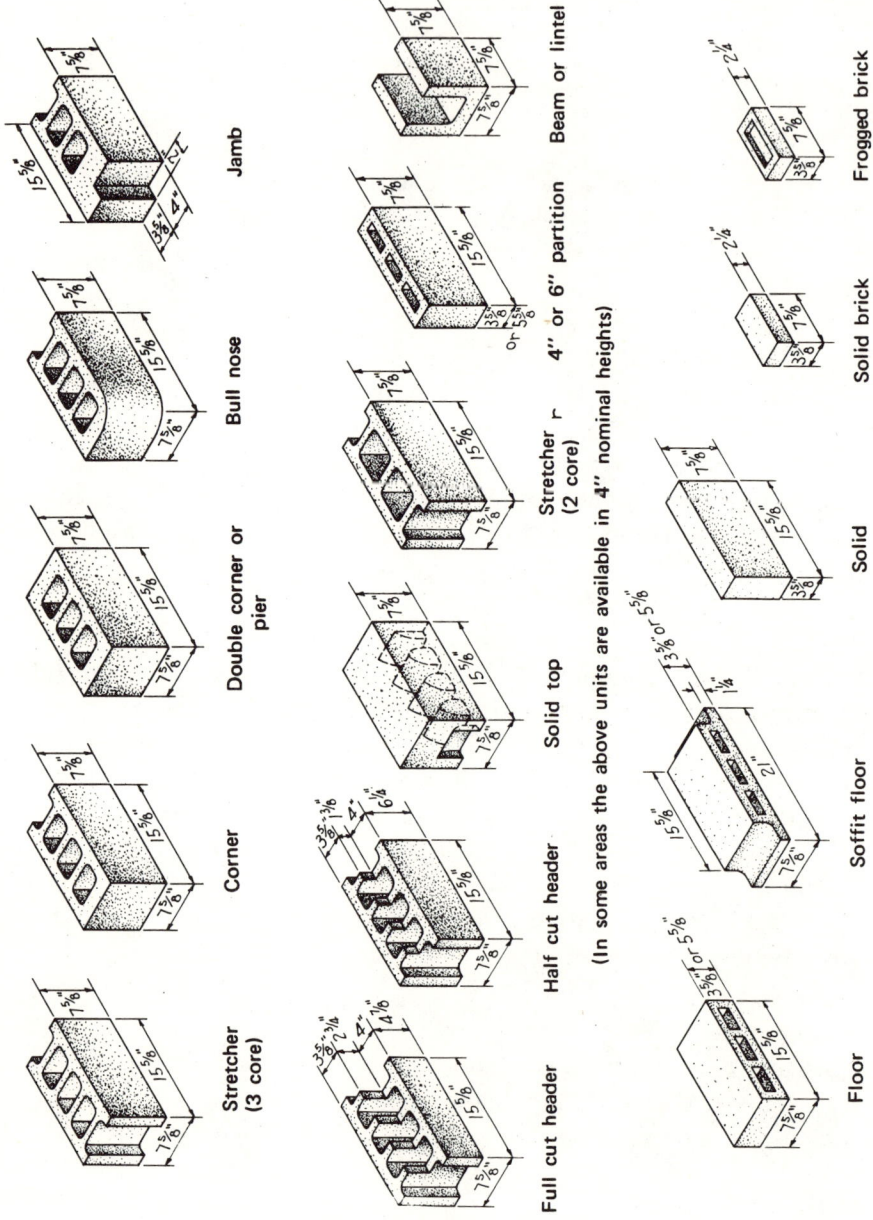

296 CONCRETE MASONRY

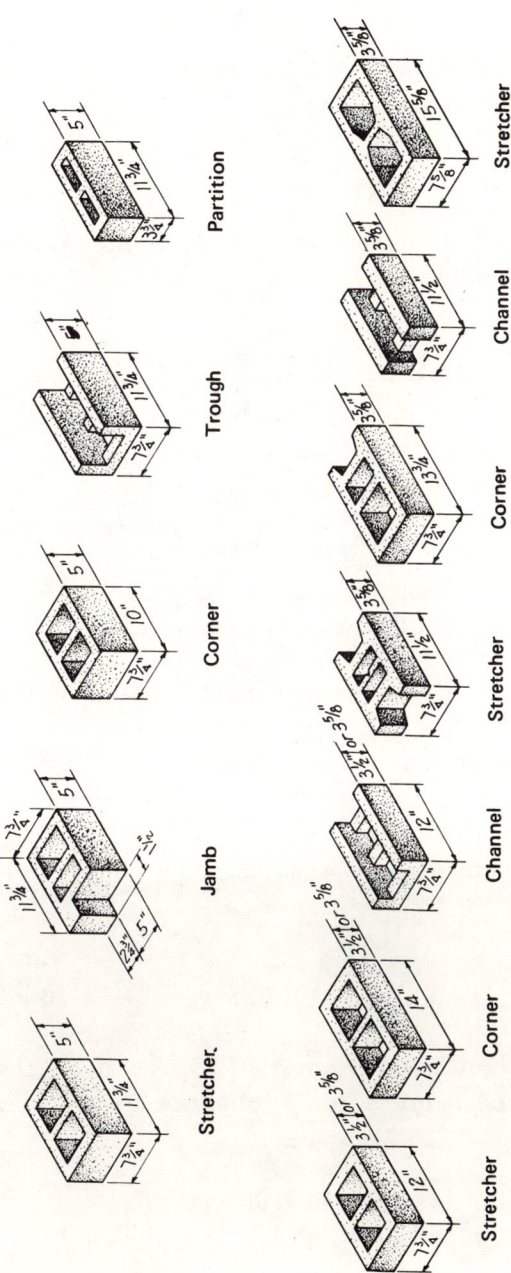

Figure 13-2 Typical shapes and sizes of concrete masonry units. Dimensions shown are actual unit sizes. A $7\frac{5}{8}''$ x $7\frac{5}{8}''$ x $15\frac{5}{8}''$ unit is commonly known as an 8"x8"x16" concrete block. Half length units are usually available for most of the units shown below. See concrete products manufacturer for shapes and sizes of units locally available.

SHAPES, SIZES, AND CONSTRUCTION DESIGN

Bullnose Corner Block

This block is used with three-core stretcher block when rounded corners are desired.

Pier or Double Corner Block

This block is designed for use in laying piers or pilasters, or for any other purpose where both ends of the block would be visible.

Special Concrete Blocks

These blocks are used in laying variously designed walls. Also available are special kinds of block used for building chimneys. They are square outside with either square or round flues.

WALL DIMENSIONS FOR EFFICIENT CONSTRUCTION

The height and length of block are usually some multiple of 8 in. If the dimensions of the building are divisible by 8 in., it is therefore possible to build without cutting a single block. Cutting block because of poor dimensioning is a waste of materials and is time consuming for the mason. Odd-size pieces detract from the appearance of the wall.

However, since it is tedious to check dimensions by converting to inches and dividing by 8, these easy-to-follow rules will save time.

Rule 1. A horizontal dimension in an even number of feet will always work out in full and half blocks.

Rule 2. A horizontal dimension in an even number of feet plus 8 in. (2 ft-8 in., 4 ft-8 in., 6 ft-8 in., etc.) will always work out in full and half block.

Rule 3. A horizontal dimension in an odd number of feet plus 4 in (1 ft-4 in., 3 ft-4 in., 5 ft-4 in., etc.) will always work out in full and half block.

Walls that have other dimensions and that are built with $8 \times 8 \times 16$-in. units will have to be cut.

It is often necessary to know how many courses of block will be required or how many blocks will be required for a course in a wall. The following rules will be helpful.

Rule 4. To find the number of courses in a wall built with 8-in. units, multiply the height of the wall in feet by $1\frac{1}{2}$. (Example: 10 ft times $1\frac{1}{2}$ equals 15 courses.)

Rule 5. To find the number of courses in a wall built with 4-in.-high units, multiply the height of the wall by 3. (Example: 10 ft times 3 equals 30 courses.)

Rule 6. To find the number of block in a course, multiply the length of the wall by ¾. (Example: 60 ft times ¾ equals 45 block.)

Table 13-A illustrates rules 1-3 worked out for 8×8×16-in. block in

Table 13–A
NOMINAL LENGTH OF
CONCRETE MASONRY WALLS BY STRETCHERS

Number of Stretchers	Nominal Length of Concrete Masonry Walls	
	Units 15⅝ in. long and half units 7⅝ in. long with ⅜ in. thick head joints	Units 11⅝ in. long and half units 5⅝ in. long with ⅜ in. thick head joints
1	1′ 4″	1′ 0″
1½	2′ 0″	1′ 6″
2	2′ 8″	2′ 6″
2½	3′ 4″	2′ 6″
3	4′ 0″	3′ 0″
3½	4′ 8″	3′ 6″
4	5′ 4″	4′ 0″
4½	6′ 0″	4′ 6″
5	6′ 8″	5′ 0″
5½	7′ 4″	5′ 6″
6	8′ 0″	6′ 0″
6½	8′ 8″	6′ 6″
7	9′ 4″	7′ 0″
7½	10′ 0″	7′ 6″
8	10′ 8″	8′ 0″
8½	11′ 4″	8′ 6″
9	12′ 0″	9′ 0″
9½	12′ 8″	9′ 6″
10	13′ 4″	10′ 0″
10½	14′ 0″	10′ 6″
11	14′ 8″	11′ 0″
11½	15′ 4″	11′ 6″
12	16′ 0″	12′ 0″
12½	16′ 8″	12′ 6″
13	17′ 4″	13′ 0″
13½	18′ 0″	13′ 6″
14	18′ 8″	14′ 0″
14½	19′ 4″	14′ 6″
15	20′ 0″	15′ 0″
20	26′ 8″	20′ 0″

Actual length of wall is measured from outside edge to outside edge of units and is equal to the nominal length minus ⅜ in. (one mortar joint)

the first two columns. The next column shows the length of walls using 6×6×12-in. block.

Table 13-B shows how rules 4 and 5 work out for block 8 in. high and 4 in. high.

Table 13-C shows sizes of window openings that can be made without cutting block. If other sizes are desired, they should be kept in multiples of 8 in.

Figure 13-3 shows a good example of what can happen with poor planning. An odd number of feet (17 ft-0 in.) do not accommodate the 8-in. block without cutting. It can be seen from the illustration that cutting odd sizes makes an unattractive wall. Adding 4 in. to the horizontal dimensions of the wall permits the use of 8×8×16-in. units and corner block without attaining a crazy-quilt look. The door and window in Figure 13-3 were also poorly planned. The shaded blocks are the ones that

Table 13–B
NOMINAL HEIGHT OF
CONCRETE MASONRY WALLS BY COURSES

Number of Courses	Nominal Height of Concrete Masonry Walls	
	Units 7⅝ in. high and ⅜ in. thick bed joint	Units 3⅝ in. high and ⅜ in. thick bed joint
1	8″	4″
2	1′ 4″	8″
3	2′ 0″	1′ 0″
4	2′ 8″	1′ 4″
5	3′ 4″	1′ 8″
6	4′ 0″	2′ 0″
7	4′ 8″	2′ 4″
8	5′ 4″	2′ 8″
9	6′ 0″	3′ 0″
10	6′ 8″	3′ 4″
15	10′ 0″	5′ 0″
20	13′ 4″	6′ 8″
25	16′ 8″	8′ 4″
30	20′ 0″	10′ 0″
35	23′ 4″	11′ 8″
40	26′ 8″	13′ 4″
45	30′ 0″	15′ 0″
50	33′ 4″	16′ 8″

For concrete masonry units 7⅝ in. and 3⅝ in. in height laid with ⅜ in. mortar joints. Height is measured from center to center of mortar joints.

Table 13-C
MODULAR CONCRETE MASONRY OPENINGS FOR WOOD WINDOW FRAMES

Types of Windows	Masonry Openings	Glass Size	Masonry Openings	Glass Size	Masonry Openings	Glass Size
Double	2'0"×3'4"	16"×12"	2'8"×3'4"	24"×12"	3'4"×3'4"	32"×12"
	4'0"	16"	4'0"	16"	4'0"	16"
	4'8"	20"	4'8"	20"	4'8"	20"
	5'4"	24"	5'4"	24"	5'4"	24"
	6'0"	28"	6'0"	28"	6'0"	28"
	6'8"	32"	6'8"	32"	6'8"	32"
	4'0"×4'0"	40"×16"	4'8"×4'0"	48"×16"	5'4"×4'0"	56"×16"
	4'8"	20"	4'8"	20"	4'8"	20"
	5'4"	24"	5'4"	24"	5'4"	24"
	6'0"	28"	6'0"	28"	6'0"	28"
	6'8"	32"	6'8"	32"	6'8"	32"
	7'4"	36"	7'4"	36"	7'4"	36"
Casement	1'4"×3'4"	8"×25"	2'0"×3'4"	16"×25"	2'8"×3'4"	24"×25"
	4'0"	33"	4'0"	33"	4'0"	33"
	4'8"	41"	4'8"	41"	4'8"	41"
	5'4"	49"	5'4"	49"	5'4"	49"
	6'0"	57"	6'0"	57"	6'0"	57"
	6'8"	65"	6'8"	65"	6'8"	65"
Basement	Two light sash		Three light sash			
	2'0"×2'0"	8"×12"	2'8"×2'0"	8"×12"		
	2'8"	20"	2'8"	20"		
	2'8"×2'0"	12"×12"				
	2'8"	20"				

Note: Modular masonry openings shown above should also be used for metal window frames. It may be necessary, however, to provide metal surrounds to fit the metal frames into the modular openings.
Masonry openings are dimensioned from jamb to jamb and from bottom of lintel to bottom of precast concrete sill. Openings of sizes other than those shown may be used by keeping the dimensions in multiples of 8 in. Glass sizes shown are for one light but any division of the sash may be used.

SHAPES, SIZES, AND CONSTRUCTION DESIGN

would have to be cut. When correctly planned, a uniform and pleasing appearance can be created. Only one block varies from the running-bond pattern and, since it is a half length, it does not have to be cut.

PATTERNS

Designers and builders have found that they can do startling things with conventional block. The stacked-bond patterns add pleasing architectural variety. Block in stacked-bond patterns are laid directly over each other, giving the wall distinct horizontal and vertical lines. Four-inch block, half as high as the standard 8×8×16-in. unit, give a long, low silhouette. Some builders have laid block in small sections of walls on the diagonal. This involves careful masonry work but does create a striking pattern. In other instances, builders have offset a few block to achieve an unusual pattern that appears in reverse on the other side of the wall.

The structural strength and stability of load-bearing walls constructed of 8×8×16-in. units laid in running bond (Fig. 13-4) have been established experimentally. Recently, tests have been carried out to determine the compressive and flexural strength of concrete masonry wall panels laid up in different patterns.

In the tests there is little difference in compressive strength of walls in which all block are laid horizontally (Fig. 13-4). The strength of walls containing diagonal or vertical block was about 75% of the standard. The strength of the individual block tested in a vertical direction (loaded on the ends) was 60 to 70% of the srtrength obtained when tested horizontally.

The strength of the mortar did not appear to have a marked effect on the compressive strength of the walls.

All nine wall patterns were tested with two strengths of mortar, and results showed that the transverse strength was primarily dependent on the bond strength obtained between mortar and units. All patterns with continuous horizontal mortar joints failed in bond between unit and mortar along one of the horizontal mortar joints at approximately midheight of the wall. The diagonal patterns also failed in bond, but the failure followed a saw-tooth line across the width of the wall, following the mortar joints. The diagonal patterns developed strength about 50% higher than those of the running bond. This could be expected because of the added bond area and the angular displacement of the tensile forces along the mortar lines.

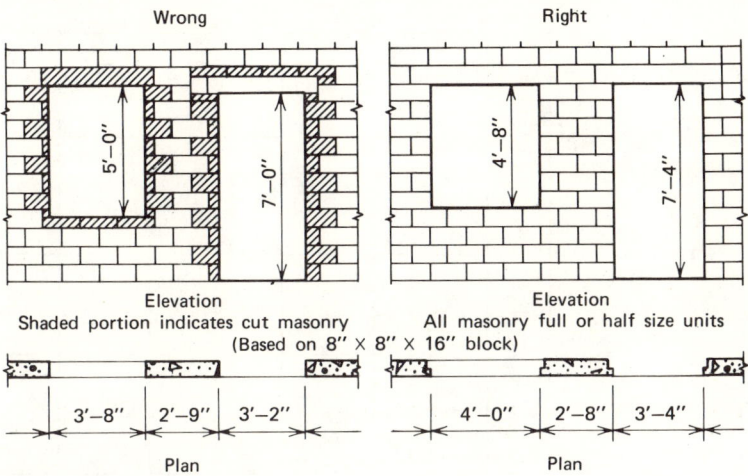

Figure 13-3 Examples of wrong and right planning of concrete masonry wall openings.

Any of the patterns tested would be satisfactory for load-bearing walls since even the weakest (Basket Weave A) tested above the safety factor. All walls tested had an adequate safety factor according to national building codes. However, the compressive strength of walls in which all blocks are laid in a horizontal position is, as expected, greater than when block is laid on the diagonal. Compressive strength of block walls depends primarily on the strength of the block. The strength of the mortar has little effect on the failure of the wall except in the case of walls with diagonal patterns that may fail in shear along the diagonal if built with weak mortar.

Although the compressive strength of the walls was not affected to any great degree by the strength of the mortar, the flexural strength of the stacked-bond patterns was entirely dependent on the bond strength of the mortar.

These tests are documented in Tables 13-D and 13-E.
Faced masonry walls usually consist of masonry facing and backing that are of different materials. A typical faced wall (Fig. 13-5) would consist of a concrete split unit, brick, or stone facing with a concrete masonry backing. The two sections of masonry are bonded together to act as one. The height limitation for an 8-in. faced wall is 35 ft.

Table 13-D
COMPRESSIVE STRENGTH OF MASONRY WALLS

Wall Pattern	Compressive load on wall at failure (lb)		Primary Failure	Compressive Strength							Safety§ Factor
	Mortar Type						Percent of Standard		Block‡ Corrected	Ratio‡ Wall to Block, Percent	
	M	S		Block*	Measured*	Adjusted†	Individual	Average			
Running bond	233,700	201,000 175,500	Block	1400 1450 1165	638 549 480	638 530 576	100 100	100	1400 1450 1165	45.6 38.9 41.2	7.5 6.5 5.6
Horizontal stacked bond	236,250	219,700	Block	1400 1450	645 601	645 580	101 105	103	1400 1450	46.1 41.4	7.6 7.1
Vertical stacked bond	161,100	164,400	Block	1400 1315	440 449	440 478	69 86	78	805ª 915ª	54.7 49.1	5.2 5.2
Diagonal basket weave	176,700 143,750	131,000	Mortar	1400 1165 1450	515 419 381	515 503 367	81 79 66	75	1100ᵇ 910ᵇ 1200ᵇ	46.8 46.1 31.8	6.1 4.9 4.5
Diagonal running bond	150,200 146,200	139,000	Mortar	1400 1165 1315	438 426 405	438 512 430	69 80 78	76	1100ᵇ 910ᵇ 1115ᵇ	39.8 46.8 36.3	5.2 5.0 4.8
Basket weave A	205,400	129,500	Block	1400 1450	562 354	562 341	88 62	74	1100ᶜ 1200ᶜ	51.1 29.5	6.6 4.2

Pattern	Gross area, psi*		Block							
Basket weave B	163,000			1450	446	430	67		1120[c]	39.8	5.2
	170,200			1450	455	438	69	77	1120[c]	40.6	5.4
	149,000		Block	1320	407	432	68		825[c]	49.4	4.8
		190,000		1315	519	551	100		1050[c]	49.4	6.1
		156,250		1315	427	454	82		1050[c]	40.7	5.0
Coursed ashlar	187,250			1450	512	554	87		1295[d]	39.6	6.0
	176,500		Block	1165	482	580	89	95	1165	41.3	5.7
		178,000		1315	486	554	100		1225[d]	39.6	5.7
		179,500		1165	491	590	106		1165	42.1	5.8
4-in. running bond	238,750		Block	1450	654	670	105	109	1365[d]	47.9	7.7
		201,250		1315	550	628	113		1225[d]	44.9	6.5

* Gross area, psi
† Adjusted to block strength of 1400 psi
‡ Corrected for position of block in wall

[a] Average unit compressive strength when tested in vertical position (as used in vertical stacked bond patterned walls)
[b] Since units are laid at a 45° angle, the assumed unit strength used is the average of the strengths in the horizontal and vertical direction

[c] Assumed unit strength = $\dfrac{f_{c1}A_1 + f_{c2}A_2}{(A_1 + A_2)}$

f_{c1} = unit strength (horizontal position)
f_{c2} = unit strength (vertical position)
$A = A_1 + A_2$ = area of weakest cross section of wall (cross section containing highest percentage of vertical units)
A_1 = that part of A containing horizontal units
A_2 = that part of A containing vertical units

[d] Unit strength of 4-in.-high units

§ Safety factor = $\dfrac{\text{Wall strength}}{\text{Wall allowable load (85 psi)}}$

SHAPES, SIZES, AND CONSTRUCTION DESIGN

Table 13-E
TRANSVERSE STRENGTH OF MASONRY WALLS

Wall Pattern	Mortar Type	Vertical Span Transverse Strength (lb per sq ft)			Com-pressive Load (85 psi)	Horizontal Span Transverse Strength					
		No Compressive Load				Unreinforced			Horizontal Reinforcement (lb per sq ft)		
		Lb Per Sq Ft	Percentage of Standard	Average[a]		Lb Per Sq Ft	Percentage of Standard	Average[a]	16-in. c-c		8-in. c-c
8-in. running bond (standard)	M	60.0	100	100	425.0	127.0 136.0 127.0	100	100	149.4		203.0
	S	34.7 32.2	100			123.0	100		149.9		202.3
Horizontal stacked bond	M	85.0	141	130	405.0	47.7	37	28	130.0		191.2
	S	39.9	120			29.2	24		131.3		190.0
Vertical stacked bond	M	60.6	101	87	357.5						
	S	24.6	74								
Diagonal basket weave	M	89.0	148	151	410.5	69.5 76.9 65.8	56 67 57	60			
	S	51.7	155								
Diagonal running bond	M	103.0	172	158	429.0						
	S	48.1	144								
Basket weave A	M	69.9	117	105	400.0						
	S	30.7	92								
Basket weave B	M	42.9 44.5 24.4	72 72 73	72							
	S										
Coursed ashlar	M	43.3 55.5 26.3 34.7	72 93 79 104	83							
	S										
4-in. running bond	M	65.0	108	101		169.4 173.0 158	129 133 128	130	160.0 193.0 186.6		196.0 194.7
	S	31.3	94								

[a] Average for Type M and S Mortars

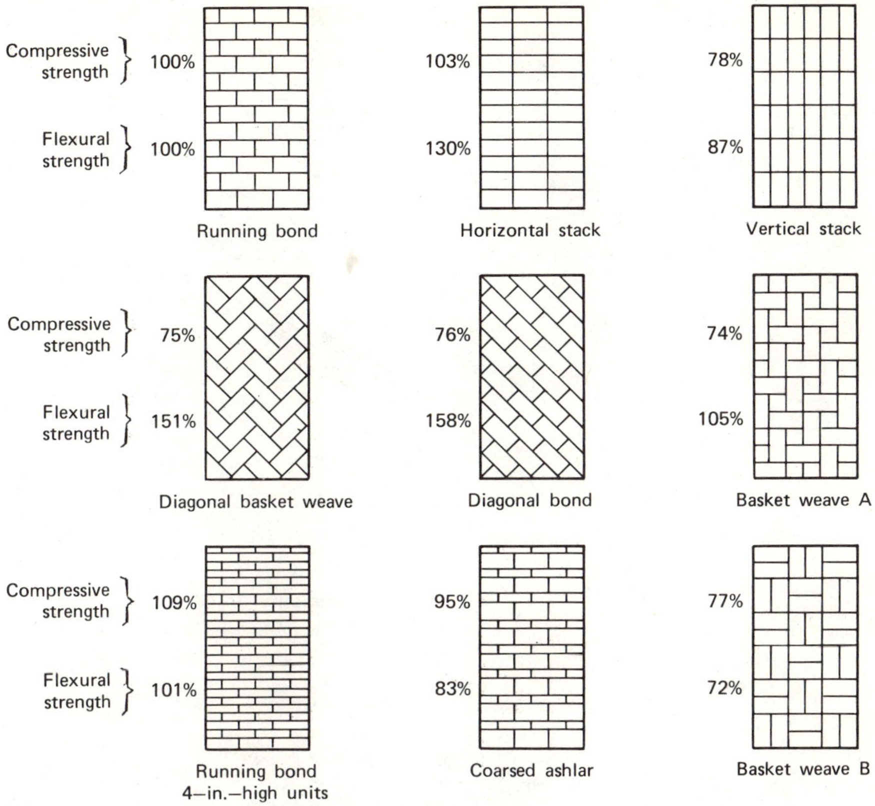

Figure 13-4 Concrete masonry patterns for structural tests.

Compressive strength: 100% — Running bond
Flexural strength: 100%

Compressive strength: 103% — Horizontal stack
Flexural strength: 130%

Compressive strength: 78% — Vertical stack
Flexural strength: 87%

Compressive strength: 75% — Diagonal basket weave
Flexural strength: 151%

Compressive strength: 76% — Diagonal bond
Flexural strength: 158%

Compressive strength: 74% — Basket weave A
Flexural strength: 105%

Compressive strength: 109% — Running bond 4–in.–high units
Flexural strength: 101%

Compressive strength: 95% — Coarsed ashlar
Flexural strength: 83%

Compressive strength: 77% — Basket weave B
Flexural strength: 72%

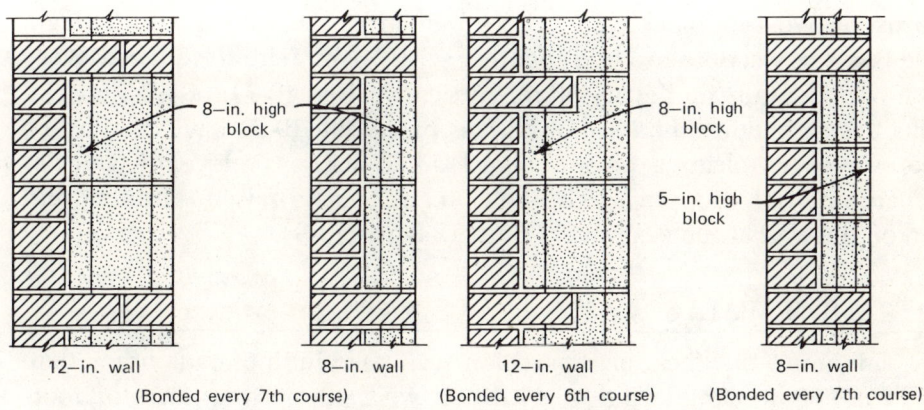

Figure 13-5 Examples of concrete masonry used as backup for brick.

12–in. wall (Bonded every 7th course) — 8–in. high block

8–in. wall

12–in. wall (Bonded every 6th course) — 8–in. high block

8–in. wall (Bonded every 7th course) — 8–in. high block, 5–in. high block

SHAPES, SIZES, AND CONSTRUCTION DESIGN

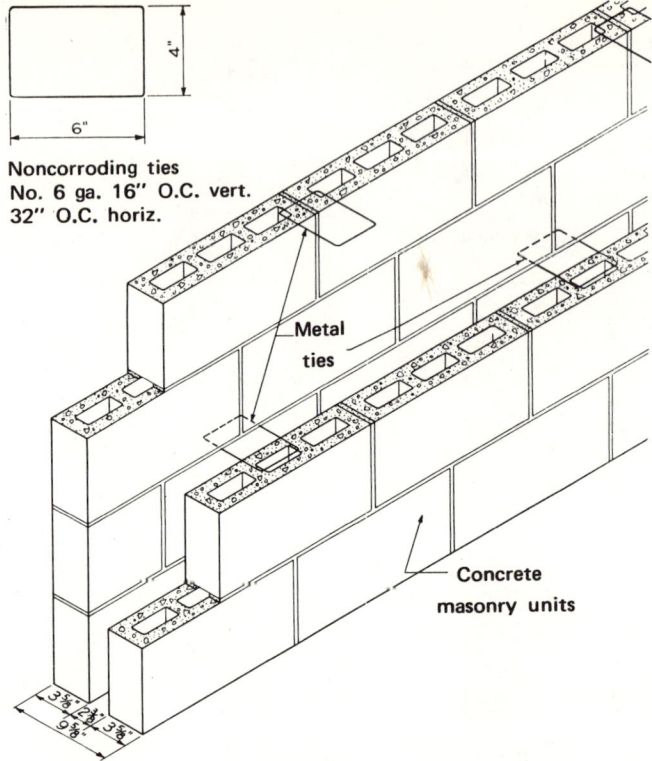

Figure 13–6 10-in. cavity wall.

CAVITY WALLS

Cavity walls (Fig. 13-6) are also used in building construction. They consist of two 4-in. block walls with a 2-in. air space, giving a 10-in. wall. The two 4-in. sections are tied together with $3/16$-in. wire ties placed every 36 in. horizontally and every 16 in. vertically (Fig. 13-7), or with continuous ties. A height limitation of 25 ft is placed on 10-in. cavity walls. To keep the cavity clean, a 1×2-in. board is laid across the level of wall ties to catch mortar droppings. The board can then be raised, cleaned, and laid on the wall at the next level (Figs. 13-8 and 13-9).

PARTITION WALLS

The minimum thickness of a partition wall carrying no loads other than its own weight should be 4 in. Partition walls supporting floors or roofs should be 8 in. thick. Partition walls should not be tied to exterior walls with a masonry bond; they must abut the face of the wall and be tied to it with steel ties (Figs. 13-10, 13-11, and 13-12). These ties should be spaced not over 4 ft apart vertically. The bends at the ends of the tiebars are embedded in cores filled with mortar or concrete.

Nonbearing block walls are tied to other walls with strips of metal lath or 1/4-in. mesh galvanized hardware cloth that is placed across the joint between the two walls. They are placed in alternate courses in the wall (Figs. 13-13 and 13-14).

Figure 13–7

Figure 13–8

Figure 13–9

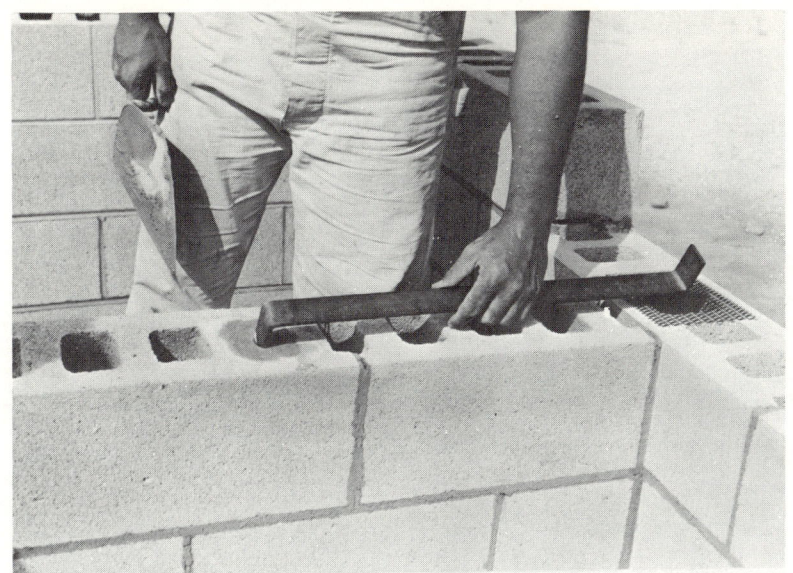

Figure 13-10

Figure 13-11

SHAPES, SIZES, AND CONSTRUCTION DESIGN 311

Figure 13-12

Figure 13-13

Figure 13–14

CHAPTER 14
MORTARS FOR CONCRETE MASONRY

Centuries ago, combinations of sand and lime were used as mortar. These combinations took months and even years to harden as the lime slowly combined with carbon dioxide from the air to form a calcium carbonate. Because it took so long for these mortars to harden and gain strength, it was necessary to use very thin joints. In many instances the joints were so thin that the masonry units were bearing on adjacent units rather than on mortar. This type of construction required an excessive amount of labor to carefully fit and place each unit.

The development of mortars that hardened and gained strength rapidly made it possible to place masonry units more rapidly. A thicker joint provided a cushion for dimensional variations in masonry units. The stronger mortars were first obtained by "sweetening" lime with a small amount of natural cement or portland cement. Later, the amount of portland cement in the mix was increased until the process involved "sweetening the cement with a small amount of lime."

The purposes of mortar are (1) to create a tight seal between concrete masonry units and to join adjacent units; (2) to make up for the slight size variations that are unavoidable in the manufacture of concrete masonry units; (3) to strengthen the structure by bonding with reinforcement, metal ties, and anchor bolts so they perform integrally with the masonry; and (4) to provide an architectural quality.

DESIRABLE PROPERTIES OF MORTARS

Consistency can be described as the degree of workability of the mortar. To attain the same degree of workability in different mortars, varying amounts of water are needed. Consistency is related to the ease of handling and placing of the mortar. To achieve the most desirable mix, the amount of water used should be the maximum consistent with good workability. Too stiff a mix is the cause of a large number of mortar failures.

Consistency can be measured with a device called a flow table. A truncated cone of mortar is placed on a standard flow table top. The table top and mortar are raised and dropped a specified number of times and from a specified height (standard tests call for a ½-in. drop, 25 times). The final diameter of the mortar is measured and the consistency is figured as the percentage of increase in mortar diameter. For example, the diameter of the original mortar is always 4 in.; if the final diameter measures 8.5 in., the increase would be 4.5 in. The percent increase would be $4.5/4.0 \times 100 = 112.5\%$.

Specifications require that laboratory test mortar have a consistency range from 100 to 115%, but on the job an average of about 120 to 130% is more common. The difference is due to the sand and water content of mortars.

Workability is the ease with which the mortar can be handled. This is related to its plasticity. The more plastic and workable a mortar is, the easier it is to mix to a uniform composition. A mortar has good workability if it handles well with a trowel. When mortar is placed on a trowel, snapped, and then turned upside down, the mortar is judged as having good workability if it sticks. It should stick to the trowel or slide from it easily and be easy to spread to a uniform thickness. Good mortar clings to the masonry units and does not run down the face to stain the block. Loss from mortar that falls from the trowel is costly. When mortar has good plasticity, it will make a continuous bond that is watertight and durable.

Water retentivity is the ability to hold water. The test for this is described in ASTM C 91, "Standard Specification for Masonry Cement." A sample of mortar is measured first for flow characteristics. A second sample of mortar is subjected to suction, remixed, and then measured for its flow characteristics. The second flow value is divided by the first value and multiplied by 100. For example, if the flow of mortar before suction was 100 and after suction was 95, the water retention value is

$$\frac{95}{100} \times 100 = 95$$

Specifications generally allow a value of 70. High water retention is desirable for high absorption units laid on hot, dry days; low water retention is desirable for low absorption laid on cold days.

Brick and block absorb water used in the mortar applied to them. Water will evaporate from mortar, too, especially in hot, dry weather. If the mortar has good retentivity, it will stiffen more slowly and allow time to adjust the block to position. Mortar that stiffens too fast forms cracks and capillaries through which water can flow. When water gets into the cracks it may freeze and expand, causing damage to the wall.

Since one purpose of mortar in concrete masonry work is to bond the units together, a measure of mortar quality is the extent and strength of the bond achieved. Incomplete bond could permit water to penetrate while complete bond makes a structure watertight and able to withstand lateral stresses. Strength is no problem if the proper mortar is chosen for the particular job (see section on mortar types, below). Good workmanship in combination with the proper mortar ensures a strong structure.

Good workmanship entails completely filled joints so that intended bonding surfaces are fully covered. Proper tooling at the proper time insures that the mortar is well compacted and in tight contact with the bonding surfaces forming the joints.

Durability of hardened mortar is usually thought of in terms of its resistance to cyclic freezing and thawing. In general, damage to mortar joints and to mortar bond by frost action has not been a problem in concrete masonry construction. In order for frost damage to occur, the hardened mortar must first be nearly water saturated. Newly placed mortar becomes less than saturated due to the absorption of some of the mixing water by the units. The saturated condition does not readily return except where there is direct mortar contact with groundwater or where water enters the masonry through inadvertent channels. Traditionally, concrete masonry in below-grade construction receives an exterior coat of ½ to ¾ in. of portland cement plaster. Above-ground masonry construction is usually painted or stuccoed for decorative purposes. These coatings afford protection against water penetration. Usually, they improve with weathering and age.

MATERIALS USED IN MORTAR

Cementitious materials hold the aggregate (in this case, sand) and bind the masonry together. After these materials are mixed with sand and water, they are called mortar. The same materials mixed with stone or gravel would be called concrete. There are several types of cementitious materials used in making mortar.

Portland Cement
ASTM C 150 covers several types of portland cement. However, Types I and III are most frequently used in portland cement-lime mortars.

Air-Entraining Cements
Air-entraining cements are sometimes used where freezing and thawing are common. However, it must be taken into consideration that they reduce the strength of the mortar.

Portland-Pozzolan Cements
Portland-pozzolan cements are made by grinding together portland cement clinker and pozzolanic material. They are used in masonry for special situations such as when masonry is in constant or intermittent contact with salt water or high-sulfate soils.

Masonry Cements
Masonry cements are mixtures marketed under various trade names. Included in their compositions are portland cement, a plasticizing material such as finely ground limestone, hydrated lime, air-entraining agents, and sometimes water-repelling agents. They have developed to provide a blend of materials that will produce a workable, durable, and economical mortar that meets requirements for presentday construction.

Natural Cements
In a few localities in the United States and elsewhere, there occur natural deposits of argillaceous limestone, called "cement rock," which contain lime, silica, and alumina in such proportions that a hydraulic cement can be produced simply by heating the raw material to a sufficiently high

temperature and grinding it. Although this natural cement resembles portland cement, it is not readily available.

Slag Cements

Slag cement is a particular type of lime-pozzolan. It is made by grinding together hydrated lime and blast-furnace slag. The slag functions as a pozzolanic material and is combined with lime to allow the materials to react chemically. They are generally light colored and low in strength.

Sand

The sand used in preparing mortar can be natural or manufactured. Manufactured sand is obtained by crushing stone, gravel, or air-cooled, blast-furnace slag. Natural sand is rounder and smoother than manufactured sand. Manufactured sand, with its characteristic sharp and angular particles, can produce mortars with workability properties that are much different from mortars made with natural sand.

To make 1 cu ft of mortar, it is necessary to use between 0.97 and 0.99 cu ft of sand or, for practical purposes, 1 cu ft, since measuring sand on the average job is not done to a tolerance of 3%. Any material constituting such a large volume of the mortar naturally will be a major determinant of the quality of the mortar. A logical question is that if 1 cu ft of mortar sand is used to make 1 cu ft of mortar, how can the water and cementitious materials fit into that cubic foot? The cement occupies most of the remaining voids between the sand and cement particles plus a small additional volume to insure lubrication of particles and good workability.

Sand is most often specified or defined by referring to a standard. The American Society for Testing and Materials (ASTM) specifies mortar sand in "Standard Specification for Aggregate for Masonry Mortar," ASTM C 144. Recommended gradations are shown in Table 14-A.

Ranges of gradation are given in Table 14-A rather than one value for each sieve size. To illustrate the variety in gradations that will pass the requirements, the coarsest allowable and finest allowable sands are shown in Table 14-B. The middle column is the mean between the two extremes and is more nearly the ideal gradation.

The values for the coarsest and the finest gradations are the extremes allowable. The best sands will fall well within these limits. Sands that are close to the extremes are not as desirable as those close to ideal gradation.

Table 14–A
RECOMMENDED SAND GRADATION

Sieve Size	Percentage Passing[a]	
	Natural Sand	Manufactured Sand
No. 4	100	100
No. 8	95 to 100	95 to 100
No. 16	70 to 100	70 to 100
No. 30	40 to 75	40 to 75
No. 50	10 to 35	20 to 40
No. 100	2 to 15	10 to 25
No. 200	—	0 to 10

[a] Additional requirements: Not more than 50% shall be retained between any two sieve sizes nor more than 25% between No. 50 and No. 100.

Table 14–B
LIMITS OF ALLOWABLE SAND GRADATION

Sieve Size	Percentage Retained[a] (natural sand)		
	Coarsest	Ideal	Finest
No. 8	5	2.5	0
No. 16	25	12.5	0
No. 30	30	27.5	25
No. 50	30	35.0	40
No. 100	8	14.0	20
No. 200	2	8.5	15

[a] Fractional percentage retained between sieves, not total percentage retained.

Good gradation reduces separation and bleeding and improves water retention and workability. Air entrainment has the same effect.

Harmful substances such as clay and lightweight particles with a specific gravity of less than 2.0 must not be present in large quantities. While these materials often will not affect the plastic properties of the mortar and may even improve workability and plasticity, they usually have a detrimental effect on the mortar's strength and durability. Table 14-C shows potentially harmful materials that may be contained in a mortar sand and includes references to ASTM tests that can be used to determine the presence of injurious amounts of such materials.

Table 14–C
TESTS FOR HARMFUL MATERIAL

Material	Effect on Mortar	ASTM Test
clay	affects workability, durability, strength, and may cause popouts	C142
lightweight materials (coal, lignite, and others)	same as above	C123
organic impurities	affects setting and hardening, may cause deterioration and discoloration	C40
silt and powdered clay	affects workability, durability, and strength	C117

Lime
Two types of hydrated lime are used for masonry purposes.

> Type N—Normal hydrated lime for masonry purposes.
>
> Type S—Special hydrated lime for masonry purposes.

Type N hydrated lime has unhydrated oxides that could hydrate, causing the mortar to grow, cracking the wall. For this reason, Type S hydrated lime is recommended.

Water
Water should be free from acids, alkalis, salts, or organic substances that might affect setting. Water can be the source of efflorescence and scum on finished masonry. Water that is potable will usually be acceptable for mixing mortar.

Admixtures
Admixtures should generally not be used where certain special properties are required in the mortar. They may reduce strength and impair durability, so their use is very controversial.

During cold weather construction, calcium chloride additions may be used to accelerate the setting time of the mortar. In all cases, calcium chloride additions should not exceed 2% by weight of the portland cement and 1% by weight of masonry cement in the mortar batch.

Color pigment additions should be mineral oxides and limited to 10% by weight of the portland cement. Carbon black additions should not exceed 3%. Only admixtures and coloring compounds of known suitability should be used.

PROPORTIONING MORTAR MATERIALS

Uniform batch yields, workability, and color will result from proper batching of mortar ingredients. Aggregate proportions are generally expressed in terms of loose volume. Experience has shown that, though volumetric measurement from batch to batch may be consistent, actual weights of sand per batch can vary due to bulking. Damp sand, for example, can vary in weight from dry sand for the same loose volume by as much as 30%. Volume-weight ratios of cement and dry hydrated lime batched in loose form will vary for similar reasons. Therefore, job-site checks of volume-weight relationships may be occasionally needed. Table 14-D gives masonry mortar proportioning recommendations. Unit weights of common mortar ingredients are as follows.

Material	*Unit Weight, pound per cubic foot*
Portland cement	94 (United States)
Blended hydraulic cement	Printed on bag
Masonry cement	Printed on bag
Hydrated lime (dry)	50 (United States)
Hydrated lime (putty)	80
Sand, damp, and loose	80 lb of dry sand

Table 14–D
MORTAR PROPORTIONS BY VOLUME

Mortar Type[a]	Parts by Volume of Portland Cement or Portland Blast-Furnace Slag Cement	Parts by Volume of Masonry Cement	Parts by Volume of Hydrated Lime or Lime Putty	Aggregate Measured in a Damp, Loose Condition
M	1	1	—	Not less than 2¼ and not more than 3 times the sum of the cements and lime used.
	1	—	¼	
S	½	1	—	
	1	—	over ¼ to ½	
N	—	1	—	
	1	—	over ½ to 1¼	
O	—	—	—	
	1	—	over 1¼ to 2½	
K	1	—	over 2½ to 4	

[a] Mortar type designation A-1, A-2, B, C, and D are the former type designations in effect prior to 1954.

MORTAR MIXING

To obtain good workability and other desirable properties, proper care in proportioning the ingredients must be followed by thorough mixing. With the possible exception of very small jobs, mortar should be machine mixed. Conventional power mixers are of either the fixed drum or rotating blade design. After all batched mortar materials are in the drum, they should be mixed for 3 to 5 min. Longer mixing times may entrain too much air in cold weather.

Batching procedures will vary with individual preferences. Experience has shown that good results can be obtained when about ¾ of the required water, ½ of the sand, and all of the cementitious materials are briefly mixed together. The balance of the sand is then charged and the remaining water is added. The amount of water added should be the maximum that can be tolerated with satisfactory workability.

When hand mixing becomes necessary, dry materials should first be mixed together by hoe, working first from one end of the mortar box and then from the other. Subsequently, ⅔ to ¾ of the required water should be placed in the box and the mixing continued as above until all the material is uniformly wet. The balance of the water is slowly added with continued mixing until the desired workability is attained.

TYPES OF MASONRY MORTARS

ASTM C 270 describes five types of masonry mortars, types M, S, N, O, and K.

Type M is a high-strength mortar recommended for concrete masonry subject to high compressive loads, severe frost action, or for masonry that will be steel reinforced to resist lateral loads from earth pressures, strong winds, or seismic motions. It is also used for below-grade structures such as manholes and catch basins.

Type S mortar is of medium-high compressive strength and is excellent for structures requiring high flexural bond strength, but subject only to normal compressive loads. This type mortar gives excellent results in walls exposed to high winds.

Type N is a medium-strength mortar that can stand severe exposure and, therefore, it is used in exposed, above-grade masonry. It is also suitable for interior walls and partitions and for concrete masonry veneers applied to frame construction.

Type O mortar is of relatively low compressive strength, but it is useful for interior nonload-bearing walls and partitions. It can be used for solid load-bearing masonry where the allowable stress does not exceed 100 psi.

Type K mortar has a very low compressive strength. It can only be used for interior nonload-bearing walls.

Table 14-F shows the minimum compressive strengths required under each of these mortar types (ASTM C 270), based on tests of 2-in. cubes.

Mortar types PM and Pl for grout and reinforced masonry are referred to in ASTM C 476. The physical requirements of these mortars are shown in Table 14-E.

Table 14–E
PHYSICAL REQUIREMENTS OF
MORTAR TYPES PM AND PL (ASTM C 476)

Compressive strength:		
Average of three 2-in. cubes, minimum psi:		
7 days[a]	1600	
28 days	2500	
Water retention:		
Flow after suction, minimum, percentage of original flow		70
Air content:		
Volume, maximum, percentage		18

[a] If the mortar fails to meet the 7-day compressive strength requirement, but meets the 28-day compressive strength requirements, it shall be acceptable.

Table 14–F
MINIMUM COMPRESSIVE STRENGTH OF
CUBES FOR MORTAR TYPES

Mortar Type	Compressive Strength at 28 days, psi (average of 3 cubes)	Water Retention Flow After 1 min. Suction
M	2500	70%
S	1800	70%
N	750	70%
O	350	70%
K	75	70%

CHAPTER 15
USING CONCRETE MASONRY

LAYING OUT THE BUILDING

The easiest, quickest, and most accurate way to lay out the corners of a building is with surveying instruments. When such instruments are not available, the right-triangle method can be used. First, a base line is established, marking out one end or side of the new building (Fig. 15-1, line A-B). Stakes are set at A and B on this line, locating two corners. A nail is driven in the top of stake A near the center. This nail locates the corner.

To lay off the corner by the right-triangle method, drive a stake at F so that it is on A-B and 6 ft from stake A. A nail is driven in the top of this stake exactly 6 ft from the nail in stake A. Stake E is driven into the ground so that it is 8 ft from stake A and 10 ft from stake F. A third nail is then driven into this stake, exactly 8 ft from the nail in stake A and 10 ft from the nail in stake F. The corner represented by the angle EAF is a right angle. The line A-E extended to D forms the second boundary line of the building, and D will represent the third corner. Other corners are laid off the same way as EAF.

After this has been done, strings are stretched above the corner stakes A, B, C, and D and carried to outside supports called "batter boards." The top of the horizontal batters should be set at first floor level or some

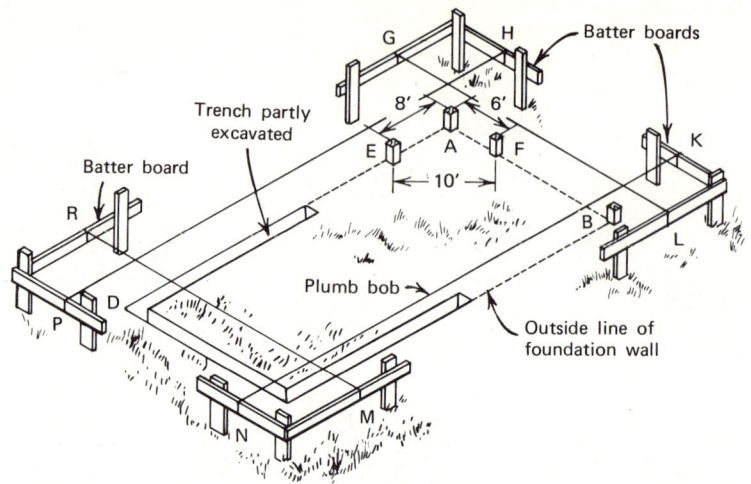

Figure 15-1 The right-triangle method of laying out boundary lines.

other convenient level. The building lines may be projected from the strings to the ground by means of a plumb bob suspended as shown in the drawing. When the batter boards have been set and the strings indicating the layout of the building are stretched between them, the corner stakes A, B, C, and D and stakes E and F are removed so that the trench may be excavated. Excavation should be measured down from the string line. Nails should be driven in the batters where the strings are fastened, so that if strings are broken or removed they can be replaced accurately. Having found the building lines, it is easy to locate piers, posts, columns, or other supports.

STORING MASONRY

Specifications limit the moisture content of concrete masonry at the time of delivery, and care must be taken to keep it dry on the job. It should be stockpiled on planks or other supports to keep it free from contact with the ground.

A quick and easy method of properly storing concrete masonry units at the job site is to place a layer on their sides to make a platform. The rest of the units are then stored on top of it. This "platform block" is used for this purpose only. Later deliveries will also be stockpiled here and, at the conclusion of the job, these units will be cleared away in the cleanup.

Protection from rain and snow is provided by covering the stockpile with a tarpaulin, kraft paper or other weathertight coverings.

A construction practice common and acceptable with most other types of masonry units must not be used with concrete masonry: This practice is wetting the units prior to laying them in the wall. The moisture content of the unit must be kept low to minimize subsequent shrinkage movements, so *do not* wet concrete masonry before using it except in extremely hot, dry weather when it may be dampened by applying water with a brush.

CONSTRUCTING FOOTINGS

Concrete footings provide a level surface for starting the wall and an increased bearing area on the soil that reduces settling. One- and two-story buildings erected on soils of average load-carrying capacity generally require a concrete footing twice as wide as the wall it supports. The depth of the footing is usually one-half its width and is equal to the thickness of the wall. Thus, for a building with an 8-in. wall, the footing is made 16 in. wide and 8 in. deep. For a building with a 12-in. wall, the footing is made 24 in. wide and 12 in. deep.

In designing footings for heavy buildings, the weight of the building and its contents are computed, and the footings are made wide and deep enough to carry the load. The carrying capacity of soil varies considerably, and this is also taken into account in designing footings. Soft clay, for example, may have a carrying capacity of 1 ton per square foot; hard, dry clay may carry 4 tons per square foot; and gravel, 6 tons per square foot.

Footings are usually placed below grade in order to be on firm soil. In areas where the ground freezes, the footings should be below frost penetration so that freezing will not heave them and cause the walls to crack.

The bottom of the footing should be flat, not rounded, so the trench (Fig. 15-3) should be flat and level on the bottom. Forms for footings are usually made of lumber 2 in. thick and are held in place with stakes driven along the outside (Fig. 15-4). Stakes are placed close enough together so that the form will not bulge when filled with concrete. Forms are removed after the concrete hardens (Fig. 15-5).

The concrete footing should be swept clean before masonry is laid to remove all loose material.

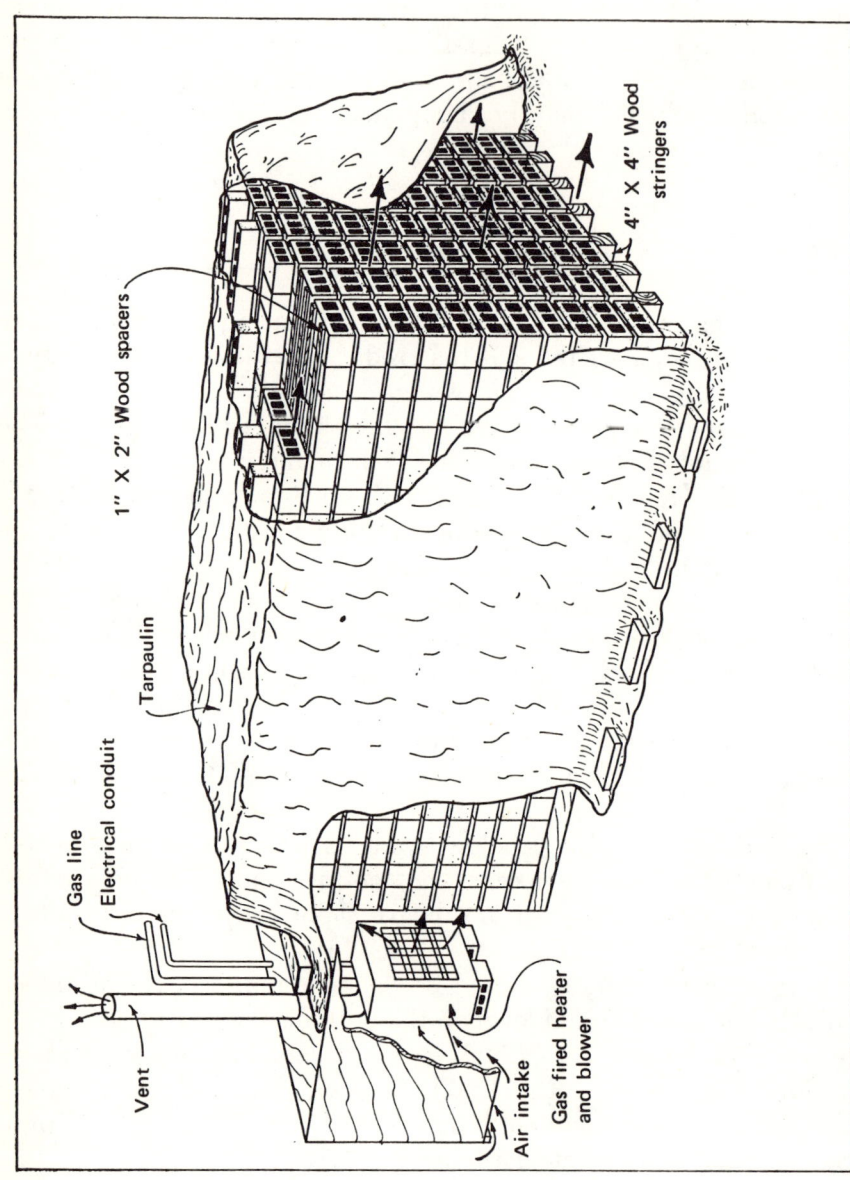

Figure 15-2 A suggested method for drying concrete block. It can be used indoors or outdoors at plant or job site.

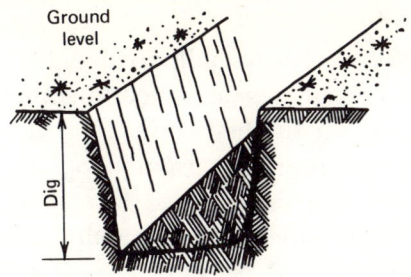

Figure 15–3 Trench for footing.

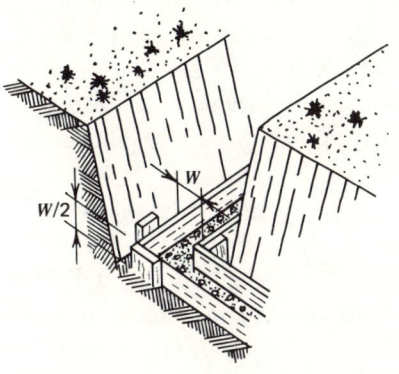

Figure 15–4 Forms for footing in place.

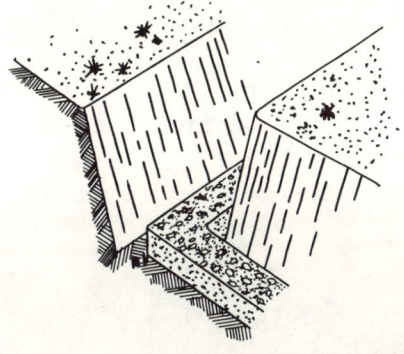

Figure 15–5 The completed footing.

LAYING BLOCK AT CORNERS

It is common practice to lay up the corners three or four courses high and then use them as guides in laying the walls. A good layout practice is to lay the block for the first course without mortar to check and make sure no units will have to be cut and no joints are too thick or thin.

Special care is taken to insure that the corners and first course are laid to correct alignment and are level and plumb. Any error will give continuing trouble in the rest of the wall. A full bed of mortar, extending the full width of the wall, is placed on the footing for the first course.

After the first course is laid on the footing, the mortar is usually applied only to the horizontal face shells of the block. This is called face-shell bedment. Block should be laid with the thicker side of the face shell in the up position, since this provides a larger mortar bedding area. Full mortar bedding adds only slightly to the strength of the wall and requires much more mortar and more of the mason's time. Face-shell bedment has the extra advantage of providing a continuous air space and of interconnecting the air spaces of the cores.

Having laid the first course of block, the mason proceeds to build up his corners or leads, using his level to check each course for alignment and a ruler to make sure he is at proper elevation.

Each block is carefully checked with a straightedge or level to make certain that the faces are all in the same vertical plane.

After the coners at each end of the wall have been laid up, a line is stretched tightly along the top outside edge of the corner block to serve as a guide. The cord is fastened to line pins, nails, or wedges that are driven into the mortar joints so that when stretched, it just touches the upper outer edges of the block laid in the corners. The block in the wall between corners are laid so they just touch the cord, which is moved up as each course is laid.

Occasionally, a corner pole is used to keep the courses level. The pole is a 1×2-in. board that is marked off exactly every 8 in. These marks locate the top of the masonry for each course (Fig. 15-7).

Corner poles are sometimes used to help the mason keep the masonry

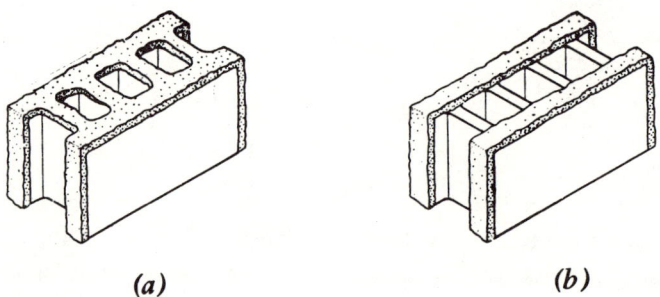

(a)   (b)

Figure 15–6 Examples of (a) full mortar bedding and (b) face-shell mortar bedding.

CONCRETE MASONRY

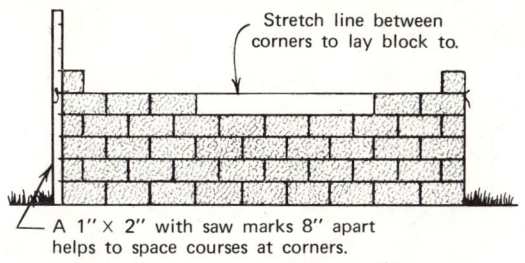

Figure 15-7 Using corners as guides to lay walls in between.

courses straight. Corner poles are free-standing poles braced to keep them plumb. A pole is set at each corner of the wall prior to construction. When this method is used, masonry is laid up a full course at a time instead of building out from the corners first.

When laying the unit, the mason keeps its top tipped slightly toward him. This enables him to place the lower edge of the unit directly over the course below. By rolling the block slightly to a vertical position and shoving it against the previously laid unit, it can be laid to line with a minimum of adjustment. This speeds the work and reduces the possibility of breaking the bond by moving the block excessively after it has been pressed into the mortar. All adjustments to final position must be made while the mortar is soft and plastic. The use of a level between corners is usually limited to checking the face of each unit to keep it lined up with the face of the wall.

APPLYING MORTAR TO BLOCK

Mortar is usually applied in two separate strips on the outer shells of the wall for the horizontal or bed joints. For the vertical or end joints, the mortar is applied only on the face shells of the block being placed.

Masons often stand the next block to be laid on end and apply mortar for the vertical joint. Sufficient mortar is put on to make sure joints will be well filled. Some masons apply mortar to the end of the block previously laid as well as to the end of the next block to be laid, to make sure that the vertical joint will be completely filled with mortar.

Three units are buttered and placed in rapid succession. Three block plus mortar joints is usually 4 ft long, the same length as a mason's level, making it convenient to adjust three units at a time. However, in hot, dry weather it may be necessary to spread only enough mortar for each

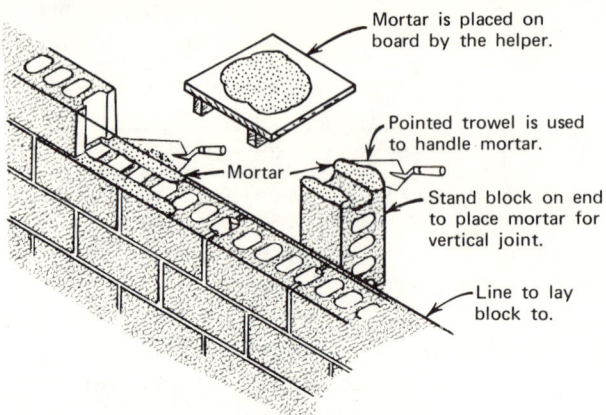

Figure 15-8 Face-shell bedding.

block as it is laid to prevent loss of moisture in the mortar through evaporation or suction.

Mortar squeezed out of the joints is carefully scraped off with a cutting motion of the trowel and applied to the other end of the block or thrown back onto the mortar board for use later. The blocks are laid to the line and are tapped with the trowel to get them straight and level (Fig. 15-9). In a well-constructed wall, mortar joints will average $3/8$ in. thick, which gives an 8-in. dimension with modular block.

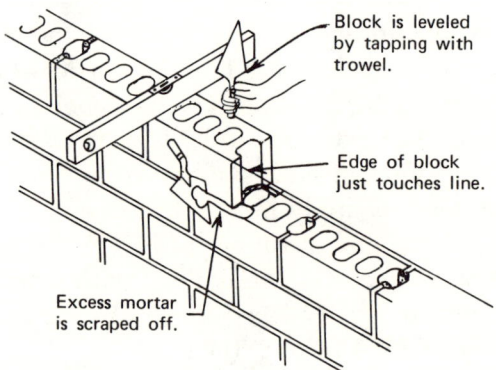

Figure 15-9 Good workmanship requires straight courses and the face of the wall plumb and true.

PLACING CLOSURE BLOCK

Although placing the final or closure block within each course appears quite simple, it is not. Some skill is required to fit the block between those already in place without knocking the mortar off the edges. All edges of the opening and all four vertical edges of the closure block are buttered with mortar, and the block is carefully lowered into place. Should the mortar be knocked off, it is necessary to remove the block and the mortar and start again.

SPECIAL TYPES OF WALLS

When masonry units of dimensions other than 8×8×16 in. are used, adjustments often have to be made so as not to break the pattern. To make a corner for a 10-in.-thick wall, a special "L" corner unit, with the corner end reduced to an 8×8-in. face, is placed in each course, thus keeping the pattern in multiples of 8. Normally, all block are placed with the thick edge of the face shell up to provide a larger base for the mortar joint, but in this case it is necessary to turn the "L" corner block over in every other course to put the 8×8-in. face on alternate sides of the corner.

This special type of corner block is not available in all areas or for all thicknesses of wall. The same thing can be accomplished for a 12-in. wall, however, with an 8×8×16-in. single corner and a concrete brick. For a 10-in. wall where no "L" corners are available, an 8×8×16-in. single corner could be used with a piece cut from the face shell of another block.

Cavity walls, which give excellent thermal results, were discussed in Chapter 13. However, something should be said about how to prevent moisture, which might collect in the cavity, from coming in contact with the bottom of the inner wall. Since flashing and weep holes are necessary, the following practices are used.

The heads of windows, doors, and other wall openings, and the bottom course of masonry immediately above any solid belt course or foundation, are flashed so that moisture entering the wall cavity will be directed toward the outside walls. Only rust-resisting metal or approved materials treated with asphalt or pitch preparations should be used for flashing.

Weep holes are placed every 2 or 3 units apart in the vertical joints of the bottom course of the outside wall immediately above any solid belt course or foundation. In no case should the weep holes be placed below grade. Weep holes can be formed by placing cotton rope (which is left in place), fiberglass, or a well-oiled rubber tubing in the mortar joint and then extracting it after the mortar has hardened. The tubing should extend up into the cavity for several inches to provide a drainage channel through any mortar droppings that might have accumulated.

Sometimes, 12-in. walls are built using a combination of 4-in.- and 8-in.-thick units. When this is done, the block are laid so the 4-in. block are all on the inside wall for one course and on the outside wall for the next course. This interlocks the block and produces a good bond. It is necessary to completely fill the joint between the units for best results.

Another type of wall in common use is a 4-in. face brick backed up with an 8-in. concrete block. In this type of construction, it is advisable to lay the block backing first so that the mason has a more substantial 8-in. block wall on which to apply the pargeting rather than the lighter 4-in. brick wall. If the 4-in. brick wall is pargeted first, the pressure of applying the mortar may tilt the brick facing enough to break the bond in the horizontal mortar joints resulting in a leaky wall.

Tying the facing and backing together is important. This can be done with metal wall ties laid in the horizontal mortar joint, or it can be done by using a header block that has one corner notched. Brick is laid edgewise, rather than lengthwise, every sixth course (Fig. 15-10). The header block must be laid with the notch up.

A brick wall with a 4-in.-block backup can be bonded by turning every

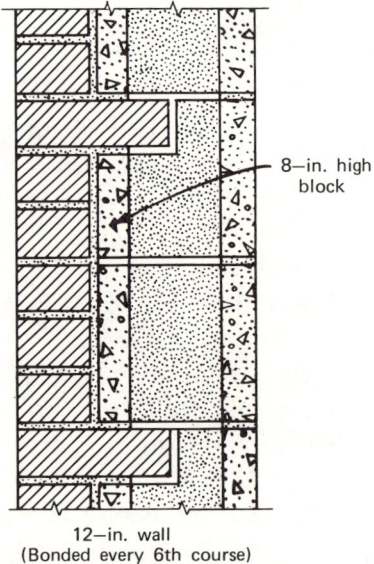

12—in. wall
(Bonded every 6th course)

Figure 15–10

334   CONCRETE MASONRY

seventh course of brick facing sideways so it extends the full 8-in. width of the wall (Fig. 15-11). A 12-in. wall with 8×8×16-in. block and brick facing can be bonded by using block for two courses, faced with six courses of brick laid end to end. The next course has the facing brick laid side by side with a lengthwise course of brick behind it (Fig. 15-12). However, local codes must be followed.

CONTROL JOINTS

Concrete masonry shrinks when drying and expands when wet. Thermal changes also affect the length of a concrete block, and foundations sometimes settle. The actual amount of increase or decrease in length of one block is very slight, but when firmly bonded together, the cumulative stresses are sometimes sufficient to crack the wall.

To relieve these stresses before they can build up, control joints are used. A control joint is a vertical joint from top to bottom of a wall. Any stresses that have concentrated in the wall will cause the control joints to open slightly. Openings in the wall, such as doors and windows, provide weakened sections where stresses are likely to accumulate. It is logical, therefore, to locate control joints at such places. When aligned with the side of a window or a door, the control joint utilizes the opening as part

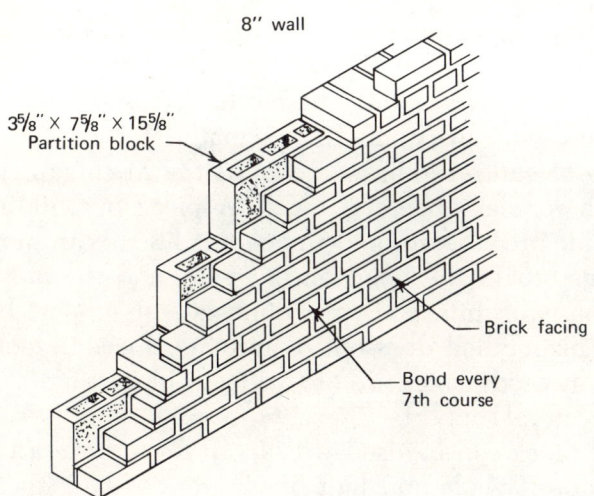

Figure 15-11 A wall with 4-in. brick facing and 4-in. concrete blocks.

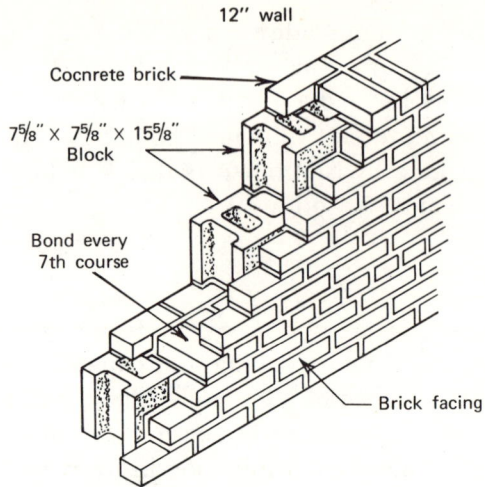

Figure 15–12 Bonding brick facing to concrete masonry by making every seventh course of brick a header course.

of the joint. Control joints must extend from the top of the foundation to the top of the wall.

Special block is used at control joints to assure lateral support between wall sections. The block has tongue-and-groove ends. It is laid up with two full block in the first course and two half block in the next. This produces a joint the full height of the wall. Mortar is used in control joints just as in any other vertical joint. The mortar is raked out to a depth of 3/4 in. and replaced with an expansive-type material.

Another type of control joint is known as the Michigan control joint. It is made with regular open-end block. A piece of building paper is placed in one side of the joint, and the core is filled with mortar or concrete. The paper breaks the bond on one side, and the mortar forms a key between the two units, thus providing lateral support between the wall sections. This method does not require any special block.

Still another method makes use of two jamb blocks and a "Z" tie. The tie provides the lateral support.

An open-end block can be used with a "Z" tie to form a control joint. Alternate courses of whole and half block are also laid for this type of joint.

Control joints should be of the built-in type with a ⅜-in. vertical joint running the full height of the wall. After the mortar has stiffened, the mason rakes out the joint to a depth of ¾ in. This recess is later filled with a calking compound that prevents water penetration in case the mortar cracks. To prevent the dry block from sucking the oils from the compound, the edges of the joint should first be primed with a coat of shellac or other sealer. Recommendations of manufacturers of calking compounds regarding priming should be followed.

When the side of a window or other opening is used as part of a control joint, it is necessary to provide a slip-plane bearing for the lintel on the control joint side. This can be done with a sheet of copper or other noncorroding metal. Thus, if the control joint opens up, the lintel can slide on the metal plate without restraining the control joint. A full bed of mortar should be placed under the ends of the lintels.

A control joint should be placed at the intersection of two walls, such as the inside and outside walls. To construct a control joint at this point and also provide lateral support to the two walls, a ¼ × ¼-in. bar, at least 24 in. long with 2-in. right-angle bends on each end, is used to tie the walls together. The tie bars are placed at least every 4 ft vertically. One end of the bar is grouted in. After the next course is laid, the other end will be grouted in. This type of joint ties the two walls together while permitting independent lateral movements. The continuous joint at the junction of the two walls is raked as is any other control joint. If it is exposed to view or to weather, it is calked.

A similar joint for a nonbearing wall is made by placing ¼-in. hardware cloth in every other course. It can be placed in one wall and left projecting until the other wall is laid. The projecting mesh is bent down into the joint of the intersecting wall as it is laid.

REINFORCEMENT

Some codes specify a reinforced concrete belt course or a bond beam. This is formed by special U-shaped or trough-shaped bond beam block in which the reinforcement bars are placed and the block is filled with concrete. Sometimes these beam block are formed with grooves in the cross web so that the webs can be knocked out. The reinforcement is placed in the bottoms of these bond beam block which are then filled with grout. Special beam block are used for the corners. They have one solid face and one face with a cutout area near the edge.

DOOR AND WINDOW FRAMES

There are several acceptable methods of building door and window frames in concrete masonry walls. One method is to set the frames in the proper position in the wall. The frames are then plumbed and carefully braced, after which the walls are built up against them on both sides. The frames are often fastened to the walls with anchor bolts passing through the frames and embedded in the mortar joints.

A more recently developed method is to build openings for doors and windows using special jamb blocks (Fig. 15-13). The frames are inserted after the wall is built. The advantage of this method is that the frames can be taken out without damaging the wall or the door.

SILLS AND LINTELS

Openings over doors and windows may be bridged by reinforced concrete beams called lintels. These are usually precast in a plant but can be cast on the job. Lintels are reinforced with steel bars placed 1½ in. from the lower side. The number and size of reinforcing rods depend on the width of the opening and the load to be carried.

Lintels can be made on the job by placing two steel angles back to back. Mortar is placed in the angles, and 4-in. units are laid in the mortar. This construction allows the block texture to be continued across the top of the door.

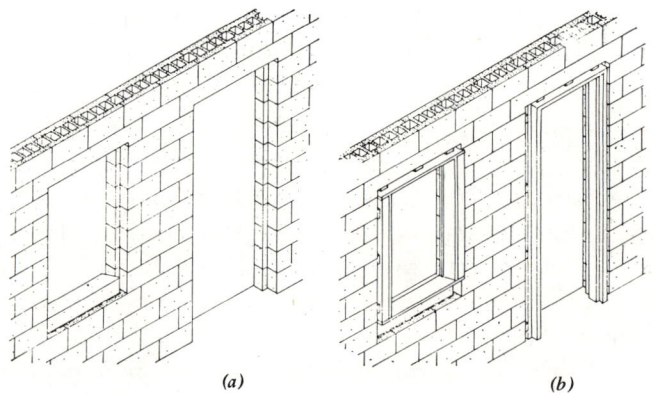

Figure 15-13 Openings for windows and doors are built first; then frames are installed. (*a*) Openings completed. (*b*) Frames installed.

Sills may also be either precast or cast in place on the wall. They are made to project about 1 in. beyond the face of the wall and are provided with a drip so rain water running down the sill will fall free and not flow down the wall.

ATTACHING FRAME ROOFS

Where a frame roof is used, anchor bolts are set in the top of the wall. A small piece of mesh is placed under the core in which the bolt is to be placed. The bolt is grouted in with mortar. Bolts usually extend down into the last two courses of masonry for good anchorage. After the mortar has set, the plate is drilled to fit over the bolts and then fastened down.

TOOLING

Often, the difference between an attractive job and an unsightly one is the tooling of the joints. Neat, straight, well-tooled joints add greatly to the appearance of a wall.

Joints are finished when the mortar has become "thumbprint" hard. The reason for tooling masonry joints is to compress the mortar between the masonry units. Tooling also makes the wall more watertight and helps locate unfilled or partially filled joints. These loosely filled areas should be repacked at the time of tooling.

The tooling instrument (Fig. 15-14) should be wider than the joint. If it is not, the mortar may be pushed in too far, disturbing the bond

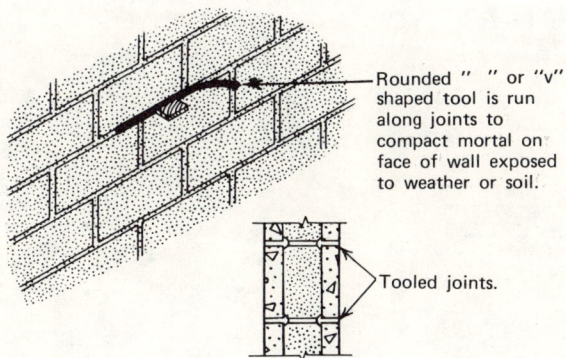

Figure 15-14 Tooling to compact the mortar in the joint.

between the units and creating ledges where water may collect. The tooling device should be 24 to 36 in. long since a long tooling instrument makes straighter, more uniform joints. For vertical joints, an "S" jointer is used.

Concave and V-shaped joints are the most widely used (Fig. 15-15). They have good weather resistance and can be used both on exterior and interior walls.

The flush joint (Fig. 15-16) is cut even with the masonry units. It is then rubbed with a wood float or a piece of burlap or carpet to give a texture similar to that of the block. The flush joint's overall weather performance is fair to poor and is most frequently used where a plaster surface is specified. It should be used only on vertical exterior joints or on interior joints. Raked joints (Fig. 15-16) are made by cutting all mortar out to a depth of approximately 1/4 in., giving distinct shadows. The raked joint leaves a ledge that can collect water. Extruded joints (Fig. 15-17) are used to give the wall a rustic effect. The mortar is not cut off when the units are laid. Raked and extruded joints should not be used on exteriors except in dry areas or where protected from rain and snow.

Figure 15-15 (a) Concave joint. (b) "V" joint.

Figure 15-16 (a) Raked mortar joint. (b) Flush mortar joint.

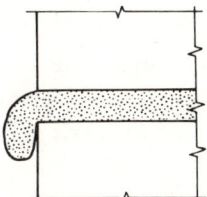

Figure 15-17 Extruded mortar joint.

COLD-WEATHER MASONRY WORK

When masonry construction is carried on during periods of freezing weather, facilities must be available for protecting fresh masonry work from frost damage for at least 48 hours. The preparation of mortar for use under these conditions is particularly important. The temperature of mortar as placed should range between 70 and 120°F. This means that the water, and sometimes the sand, may need to be heated.

Mixing water may be heated in 55-gallon steel drums supported over a fire. If the size of the job warrants more sophisticated water heaters, they can be obtained from equipment suppliers, often on a rental basis. In any case, the temperature of the water, as added at the mixer, should not exceed 160°F.

Sand is usually heated on the job by heaping a quantity over and about a steel pipe approximately 30 in. in diameter, inside of which is maintained a moderately hot fire. The sand should be turned over at intervals to insure uniform warming (to 50°F) and to avoid excessive drying or scorching of sand.

Calcium chloride is sometimes used in cold-weather work to accelerate the rate of mortar strength gain and to shorten the necessary protection time. The use of calcium chloride in masonry that will incorporate joint reinforcement, metal ties, and fasteners in contact with the mortar is not recommended. If calcium chloride is used, it should always be added in solution (never as dry crystals), and its weight should not exceed 2% of the weight of the portland cement or 1% of the masonry cement component. For convenience in batching calcium chloride, the solution can be designed so that one or two measures of convenient volume will serve the requirements of one mortar batch. For example, from a solution prepared by dissolving 100 lb of flake calcium chloride in 25 gal of water, 1 qt added to the mortar mix is equivalent to 1 lb of calcium chloride salt.

In portland cement mortar mixtures, high early-strength development can be obtained through the use of Type III (high-early-strength) portland cement.

Under no circumstances should calcium chloride or other admixtures be used as an "antifreeze" in mortar, because the required concentrations of the admixture are so large for any effective protection that mortar strength, durability, and other desirable properties would be seriously impaired.

The masonry units must be dry and free of snow or ice. If the temperature at the time of placement is below 20°F, the units should be heated until they are at least 50°F. With careful planning and workmanship, excellent masonry work can be accomplished in cold weather as well as warm.

RETEMPERING MORTAR

Mortar that has been mixed, but not used immediately, tends to dry out and stiffen. The loss of water by absorption and evaporation on a dry, windy day can be reduced by wetting the mortarboard and covering the mortar in the mortar boxes or wheelbarrows.

If it is necessary to restore workability, mortar may be retempered by thorough remixing and by adding water. This can be done within an hour or two after mixing without damage to the mortar. Water should be added slowly and sparingly until a workable consistency is achieved. Although the addition of water may slightly reduce the strength, the effect on the wall is preferable to that of dry, stiff mortar.

If mortar stiffens because of hydration (setting), it should be discarded. Since it is difficult to determine by sight or touch whether mortar stiffening is due to evaporation or hydration, the most practical method of determining the suitability of mortar is on the basis of time elapsed after mixing. Mortar should be used within 2½ hours after mixing; otherwise, it should be discarded.

SECTION FIVE
CONCRETE PIPE

CHAPTER 16
THE MANUFACTURING OF CONCRETE PIPE

Concrete pipe is used throughout the world for irrigation, drainage, sewerage projects, culverts and pressure transmission, and distribution lines. It is made of either plain, reinforced, or prestressed concrete of high strength. Some of the processes used to manufacture concrete pipe include vibration, centrifugation, tamping, and the packerhead process. The concrete pipe industry is in three divisions: (1) agricultural pipe; (2) sewer and culvert pipe; and (3) pressure water pipe. Few manufacturers make all three types of pipe, but many produce pipe for both agricultural uses and for sewers and culverts.

HISTORY OF CONCRETE PIPE

Natural cement concrete was used in building the *Cloaca Maxima*, the main sewer in ancient Rome, in 200 B.C., portions of which are still in service today. A rammed concrete waterline constructed by the Romans in the year A.D. 80 delivered water from the Eiffel Mountains to *Colonia Agrippina* (Cologne, Germany) up until 1928—a period of 1848 years.

The American concrete pipe industry has grown with the country. Concrete pipe sewers were developed in the 1840s. Concrete drain tile and storm sewers also were first used in the 1840s followed by a great period of expansion in the Midwest in the 1890s and 1900s. Concrete irrigation pipe was used in the West shortly after the Gold Rush in 1849. Railways pioneered the development of concrete pipe culvert following the Civil War. Highway development from 1915 to the present carried on this growth of concrete pipe highway culverts. Concrete pressure water pipelines have been designed and built extensively during the last 50 years because of the increasing demands for adequate supplies of drinking water and the recognition of the maintenance-free service of properly made and installed concrete pipe.

1800s

In 1840, concrete drain tile was produced in New England and used for the drainage of farms. It was generally not reinforced, no smaller than 4 in. in diameter and 1 ft in length, nor larger than 24 in. in diameter and 2 ft in length. The relatively dry mix was hand-tamped into crude forms. The cement used in this pipe was either imported from Europe or produced in hand-operated kilns of the most primitive type.

Nashua, New Hampshire, constructed the first concrete pipe sanitary sewer system using concrete drain tile with a bell-and-spigot end. In 1849, the first nonreinforced concrete pipe irrigation lines were made in California by tamping a dry mix into forms by hand.

The first nonreinforced concrete railway culverts were built in 1865,

Figure 16–1 The pipe on the right removed from a St. Louis, Mo., sanitary sewer in 1962 after being in service since 1868. A new pipe section is on the left.

when the railroads pushed west across the prairies and plains at the close of the Civil War. These culvert pipes were generally less than 36 in. in diameter and not more than 2 ft in length with bell-and-spigot ends. They also were made by hand-tamping.

1900s
In 1906, the first reinforced concrete railway culvert pipe was made at Red Oak, Iowa. Circular rings of reinforcing rods were added as the dry mix was hand-tamped.

Ames, Iowa, was the scene in 1908 where a group of concrete drain tile manufacturers met and formed an association that later developed into the American Concrete Pipe Association. From this organization grew the American Concrete Pressure Pipe Association and the American Concrete Agricultural Pipe Association. (The American Concrete Agricultural Pipe Association was dissolved in 1962 and became a division of the American Concrete Pipe Association.)

In 1909, the first low-head concrete pressure lines were built mainly for irrigation purposes. They were not very successful because they leaked at the joints. Between 1910 and 1920, joint design improved. Reinforced concrete culvert and sewer pipes came into use in many areas of the United States after 1910, when construction of highway and sanitary sewer systems began to accelerate.

Since 1920, there has been much research in the use of reinforcement in pipe, the use of gaskets for pressure and nonpressure joints, and the development of coatings and linings for concrete sewers carrying corrosive effluent.

DEVELOPMENT OF CONCRETE PIPE-MAKING EQUIPMENT IN THE UNITED STATES

1800s
From 1840 until about 1900, all concrete pipe in the United States was made by hand-tamping a dry mix into forms. The limited amount of large pipe that was manufactured was wet cast.

1900s
Shortly after 1900, in the Northwest, the first tamper-type machines were developed. The core form for these machines was generally fixed, with the outer form revolving while dry-mix concrete was mechanically tamped into the wall space between the core form and outer form by wooden tamper sticks, striking the dry mix with rapid, heavy blows.

Using the tamper principle with both the core and the outer form revolving, butt-end concrete drain and culvert pipe were made in 1906. Steel tamper sticks were used in this process. In the same year, steel wire reinforcement for concrete pipe over 24 in. in diameter was used extensively. Soon all concrete pipe with an internal diameter of over 24 in. used steel reinforcement.

In 1916 and 1917, the packerhead process was developed. This process uses a stationary outside form. The dry mix is packed against the form by a revolving shoe, or packerhead, as it travels upward on a heavy shaft.

In 1918 and 1919 double tampers were used, thus making it possible to reinforce pipe and tamp on both sides of the reinforcing rods. Several other tamper machines were developed after this time.

In 1920, W. R. Hume of Melbourne, Australia, toured America in the interest of establishing his method of making concrete pipe by the centrifugal method. He had been using his method in Australia since 1912 and had established Hume plants there as well as in New Zealand, India, southeast Asia, and Peru. Hume plants were set up in the United States in several locations, including Los Angeles, Detroit, Boston, and Dallas.

The "Mohr-Buchanan" centrifugal process from Scotland was introduced simultaneously in Montreal and Los Angeles in 1922. This process involved coating the inside of a solid form with a layer of wax. In the curing process the wax melts, runs off, and is collected to be used again. The outside form is then stripped in one piece.

About the same time in various parts of the country, several other manufacturers of pipe developed centrifugal machines. In the 1920s, a thin steel cylinder was embedded in the wall of reinforced concrete pipe to be used for large-diameter pressure water lines.

The development of prestressed concrete pressure pipe began in the 1930s, when concrete pipe was first wrapped with high-tensile wires to induce compression in the concrete to prevent it from cracking under the internal pressure.

In 1942, a fuel-oil pipeline for the United States Navy at Norfolk, Virginia, was constructed to withstand 300 lb per square inch working pressure. A thin-steel cylinder with steel rings welded to each end to form the joint section was lined with concrete that was placed centrifugally. The outside was wrapped with high-tensile steel wire, which in turn was protected with a cement-sand mortar placed by the brush-coat method. Since then, this type of pipe, generally known as "prestressed concrete cylinder pipe," has been used extensively in the United States for water supply lines. A later version for large diameters is made by embedding the cylinder in concrete and wrapping the outside concrete with the high-tensile wire, which in turn is protected with a cement-mortar coating or with a cast-concrete coating.

The "roller-suspension" process in which the circular outside form is suspended horizontally on a heavy roller, was established in 1947–1948. When the roller is rotated, the form revolves around the roller, and dry-mix concrete is rolled into place against the outer form by the weight of the form as it compresses the dry mix between the form and the rotating roller. This process produces a more homogeneous wall than the centrifugal process.

Electric or pneumatic vibrators, usually employing low-amplitude, high-frequency vibration, have been widely used to consolidate plastic concrete for large-diameter cast pipe.

High-amplitude, low-frequency vibrators are sometimes used in combination with pressure applied to the green concrete after the form is filled. A nonplastic (dry) mix is used. This type of process is generally referred to as the "dry cast" or "vibration" process.

DIVISIONS IN THE INDUSTRY

Agricultural and Farm Drainage

The agricultural and farm drainage concrete pipe are nonreinforced concrete pipe 24 in. and under in internal diameter. Generally, the irrigation pipe has a tongue-and-groove joint or a modified tongue-and-groove joint sealed with cement mortar or rubber gaskets.

The farm drainage pipe is either butt-ended or has tongue-and-groove ends. The joints are left open to pick up the drainage water. Porous concrete pipe is rarely used, but some perforated concrete drain pipe is used.

Sanitary, Storm Sewer, and Culvert Pipe

Sanitary, storm sewer, and culvert pipe come in many sizes. The 24 in. and smaller sizes may be nonreinforced. Larger sizes up to 144 in. and more in diameter are reinforced. Rubber gaskets of various designs are used in the joint for sanitary sewer construction. For culverts, a rubber gasket or cement mortar may be used.

Large reinforced pipe is being made in new shapes for use in culverts as storm-drain and sewer pipe. Arch pipe is slightly flattened on the bottom and arched on top. Elliptical pipe can be laid either on edge so that its greatest dimension is vertical, or on its widest side so that the greatest dimension is horizontal.

Pressure Pipe

Concrete pressure water pipe is concrete pipe in which the barrel and the joint are designed to resist internal pressure. It is used for drinking-water supply systems, for transmission of irrigation water, for pressure sewer force mains, for the overflow structures of dams, for ocean outfall sewers, for irrigation and sewer siphons under roads, railroads and water courses, and in general, for conveyance of any noncorrosive fluid under pressure.

MATERIALS USED IN CONCRETE PIPE

Cement

Portland cement, Types I and II, are the cements most used in pipe manufacture. Low-heat cement (Type IV) is not needed in pipe manufacture, and sulfate-resistant cement (Type V) would be employed only on special contracts as required by the job specifications. Type III cement (high-early-strength portland cement) might be valuable under certain circumstances to speed up production, but the most widely used cements are Types I and II. Availability depends on the local market.

The cement industry manufactures Types IA, IIA, and IIIA cements that contain an air-entraining agent. The air-entraining capabilities of these cements vary with the type and richness of mixture, gradation of aggregate, type of mixer employed, length of mixing time, temperature, and presence of organic matter, for example. Therefore, it is recommended that an air-entraining agent be added to the concrete mix when air entrainment is desired. Air-entraining agents should never be used in centrifugally spun pipes, and their use with no-slump or dry-mix concrete is extremely limited.

Slag cements, which are manufactured from blast-furnace slag in combination with normal portland cement clinker, often exhibit very satisfactory quality, particularly in concrete products that are subjected to steam curing. The selection of cement should be based upon actual performance tests of the strength and durability of concrete cylinders as well as strength tests of the concrete pipe itself.

Aggregates

Aggregates for use in concrete pipe should be clean, durable, strong, and well graded. The organic tests should be made for fine as well as coarse aggregate. Silt and gradation tests should also be made. Aggregate strength can best be determined by comparing concrete mixtures using various aggregate sources. If results of such comparisons are not available, tests for durability and inherent strength of aggregates should be made to determine an adequate source.

After selection of the source, the remaining qualitative tests for cleanliness and gradation should be made periodically.

Fine Aggregate. Perhaps the most important variable in aggregates is their gradation. This affects ease of manufacture of any kind of pipe, as well as the end qualities, such as economy, strength, and watertightness, for instance. The ASTM specifications for sand gradation tolerances are necessarily wide in order to cover the entire United States. In any given area, the tolerances should be narrowed within the extremes shown in the ASTM chart below.

In Table 16-A the information at the right gives three typical specifications for gradation based on material retained on each sieve. The "Old" specification is an example of an earlier ASTM tolerance; the next column presents the current ASTM Sand Specification C 33-71; and the last column is an improved specification with even tighter tolerances, applying to a given locality such as New England. It is important in any locality to select a sand having a gradation well within the permissible extremes of ASTM in order to provide proper attributes in the concrete mixture. Once having designed mixtures for an established gradation, the mix characteristics can be more easily maintained by maintaining constancy of gradation. The indicated "better" gradation is proper for all types of concrete pipe mixtures. It may not be possible to establish a basic gradation as desirable as this, but it will be possible to establish some basic gradation and design a proper mixture around it.

Table 16-A

Sieve Number	Clear Opening	ASTM Specifications		
		Old	C33-71	Ideal
⅜ in.	0.375 in	0%	0%	0%
#4	0.187	0–5	0–5	2–5
8	0.0937		0–20	10–20
16	0.0469	20–55	15–50	25–40
30	0.0232		40–75	45–65
50	0.0117	70–90	70–90	78–86
100	0.0059	90–98	90–98	93–97
Fineness Modulus				2.85
Maximum between sieves—			45%	30%

Table 16–B

	Job Sand			Ottawa (20–30) Sand		
Sieve Number	Individual Grams	Individual Percent	Cumulative Percent	Individual Grams	Individual Percent	Cumulative Percent
⅜ in.	0	0	0	0	0	0
#4	39	4	4	0	0	0
8	134	13	17	0	0	0
16	176	18	35	0	0	0
30	304	30	65	1000	100	100
50	188	19	84	0	0	100
100	108	11	95	0	0	100
Pan	51	5	—	0	0	—
Total	1000	100	300	1000	100	300

The fineness modulus serves a useful purpose only when the gradation has reasonable uniformity. The following tabulation is a case in point.

In the above tabulation if 1000 grams of dry typical job sand were sieved through the required testing sieves, the individual grams retained on each sieve might be as shown in the second column. The corresponding individual percentage retained would be shown in the third column. Cumulatively, the percentages retained on each successive sieve would be as shown in the fourth column.

By definition, the fineness modulus is computed by adding the cumulative percentages retained on the sieves shown and dividing by 100. In this case, the fineness modulus of the well-graded sand is 3.00 (300 ÷ 100).

Standard Ottawa sand has grains of pure silica essentially of one size, passing the number 20 sieve, and retained on the number 30 sieve. Consequently, a gradation of this type sand would be as shown in the last three columns of the above tabulation. One hundred percent of the sand would pass the number 16 sieve and 100 percent would be retained on the number 30 sieve. The fineness modulus in this case would also be 3.00. A very poorly graded sand can have the same fineness modulus as an ideally graded sand.

Poorly graded aggregates can be blended to improve the gradation. To aid in this blending, care should be exercised in describing gradation of aggregate. For example, a sand should not be merely described as being too fine, but rather either as having an excess of fines or a shortage of coarse sizes. A coarse sand should be described rather as having an excess of coarse sizes or a shortage of fines.

In Table 16-C, the sieve size and the required tolerances are shown for the ideal gradation on a simple job. The third column shows the grada-

Table 16–C

Sieve Size	Ideal	Available Sand	Masonry Sand	60 Percent Available Sand	40 Percent Fine Sand	Blend
⅜ in.	0%	0%		0	0	0%
#4	2–5	5	0%	3	0	3
8	10–20	25	2	15	1–	16
16	20–40	46	5	28–	2	30
30	45–70	73	36	44–	14+	58
50	79–86	93	68	56–	27+	83
100	93–97	99	90	59+	36	95
Fineness Modulus		2.85	3.41	2.01	2.05	0.80	2.85

tion of an available sand. The individual sieve retentions and the fineness modulus show that the sand is lacking in fines. A fine masonry sand was available for purposes of blending (fourth column). The following formula was used to design a blend of the two sands.

$$p = \frac{C-M}{C-F}, \text{ where}$$

p = % of fine sand required by blend
C = fineness modulus of coarse sand
F = fineness modulus of fine sand
M = fineness modulus of desired blend

Substituting into this formula the data from Table 16-C, we obtain the following.

$$p = \frac{C-M}{C-F} = \frac{3.41-2.85}{3.41-2.01} = \frac{0.56}{1.40} = 0.40$$

This answer of 0.40 signifies that a fineness modulus of 2.85 will be obtained using 40% by weight of the fine sand and 60% by weight of the coarser sand.

We have obtained a final blend having the desired trial fineness modulus that falls well within the tolerances established for the gradation on each sieve. In the event that the sieve tolerances were not met, either a new sand would have to be investigated for blending or a new trial would have to be made using different percentages.

After determining a proper blend, it should be maintained batch to batch. This can best be done by storing the coarse and fine sands in separate bins and combining them by weight.

Coarse Aggregate. Coarse aggregate is comprised of gravel, crushed stone, rock or other artificial, or manufactured aggregate. The basic rule for gradation of coarse aggregate is similar to that for fine aggregate.

The following tabulation indicates a variety of sieve tolerances for coarse aggregate for both ASTM and "better" gradations.

The chart indicates that specific or "ideal" gradations should be established with narrower tolerances than the permissible ASTM extremes. The indicated gradations generally apply to cast concrete pipe. Modifications may be made for other types of manufacture, including spun, tamped, packerhead, and roller suspension, for example. Depending on the process, the aggregate shape, and other variables, the manufacturer may elect (based upon experience) to adjust tolerances to produce a particular result. The rules for blending and quality control of coarse aggregate are the same as those for fine aggregate.

Water

Water must be clean and free from contaminants that would adversely affect the strength of the concrete. In general, potable water is suitable. Water from wells and streams in arid, western states often contains chlorides and sulfates. Sulfate concentrations should not be above 1% nor chloride concentrations above 2%.

With a given mixture of concrete, it takes an increase of approximately 3% of the net mixing water to increase the slump by 1 in. The increase in slump by the use of mixing water causes a reduction in strength of approximately 300 psi for each inch of increase in slump. Obviously, the use of excessive water is to be avoided.

Admixtures

Accelerators. Calcium chloride ($CaCl_2$) is generally used as an accelerator to speed up setting and hardening and, in the presence of proper

Table 16–D

Sieve Size	Clear Opening	1 in. Size ASTM	¾ in. Size		⅝ in. Ideal	½ in. ASTM	Pea
			ASTM	Ideal			
1½ in.	1.500 in.	0%					
1	1.000	0–5	0%	0%			
¾	0.750		0–10	4–8	0%	0%	Special
½	0.500	40–75		40–60	25–30	0–10	
⅜	0.375		45–80	65–75	45–65	30–60	
#4	0.187	90+	90+	94–97	85–95	85+	

moist curing, it will permit a shortening of the total curing time in the kiln. Calcium chloride should not exceed 2% by weight of cement and should never be used with prestressed pipe. Normally, a solution is mixed to provide one pound of calcium chloride per quart of aqueous solution.

Air-Entraining Agents. Air-entraining additives have been found useful in making cast pipe. They aid workability, reduce bleeding, and improve durability under some conditions of exposure. They are sometimes helpful as lubricants in the manufacture of pipe from nonplastic mixtures. Here the employment of more than the normally recommended amounts will impart a slipperiness to the mix that creates an increase in density and an increase in strength of the product. In general, for machine-made pipe, use of the air-entraining agent becomes effective when the original mixtures lack the characteristics to produce good density, interlocking of the aggregate particles, and ability of the pipe to stand after stripping. Air entrainment should never be employed in spun pipe. Such air entrainment migrates toward the surface during the spinning process and creates a froth that induces wide-spread surface imperfections. Furthermore, the froth becomes a waste product which is difficult to discharge. In tamped pipe, air entrainment should be used sparingly because too much causes a spongy mix in which the tampers may bury themselves.

Careful control of the mix design and proper air content are necessary when using air entrainment for wet cast pipe. Normally about ½ of the manufacturer's recommended amount of air-entraining admixture will produce a reasonable air content in the concrete. The resultant air content must be controlled between 3.5 and 5.0% if predictable quality is to result and adequate strengths are to be assured.

Pozzolans. Pozzolans are either treated natural minerals or fly ash (a residue from the combustion of coal). Fly ash must have the proper chemical composition and fineness to be effectively used in concrete. The carbon content should not be less than 7% nor more than 12%. The specific surface area should be at least 2800 and preferably over 3000 sq centimeters per gram. ASTM C 350-57T defines the requirements for pozzolans.

Pozzolans are used in concrete to reduce cement-aggregate reaction in localities where this is a problem. In the pipe industry, pozzolans are at times useful in improving workability and reducing wear on casting machinery. This type of admixture should not be employed without careful testing.

MANUFACTURING METHODS

Cast and Vibrated Pipe

Cast pipe is made by the conventional method of preparing the mix and placing it into oiled steel forms of the desired sizes. Cast pipe is 24 in. in diameter or larger. The concrete used should generally have a slump not in excess of 3 in. Entrained air in cast concrete should never exceed 2.5% of total volume.

Concrete should be slowly and evenly distributed around the pipe forms. Vibration should continue while the concrete is entering the forms and as long as bubbles of entrapped air emerge from the concrete. As soon as the bubbles stop, the vibration should cease. Overvibration will cause segregation of the aggregates, resulting in poor concrete in the top of the mold.

The forms generally have external vibrators located $1/4$ and $3/4$ of the way up the sides on long molds and in the center of molds 8 ft or shorter. For long pipe sections, it may be wise to discontinue lower vibrators when the level of concrete has risen well above them.

If pipe, after stripping, shows shrinkage cracks on the top, revibration of subsequently poured pipe at 20 to 30 min after pouring will eliminate or reduce this. These cracks are usually in the upper portion of the pipe because the lower portion is revibrated during placement of successive lifts and initial shrinkage or additional increments of consolidation are cumulative at the top of the pipe. Revibration should be done just before concrete becomes so stiff that plasticity cannot be restored by vibration.

Forms must be absolutely tight or sand and gravel streaks will appear where mortar leaks occur. The gates closing the form cylinders and the joint between forms and the base ring are possible locations for leaks. If tight gaskets are not used in form joints, 2-in. cloth tape can be used to seal joints with base rings.

"Packerhead" Process

Concrete pipe in a wide range of sizes is commonly made by the "packerhead" method. This method derives its name from the revolving packerhead or shoe on the machine that packs the nonplastic mix into place in the pipe wall. At the start of the process the packerhead is at the bottom of a vertical form that makes up the outside form of the finished pipe. As the packerhead revolves at high speed, the nonplastic mix is fed into the form and the packerhead is raised while revolving. The mix is packed into place under pressure by the packerhead at the same time the packerhead produces a smooth finish. The mix is so dry and the force with which it is packed against the outside is so great that the freshly made pipe section can be stripped immediately and the green concrete pipe will stand vertically during curing and hardening. On some machines a vibrator is used briefly at the beginning of the process to insure a dense, well-formed section on the bell end of the pipe.

Tamped Pipe

Forms for machine-tamped pipe are similar to those used in the packerhead process—split cylinders can be easily removed without injury to the pipe as soon as the pipe is finished. In the tamped pipe process, the nonplastic mix is tamped into the form by blows from hard wood or steel tampers fitted with metal shoes and striking at the rate of 500 to 600 times a minute. The inside core generally remains stationary while the outside form revolves. As it revolves, the dry mix is fed into the space between the outside form and the core. The tampers tamp the mix into place. Ingenious devices maintain uniform pressure in the tamping blows and, at the same time, allow the tampers to travel upward as the pipe walls are built up. A uniform amount of mix is fed into the revolving form while it is being subjected to this rapid, heavy pounding, and in this way the pipe is built up until finished. Upon completion, the machine withdraws the core or inside form, and the pipe is ready to be taken to the curing room where the outside form is stripped. The outside form and base rings are so designed that when the form latches are loosened, the weight of the form causes it to drop free from the pipe and allows the form to be removed without injury to the newly made pipe.

The blows of the tamper pound the mix into the desired density while the troweling action of the concrete, turning around the stationary core, adds a smooth-finish surface to the barrel of the pipe.

The best table speed for tamping machines is usually the fastest speed at which the table will run without "burning" the pipe. Average speeds on 4-ft lengths may range from 40 rpm for small diameters to 10 rpm for larger diameters. The speed of the table must be coordinated with the feed rate of the concrete. Unless the two are in balance, poor results will ensue.

Reinforcing steel, when used, must be carefully centered in the pipe wall or such undesirable results as broken tamp sticks, injury to operators, damage to machines or equipment, and exposed steel causing rejection of pipe, may result. Cages that are improperly formed or that have been sprung or misshapen should be corrected or else not used.

Centrifugal Casting

Centrifugal concrete pipe is manufactured in horizontal forms rotating at high speed in such a manner that the centrifugal force compacts the concrete both during and after the time it is being placed. The object of this process is to force the excess water to the inside where it can be drawn off by brushing or troweling. Enough water is put into the mix to make it workable although it should not have a slump of more than 1½ in. Some centrifugal machines have knocker wheels or vibrator wheels that vibrate the concrete as the form is rotated. Some machines have a roller in the form to compress the green concrete as the form is filled. During charging, the form is rotated fast enough to hold the concrete against the form—usually about 5 Gs. This speed is maintained while the charger places the concrete. After filling, the speed is increased very slowly until it is sufficient to achieve the desired consolidation, and then it is held there for the period of time required. Timing and speed involved in this phase are critical. Rapid acceleration can be damaging; too slow an acceleration may result in unnecessary costly power demands.

When most of the water is forced out of the mix, the pipe is slowed down so that water can be removed with troweling or brushing. Speed is picked up again so that additional water, if any, will be forced out. This water is removed, and then final troweling is done before the pipe is taken to be cured.

Any deleterious material in the aggregate will likely be forced to the center where it will project through the surface. Attempts to dig it out can leave pockmarks that may require patching. One of the problems that sometimes occurs as a result of bringing the pipe to consolidating speed too quickly is the trapping of fines within the wall, forming an impervious layer through which additional free water cannot pass. When the pipe is laid down in the curing area, the water collects at the bottom of the pipe under this layer and forms a blister. Proper acceleration and timing during spinning will prevent this.

CURING PROCESSES

Curing is of the utmost importance, and no matter what method of curing is chosen, it must be such that the cement is properly hydrated. There is little agreement among pipe producers about which processes and methods are best since conditions vary in different parts of the country, making standard procedures impossible. There are, however, six general methods of curing in use: (1) low pressure steam curing (100 to

160°F), (2) water spray curing, (3) saturated curing (with wet burlap, for instance), (4) conventional water curing, (5) membrane curing, and (6) air curing. The economical production of pipe requires that curing be accelerated, usually by steam. Air curing is in all cases supplemental to one of the other methods and is not used alone.

TESTS

ASTM C 497 describes tests on concrete pipe that are used to evaluate the properties provided for in the various specifications for different types of pipe. The tests described are (1) external strength test, (2) absorption test, (3) hydrostatic test, and (4) permeability test. Preparation for testing requires that the pipe not be exposed to a temperature below 40°F for 24 hours prior to the test and that the pipe be free from visible moisture.

External Strength Test

This test is referred to as the three-edge bearing test. The machine used is designed to exert a crushing force the full length of the pipe barrel. The load must be applied uniformly at a rate of not less than 1500 nor more than 2500 lb per minute per lineal foot of pipe.

For this test, the specimen stands on two parallel strips extending the full length of the barrel. The space between the strips may not be less than 1 in. nor more than 1 in. per foot of diameter. The load applied by the top beam must also extend the full length of the barrel. The lower bearings are of hard wood or hard rubber with rounded inside corners. They are fastened to a solid base. The upper bearing should be of hard wood backed with a steel beam. It may have a hard rubber strip attached. The load is applied at the center of the overall length of the pipe. The specimen is placed on the two bottom bearing strips and the point diametrically opposite. The top bearing beam is centered at this mark. Load is applied until either the formation of a 0.01-in. crack width or ultimate strength load, as specified, has been reached.

The "0.01-in. crack load" is the maximum load that can be applied to the pipe before a crack having a length of 1 ft and a width of 0.01 in. is formed. The crushing strength in pounds per lineal foot is calculated by dividing the total load on the specimen by the nominal laying length. In some machines the total load will include the deadweight of the top bearing plus the load applied by the loading apparatus.

Table 16-E gives physical and dimensional requirements for Class I pipe.

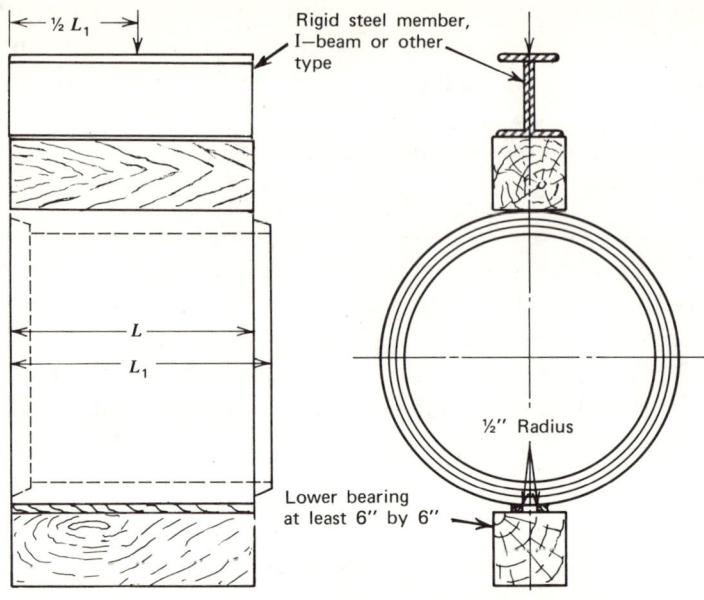

Figure 16-2 Three-edge bearings for pipe. This figure illustrates only the method of applying the load to the pipe and does not illustrate the various types of joints allowable.

Specifications may call for the three-edge bearing test to be calculated in D-load. *D-load* is the test load expressed in pounds per linear foot per foot of internal diameter.

ASTM C 76 (only one of many specifications for pipe manufacture) designates five classes of pipe. These classes are indicated by Roman numerals ranging from Class I through Class V, Class V being the strongest. Reinforced concrete sewer pipe is manufactured in sizes ranging from 12 in. to 108 in., and the supporting strength of all sizes of pipe within a given class is the same. The strength is indicated as D-load.

The advantage of the D-load concept is that all sizes of pipe within a given class generally will support the same height of fill above the top of the pipe.

There are two D-loads indicated for each class of reinforced concrete pipe. The lesser D-load is the D-load that will induce a 0.01 in. crack in the pipe. At this load, theoretically, the concrete is no longer in compression and the reinforcing steel in the pipe is beginning to assume tensile stresses. It does not mean that the pipe has failed.

The second D-load is the ultimate load—the load that will cause the pipe to fail.

In the first column of Table 16-E, the internal diameter of the first pipe listed is 60 in., which equals 5 ft. Multiplying the 5 ft by 800 (D-load to produce a 0.01-in. crack) shows that it will take 4000 psi to make a 0.01-in. crack in the pipe. Under 4000 psi, the pipe does not come up to specifications. To find the amount of pressure needed to produce the ultimate load, multiply 5 ft by 1200; thus it will take 6000 psi to bring the pipe to its ultimate load.

Boiling Absorption Test

Specimens to be tested must be dried in a ventilated, mechanical convection oven at 221 to 239°F until two successive weighings at intervals of 2 hours show an increment of loss not greater than 0.1% of the original weight of the specimen. The dried specimens are put in a covered receptacle with clean water at room temperature. The water is then heated to boiling in not more than 2 hours, boiled continuously for 5 hours, and allowed to cool for not less than 14 hours. When cooled, the specimens are removed and drained for not more than 1 min. They are dried with a cloth or a paper towel and weighed immediately. The increase in weight of the boiled specimen is the absorption of the specimen. It is expressed as a percentage of the dry weight.

Ten-Minute Soaking Absorption Test

The specimens are dried and weighed as above and are then immersed in clear water for 10 min. at room temperature. They are then removed, towel dried, and weighed as above. The percentage of absorption is calculated from the differences in weight.

Hydrostatic Test

In the hydrostatic test the pipe is filled with water, all air being excluded. A pressure gage is used and the water pressure is brought up to 10 psi in about 1 min. It is held there for 10 min., after which it is increased at a uniform rate to the pressure specified for the type of pipe being tested.

Permeability Test

A section of pipe with the spigot end down is placed on a soft rubber mat and filled with water to the level of the base of the bell. In 15 min. the pipe is inspected to see if it leaks.

Table 16-E
DESIGN REQUIREMENTS FOR FIRST CLASS REINFORCED CONCRETE PIPE[a]

	Minimum Reinforcement (Square Inch Per Linear Foot of Pipe Barrel)								
	Wall A					Wall B			
		Concrete Strength (4000 psi)					Concrete Strength (4000 psi)		
Internal Diameter of Pipe (in.)	Minimum Wall Thickness (in.)	Circular Reinforcement in Circular Pipe		Elliptical Reinforcement in Circular Pipe and Circular Reinforcement in Elliptical Pipe		Minimum Wall Thickness (in.)	Circular Reinforcement in Circular Pipe		Elliptical Reinforcement in Circular Pipe and Circular Reinforcement in Elliptical Pipe
		Inner Cage	Outer Cage				Inner Cage	Outer Cage	
60	5	0.25	0.19	0.28		6	0.21	0.16	0.23
66	5½	0.30	0.22	0.33		6½	0.25	0.19	0.28
72	6	0.35	0.26	0.39		7	0.29	0.22	0.32
78	6½	0.40	0.30	0.44		7½	0.32	0.24	0.36
84	7	0.45	0.34	0.50		8	0.37	0.28	0.41
90	7½	0.49	0.36	0.54		8½	0.41	0.31	0.46
96	8	0.54	0.40	0.60		9	0.46	0.35	0.51

Table 16–E (*continued*)

	Concrete Strength (5000 psi)					
102	8½	0.63	0.48	Inner Circular .... 0.22 Plus Elliptical 0.48	9½	0.54	0.41	Plus Elliptical .... 0.41 Inner Circular 0.20
108	9	0.68	0.51	Inner Circular .... 0.25 Plus Elliptical 0.51	10	0.61	0.46	Inner Circular .... 0.22 Plus Elliptical 0.46

Steel areas may be interpolated between those shown for variations in diameter, loading, or wall thickness. Pipe over 96 in. in diameter shall have two circular cages or an inner circular plus one elliptical cage of the minimum steel areas shown.

The elliptical steel in circular pipe or the circular steel in elliptical pipe must be held in place by means of holding rods or chairs or other positive means throughout the entire length of the pipe and throughout the entire casting operation and remain in place until the concrete has taken its initial set.

[a] The strength test requirements in pounds per linear foot of pipe under the three-edge bearing method shall be either the D-load (test load expressed in pounds per linear foot per foot of diameter) to produce a 0.01-in. crack, or the D-loads to produce the 0.01-in. crack and the ultimate load as specified below, multiplied by the internal diameter of the pipe in feet.

D-load to produce a 0.01-in. crack   800
D-load to produce the ultimate load   1200

Test loads for sand-bearing tests shall be one and one-half times those specified for the three-edge bearing test.
The minimum compressive strength of the concrete at the time of acceptance shall be as shown in this table.

CHAPTER 17
CONCRETE PIPE CONSTRUCTION

The fundamental principles for structural design of a sewer are based on studies made several decades ago at Iowa State College by Dean Marston and Professors Schlick and Spangler. Professors Schlick and Spangler also helped develop the design principles and conducted studies on loads. Further research is currently being carried on by universities, state highway departments, bureaus of public roads, and private agencies. The prime objective of Dean Marston's investigations was to study the load-carrying capacity of concrete pipe under various depths of backfill and various methods of bedding. As a result of these studies, it was established that the load on a pipe in a trench is influenced by four conditions.

1. The height of fill over the pipe.
2. The width of the trench at or near the top of the pipe.
3. The weight of the soil in pounds per cubic foot.
4. The character of the soil.

Concrete sewers are load-carrying structures designed to support specific loads. Occasionally, surface loads are important, but usually the load to be considered is that imparted by the weight of the backfill material over the pipe. The amount of this load depends not only on the height of the backfill over the pipe, but also on the width of the trench at the top of the pipe and the weight and character of the soil.

When a trench is narrow and deep, the amount of load on the pipe is significantly influenced by the frictional resistance of the trench walls as the backfill settles into position.

The width of the trench at the top of the pipe is of utmost importance in keeping the load on the sewer pipe at a minimum. The width of the trench should be kept as small as possible while still maintaining adequate working space. In no case should the trench be cut any wider than is called for in the plans and specifications. An increase in trench width over that assumed in design calculations causes the pipe to be overloaded. It is best to have the job specifications state the maximum trench widths permitted.

Trench banks can be sloped above the level of the top of the pipe without increasing earth loads over that of a trench with vertical sides. This practice of utilizing a subditch to prevent banks from caving is quite common where sheeting is not required by surface conditions. It also minimizes the load on the pipe.

LOADING FORMULA

Stated in the simplest manner, the load on a pipe equals the weight of the backfill minus the frictional support that the fill gains from the trench walls. Marston expressed this as a formula.

$W_c = C_d w B_d^2$,
    where W_c is the load in pounds per linear foot of pipe
              C_d is a load coefficient (Table 17-A)
                w is the unit weight of backfill in pounds per cubic foot
                B_d is the width of the trench at the top of the pipe in feet

C_d depends on the ratio of the height of backfill to the width of the trench and on the coefficient of friction of the soil. For convenience, C_d is usually obtained from a graph or a table (Table 17-A).

THREE-EDGE BEARING STRENGTH

The load that a pipe can withstand influenced by these two conditions: (1) the inherent strength of the pipe and (2) the manner in which the pipe is laid and embedded at the bottom of the trench. The first factor is usually referred to as the "three-edge bearing strength" and can be determined by the test of that name. This test is more severe than conditions likely to be encountered in any field installation since the loads are applied as concentrated loads at the top and bottom of the pipe. The load-carrying capacity of the pipe, in actual use, will be greater than the capacity of the pipe as measured by the three-edge bearing test because the vertical load is distributed over the width of the pipe and the pipe is uniformly supported.

FIELD SUPPORTING STRENGTH—EFFECT OF BEDDING TYPE

The effect of bedding was studied at Iowa State College to determine the ratio of the field supporting strength to the three-edge bearing strength with various types of embedment. This ratio became known as the load factor. As a result of these studies, various types of embedments were classified, and corresponding load factors were recommended for each (Fig. 17-1):

Class A, or concrete cradle, with load factor = 3.0
Class B, or first class, with load factor = 1.9
Class C, or ordinary, with load factor = 1.5
Class D, or impermissible, with load factor = 1.1

Concrete Cradle Bedding, Class A. This is a method of bedding a pipe in which the lower part of the pipe is bedded in plain or reinforced concrete. Load factors vary from 2.2 to 3.4.

First Class Bedding, Class B. This is a method of bedding a pipe in which fine granular materials in an earth foundation are shaped to fit the lower part of the pipe for at least 10% of its overall height. The earth fill material is thoroughly rammed and tamped in layers not more than 6 in. deep around the conduit for the remainder of the lower 30% of its height. Where the foundation is rock, the pipes are bedded on an earth cushion with a thickness of ½ in. per foot of height of fill, limited to ¾ of the nominal diameter of the pipe. A minimum thickness of 6 in. is required with the earth foundation carefully shaped around the conduit. This type of bedding has a load factor of 1.9.

Ordinary Bedding, Class C. This is a method in which the pipe is laid with care in an earth foundation shaped to fit the lower part of the conduit with reasonable closeness for at least 10% of its overall height. The rest of the pipe is surrounded by granular materials that are shoveled to

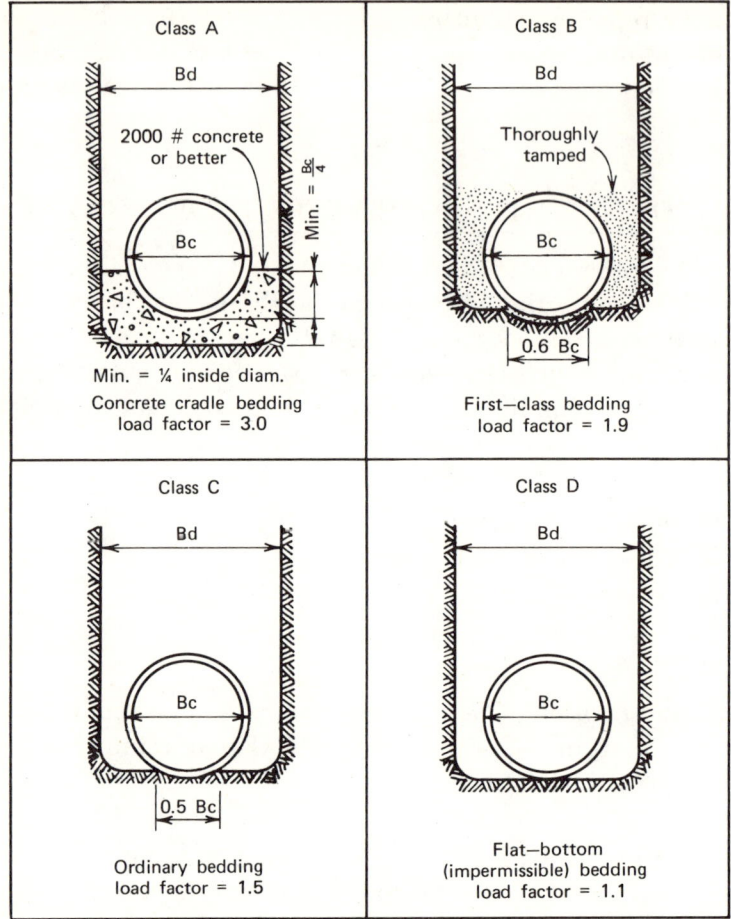

Figure 17-1 Types of trench bedding.

fill spaces under and next to the pipe. Where the foundation is rock, an earth fill is placed under the pipe for not less than ½ in. per foot of height of fill. Again, a minimum thickness of 6 in. is allowed. The load factor corresponding to this type of bedding is 1.5.

Impermissible Bedding, Class D. This is a method in which little or no care is exercised to shape the foundation to fit the lower part of the pipe or to compact under pipe haunches. Pipe installed by this method has a load factor of only 1.1 and its use is not recommended.

Figure 17-2 shows the permissible loads on trench conduits with dif-

Table 17-A
LOAD COEFFICIENTS FOR CALCULATING TRENCH LOADS
VALUES OF C_d IN THE MARSTON FORMULA $W_c = C_d w B_d^2$

Ratio of Depth to Width[a] $\frac{H}{B_d}$	Values of C_d for:			
	Granular Materials Without Cohesion $w = 100$ lb/cu ft $K = K' = 0.1924$	Sand and Gravel $w = 120$ lb/cu ft $K = K' = 0.165$	Saturated Top Soil $w = 110$ lb/cu ft $K = K' = 0.150$	Saturated Clay $w = 130$ lb/cu ft $K = K' = 0.110$
0.25	0.24	0.24	0.24	0.24
0.50	0.45	0.46	0.46	0.47
0.75	0.65	0.66	0.67	0.69
1.00	0.83	0.85	0.86	0.90
1.25	0.99	1.02	1.04	1.09
1.50	1.14	1.18	1.21	1.28
1.75	1.27	1.33	1.36	1.45
2.00	1.39	1.46	1.50	1.62
2.25	1.51	1.59	1.64	1.77
2.50	1.61	1.70	1.76	1.92
2.75	1.70	1.81	1.87	2.06
3.00	1.78	1.90	1.98	2.20
3.25	1.85	1.99	2.08	2.32
3.50	1.92	2.08	2.17	2.44
3.75	1.98	2.15	2.25	2.55
4.00	2.04	2.22	2.33	2.66
4.25	2.09	2.28	2.40	2.76
4.50	2.14	2.34	2.47	2.86
4.75	2.18	2.40	2.53	2.95
5.00	2.22	2.45	2.59	3.03
5.50	2.29	2.54	2.69	3.19
6.00	2.34	2.61	2.78	3.33
6.50	2.39	2.68	2.86	3.46
7.00	2.42	2.73	2.93	3.57
7.50	2.45	2.78	2.98	3.67
8.00	2.48	2.81	3.03	3.76
8.50	2.50	2.85	3.07	3.84
9.00	2.52	2.87	3.11	3.92
9.50	2.53	2.90	3.14	3.98
10.00	2.54	2.92	3.17	4.04
11.00	2.56	2.95	3.21	4.14
12.00	2.57	2.97	3.24	4.22
13.00	2.58	2.99	3.27	4.29
14.00	2.59	3.00	3.28	4.34
15.00	2.59	3.01	3.30	4.38
16.00	2.59	3.01	3.31	4.41
17.00	2.59	3.02	3.31	4.44
18.00	2.60	3.02	3.32	4.46
20.00	2.60	3.03	3.33	4.49

[a] Depth of trench is to the top of the conduit.

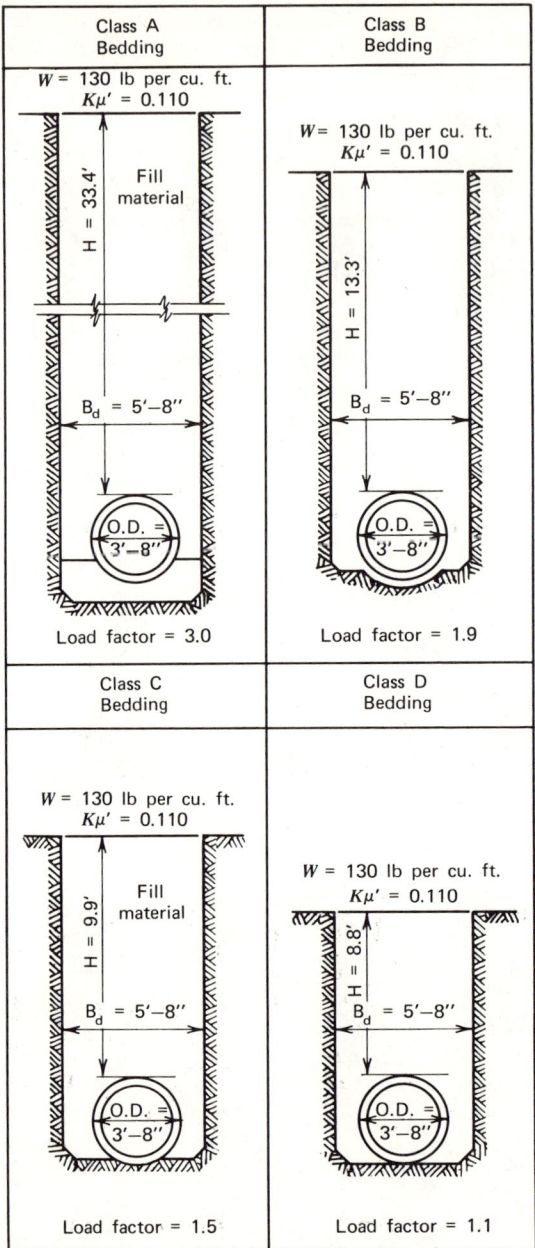

Figure 17-2 Permissible loads on trench conduits with various types of bedding, ASTM C76-59T, Class III, reinforced concrete culvert, storm drain and sewer pipe.

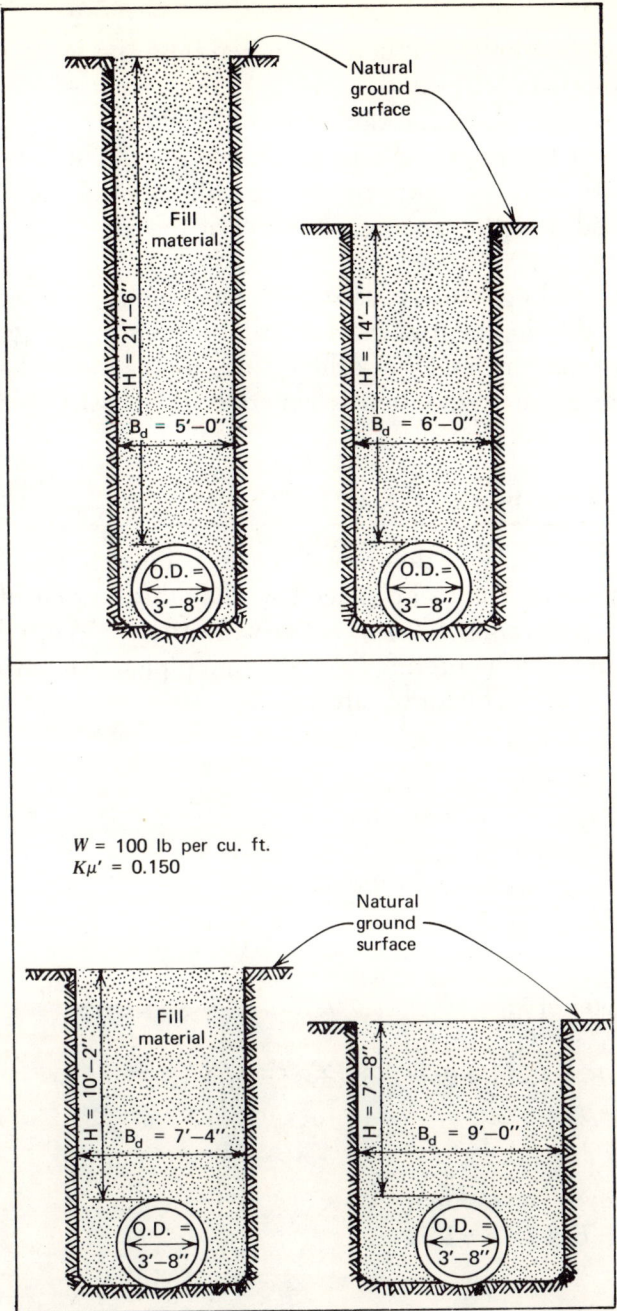

Figure 17-3 Typical variation in permissible depth of trench conduits with increase in width of trench, ASTM C76-59T, Class III, 36-inch reinforced concrete culvert, storm drain, and sewer pipe, Class C bedding.

ferent types of bedding. The better the bedding, the more load from fill material the pipe can withstand.

Figure 17-3 shows how the same Class C bedding can support various trench widths if the depth is kept in proper ratio. The allowable depth of trench varies with (1) the weight of the fill material, (2) the values of C_d, (3) the width of the trench at the top of the pipe, and (4) the type of bedding.

The above beddings were constructed when labor costs were relatively low. With the development of mechanical methods for subgrade preparation, pipe installation, and backfilling and compaction, the flat-bottom trench with granular foundation is generally the most practical method.

PROBLEM—LOADS IN A NARROW TRENCH

As an exercise, we will determine the load on a 15-in.-diameter concrete sewer pipe that is to be placed so that the bottom of the pipe is 15.5 ft below the ground level in a 3-ft-wide trench. We will also determine the adequacy of nonreinforced concrete pipe to carry the load using Class B bedding and a factor of safety of 1.5.

Conditions
  Pipe size: 15-in. inside diameter
  Depth to bottom of pipe: 15.5 ft
  Unit weight of backfill (w): 120 lb/cu ft
  Trench width (B_d): 3 ft

Outside Diameter of Pipe
  Wall thickness (t) = 1⅝ in.
  Therefore, B_c = 15 in. + $2t$ = 15 in. + 3.25 in. = 18.25 in., or
  B_c = 1.52 ft

Height of Fill over Top of Pipe
  H = 15.5 ft − B_c = 15.5 − 1.52
  H = 14.0 ft

Determination of C_d
  H/B_d = 14.0/3.0 = 4.67
  Therefore, from Table 17-A, C_d = 2.38

Trench Load
 $W_c = C_d w B_d^2$
 $C_d = 2.38$
 $w = 120 \text{ lb/ft}^3$
 $B_d = 3.0 \text{ ft}$
 $W_c = 2.38(120)(3)^2 = 2.38(120)(9)$
 $W_c = 2570 \text{ lb/ft}$

Selection of Pipe

Bedding	class B
Load factor (LF)	1.9
Factor of safety (FS)	1.5
Required pipe strength	$(W_c/LF)(FS) = (2570/1.9)(1.5) =$ 2030 lb/ft
Specification	ASTM C.14
Type	extra strength
ASTM minimum strength	2750 lb/ft

SAFETY

Often in unstable soils, it is necessary to resort to some type of temporary structure to prevent wall cave-ins. Such cave-ins are dangerous to the workmen and have resulted in many deaths over the years. Aside from

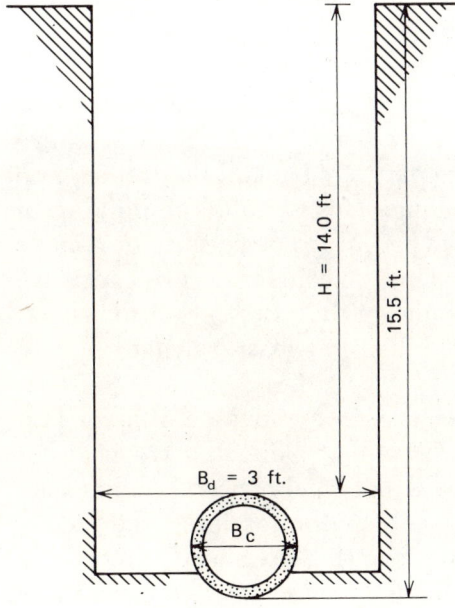

Figure 17-4

the hazard to the men in the trench, a cave-in costs the contractor money by slowing his operation. It also widens the trench, causes greater loads on the pipe, and may undermine adjacent structures. For this reason, bracings of various types are used to support trench walls.

The most widely used bracings are vertical sheeting made of wood timbers. This sheeting may be either open or closed. *Open sheeting* is the installation of vertical planks in pairs on opposite sides of the trench braced by cross members that span the trench. Subsequent pairs of planks are spaced at 2- or 3-ft intervals along the trench. The spacing is determined by the depth of trench, the type of soil, and moisture conditions. Following a ground-soaking rain, it may be necessary to space sheeting closer together than would be the case under dry conditions.

With *closed sheeting* there is no space left between the sets of planks. As a result, all of the gaps in the sheeting are closed and a solid timber wall results. Closed sheetings are usually provided for either extremely unstable soil conditions or when the project is so large as to require an unusual length of time between initial trenching and backfilling. These long periods pose the possibility of storms or other events causing a cave-in. Closed sheeting may be timber or may be steel sheet piling, depending on the size of the project.

One of the problems of sheeting is the change in loading on the pipe when it is removed. The sheeting must stay in place long enough to serve its purpose, but if the sheeting is to be removed, it must be removed as soon as possible after the backfilling operation has been started. The backfill load on the pipe is reduced by friction with the trench wall. If the backfilling operation is completed before the sheeting is pulled, the extraction of the sheeting causes the entire load of backfill to be transmitted to the pipe. It may be possible to partially backfill the trench, pull the sheeting, and then complete backfilling. This permits the buildup of friction along the trench wall. In cases where it is not feasible to remove the sheeting, it is simply left in place. This, of course, is very expensive and is avoided where possible. Another alternative is to cut off the sheeting after the backfill is partly completed. This, at least, salvages part of the timber for future use.

Another method of bracing is to construct a movable shield. This shield, sometimes called a *trench box*, is pulled along the trench after each length of pipe is placed. It may be either steel or wood and is used more for the protection of the workmen than for retaining the trench wall. However, it is just as important to move the box ahead carefully as it is to withdraw sheeting carefully. This is because the previously laid

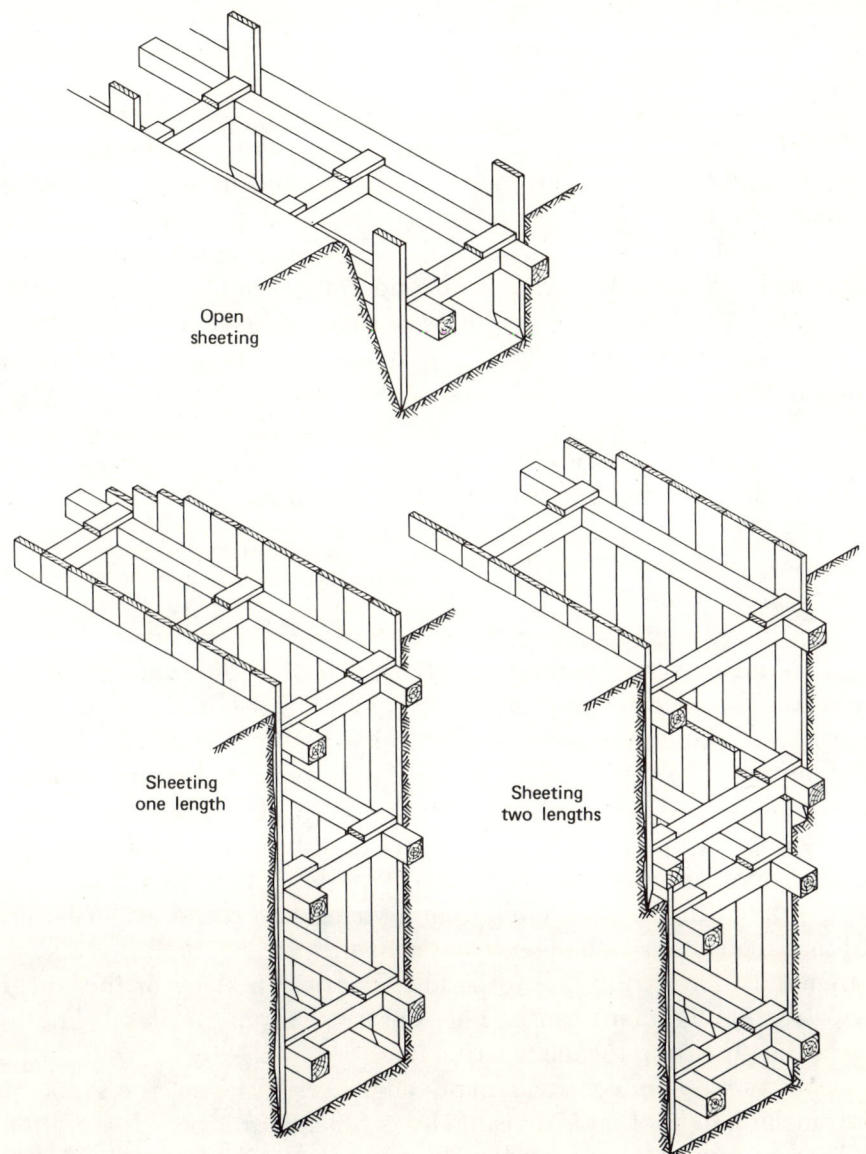

Figure 17-5 Typical trench sheeting.

pipe may become separated at the joints if the backfill material becomes wedged between the pipe and the box shield.

BEDDING

As the final shaping of the trench is being done, care must be exercised to see that the bottom of the trench is both firm and uniform. Soft spots in the trench bottom will allow unequal settlement of the pipe.

A sag will tend to open the joints of the adjoining lengths of pipe. Such openings in the line could permit infiltration of ground water. This infiltration could carry with it silt and sand particles that could cause blockage. In extreme cases the backfill material can be seriously depleted by such infiltration.

Extreme care should be taken to assure that the class of bedding specified is actually achieved in the trench. If Class B bedding is specified and Class D bedding is actually achieved, the result might be a costly maintenance problem.

PIPE LAYING

The work of laying pipe should proceed upgrade. The pipe should, whenever possible, be placed with the bell or groove end pointing upstream. As it is placed, the pipe should be checked for both horizontal and vertical alignment. Gravity systems are designed with a specific slope in order to provide adequate velocity and discharge. If this specified minimum grade is not maintained, the velocity and discharge characteristics of the line may be impeded. The proper grade can be maintained by measuring down to the pipe joint from a string line set between batter boards that span the trench. These batter boards are referenced at elevations that cause the string line to parallel the design slope of the sewer line. Horizontal alignment can be checked by dropping a plumb line to the top of the pipe from this same string line.

As a construction convenience, many engineers try to set the grade of a sewer line at a figure that is divisible by 4. Since batter boards are often set at 25-ft intervals and the grade of a sewer is specified in feet per 100 ft, a sewer grade that is divisible by 4 facilitates the setting of batter boards to the proper elevation because the increment of slope between batter boards then will be a whole number of feet.

Care should be exercised in handling the pipe during placement. If the pipe sections are too heavy to be placed by hand, power equipment should be provided. Of course, no pipe should ever be dropped into the trench. Any pipe with gross defects or cracks should be rejected. Damage to the ends of the pipe that will induce undue roughness to the interior of the line or preclude the making of a proper joint is cause for rejection.

Special attention should be given to the construction of the joints. Poor joints allow seepage out of the line and will allow ground water and roots to enter the line.

BACKFILLING

After the pipe has been placed, the grade and alignment checked, the proper bedding assured, and the joints made watertight, the backfilling has to be done. Proper backfilling procedures are as important as any other in the construction operation.

The placement of select material around the pipe and for a minimum distance of 1 ft above the top of the pipe is a necessity for proper construction. This initial backfill should also be thoroughly compacted in lifts not exceeding 6 in. in depth. This assures a minimum of lateral move-

Figure 17-6 Compacting backfill around 84-in. reinforced concrete pipeline with vibrator rollers. (U.S. Bureau of Reclamation photograph.)

ment of the pipe or settlement of the backfill adjacent to the pipe. This initial backfill also serves as a cushion to absorb loads caused by the dumping of backfill into the trench.

The remainder of the backfill need not be hand placed. However, in order to avoid excessive settlement of the backfill, it is often wise to continue backfilling procedures in lifts of 6 to 12 in., with adequate compaction between lifts. This is especially necessary where sewer trenches are located beneath roadways.

Water flooding is often employed to gain compaction in trench backfills, but it is not completely successful except in sandy soils. An inherent danger in water flooding is the fact that the water might flow to only one side of the line. The resulting pressure might cause displacement of the pipe. Care should also be taken to see that the line is not floated.

Large boulders and all organic matter should, naturally, be removed from trench backfills. Care must be taken when operating heavy equipment on or adjacent to the trench to avoid a cave-in.

Soil-cement provides a firm, uniform foundation and can be placed quickly from the top of the trench and left to cure. This eliminates tedious tamping procedures and saves time, labor, and money.

A sandy soil mixed with about 280 lb of cement per cubic yard plus enough water to make it fluid has been successful. This produces a bedding between Classes A and B in quality.

It is well to remember that no conduit construction project is complete until a good, well-compacted backfill is in place. Poor backfilling is at the very least a maintenance problem since natural settlement will take place and the trenches will have to be filled, and, at worst, excessive settlement can cause pipe displacement and even possible distress of the line.

JOINTS

Joints can be the downfall of a conduit system. To be effective, joints must be watertight and root-tight. No water should leak out and nothing should be able to grow or seep in. Good workmanship and quality materials are essential to satisfactory jointing.

For example, when mortar is being applied to joints, it has a tendency to dry out quickly unless the ends of the pipe are covered. Covering pipe ends prevents the circulation of air and slows the drying process.

Some of the joints that have been tested by time and proven dependable in gravity-flow concrete sewer pipe are discussed below. In each case the value of the joint depends on good workmanship and materials.

Types
Bell-and-Spigot (or bell end). Pipe with bell-and-spigot joints has one end flared out in a bell shape and the other straight. The straight end serves as a spigot and fits into the bell end. In larger sizes of pipe the bell end becomes heavy and cumbersome.

Tongue-and-Groove. In this type of joint, a groove or recess is built into one end of the pipe with a corresponding tongue on the opposite end. When two pipes are joined, the tongue enters the groove of the adjoining pipe to form an overlapping connection. The tongue may be longer, shorter, or equal in length to the groove. The tongue-and-groove joint is used extensively in irrigation, sewer, culvert and drainage work, and in larger sizes of pipe.

Modified Tongue-and-Groove. The basic joint in this type design is the tongue-and-groove. In order to provide more thickness of concrete and, therefore, more strength, a modified bell projects beyond the barrel of the pipe.

Concrete Collar. This is a prefabricated or field-cast, reinforced-concrete collar attached to one end of the pipe section with half of the collar protruding so that the end of the next section can be slipped into the collar. Sometimes this joint is used on straight butt-ended pipe and, at other times, it is used in conjunction with tongue-and-groove pipe.

Jointing Materials
There are three main types of jointing materials used in gravity-flow concrete pipe construction.

1. Compression-type rubber-ring gasket.
2. Cement mortar.
3. Jointing compounds.

Compression-Type Rubber-Ring Gasket. This is now widely and successfully used on concrete pipe sewer lines. The ends of the pipe are manufactured to close tolerances so that there is the correct space left for the single ring of elastomer. This ring is used with all types and sizes of circular pipe. It provides an excellent watertight joint, completely root-tight and yet having a certain amount of flexibility—enough to allow for some settlement movement in the line.

The gasket is slipped over the tongue before the section of pipe is laid in the ground. The tongue of the pipe, with the rubber ring on it, is then inserted into the groove end of the pipe in place.

Cement Mortar. The material first used on concrete pipe sewer joints is probably cement mortar. When properly made, it makes a rigid joint that can be watertight and root-tight.

Oakum-cement mortar is another jointing material. Oakum is a closely twisted fiber of hemp or jute that is bound together by coal tar. Mortar is applied, making a particularly root-tight joint even after slight movement from settlement.

Jointing Compounds. There are numerous jointing compounds on the market for use with bell-and-spigot, tongue-and-groove, or modified tongue-and-groove joints. Some have been successful and others have failed completely. Care with installation, attention, to the proportions of the compound and to the temperature at the time of installation, care in handling, and dryness of the pipe, for example, contribute to the success of these jointing compounds.

These compounds are combinations of coal tar and mineral fiber, bitumastic, creosote, sulfursilica, asphalt, coumarone indene, various resins and gums, and a variety of patented compounds. They are used with or without oakum or jute fillers.

Joint Construction

Joints in concrete pipe are of two general types: (1) rigid joints such as neat cement, cement mortar, and cement grout; and (2) flexible joints, such as bituminous materials and rubber. Sometimes a combination of these materials is used. A good joint material should bond well to the pipe (except rubber gaskets) and, where required, it should provide some flexibility.

Flexible joints should be used where unstable ground conditions are encountered. However, even with the most elastic joint material, there is a limit to the amount of displacement a pipe line will withstand without excessive leaking. In unstable soils it is always better to stabilize the sewer line by the use of piles or stabilized foundation rather than to rely entirely on a flexible joint to hold the sewer pipe together.

Cement Joints. Cement joints, when properly made, are watertight, resisttant to root penetration, and durable. When pipe lines are constructed on unstable foundations, the rigidity of this type of joint might be objectionable.

Neat portland cement paste, made with a water-cement ratio of about 0.12 to 0.16, has been used for years for calking cast-iron pipe as well as for calking bell-and-spigot pipe. Forms or molds are used to retain the grout during the placing and hardening. Grout is easy to handle but likely to develop shrinkage cracks during hardening because of its high cement content.

Mortar made with a mixture of 1 part portland cement to 2 parts sand and having a water-cement ratio of 0.6 to 0.65 also makes a satisfactory material for sealing joints. The grout is usually poured or pumped into the joint space and is retained by molds or runners (called "diapers") around the pipe. Cement mortar of a stiffer consistency may be used for the lower and upper portions of this joint.

Some engineers require that for a bell-and-spigot pipe the spigot end be centered in the bell end by means of a packing gasket of twisted oakum or hemp of proper thickness and sufficient length to pass around the pipe and lap at the top. After the pipe has been properly placed and bedded to line and grade, this gasket is calked into the joint space and

Figure 17-7 Constructing diaper joints in 42-in. pipeline. Diaper joints are hand finished at the top of the pipe. (U.S. Bureau of Reclamation photograph.)

the remainder of the space filled with cement mortar. The oakum or hemp gasket should be impregnated with neat-cement grout.

Bands of mortar are often placed around the exterior of the pipe at a joint. They usually are about 5 in. or more in width and not less than 1 in. thick for the smaller-sized pipe, increasing in thickness somewhat as the pipe size increases. The exterior of the pipe should be thoroughly cleaned and wetted just before the band is placed. The consistency of the banding mortar should be slightly stiffer than that used in the joint. As the mortar band is brought up from the bottom around the outside of the pipe on either side, backfill material can be placed against it to prevent its sagging. Another technique is to place a strip of cheesecloth under the pipe, bringing the ends up, around, and over the mortar band as it is placed.

Bituminous Joints. Bituminous joints consist of a tar or asphalt to which an organic filler is sometimes added. Some of these can be applied cold; others must be heated and poured into the joint space with the aid of a runner placed around the pipe.

Rubber Gasket Joints. To keep infiltration to a minimum and yet provide a joint that is flexible and easy to put together, consulting engineers have turned more and more to rubber-gasket joints for sewers. Properly compounded rubber gaskets that are continually under compression and away from sunlight are very durable and can provide a tight joint. Installation is simple, usually requiring fewer men than other types of jointing.

Rubber gaskets are made in varying shapes by extrusion or molding into a continuous ring. Flat gaskets have projections on one or both sides that become compressed when the pipe ends are fitted together. An adhesive is used on the joint when flat gaskets are used. Round or almost-round gaskets fit in the space between overlapping surfaces of the joints. A ring is slipped over the tongue end of the pipe section before it is laid in the trench. The tongue is lubricated with vegetable soap and inserted into the groove end of the pipe in place. To form satisfactory joints, the pipe must be manufactured to close tolerances.

Jointing Compound Installation. Instructions for the installation and laying of concrete pipe using jointing compounds should be obtained from the manufacturer of the compound.

SPECIAL CONSTRUCTION PROBLEMS

Quite often, problems will be encountered during a pipe-laying project that call for special types or methods of construction. Some of these are high ground-water tables, rocks in the excavation, tunnel construction, jacking, and laying pipe in unstable material.

High Ground-Water Table

If the ground-water table is so high that it is impossible by ordinary methods to maintain a dry trench, there are several ways to solve the problem. Some of these are:

1. Excavation.
2. Installing subdrains.
3. Pumping.
4. Well points.

One solution is to excavate a subtrench adjacent to the trench where the pipe will be bedded. This will lower the water table sufficiently to allow proper construction of the line. However, this procedure necessitates a trench that is wider than the design width, and it may induce excessive loads on the pipe.

Another method of draining a trench is to install subdrains beneath the bedding for the sewer line. This procedure is quite expensive since it necessitates the construction of two sewer lines—one on top of the other. Also, there is the problem of disposing of the water at the lower end of the subdrain.

A very common method of disposing of unwanted water is to pump it from sumps in the trench bottom. However, this procedure cannot be expected to handle large quantities of water.

The most satisfactory method of draining a trench is by the use of well points adjacent to the trench. In this procedure, numerous well points (usually 2 to 3 in. in diameter) are driven at $2\frac{1}{2}$-ft intervals adjacent to the trench and then coupled with a pumping device. The well points should be of sufficient depth and adequate spacing so that the water table will be lowered below the trench bottom when the pumping equipment is operating. This method of de-watering a trench is quite expensive, but it is frequently justified in areas where the water table is extremely high or the trench is extremely deep.

Figure 17-8 Well points are used to maintain a dry excavation for the installation of horizontal pipe. Pump, header, valves and swing joint connections are shown.

Rock

When a contractor encounters rock in the course of excavating the project, there is only one thing to do—overexcavate. This overexcavated material is later replaced with select compactible material that will provide an adequate foundation for the pipe. The depth to which this cushioning should be carried is a matter of judgment. However, one standard, which seems logical, is to provide for a minimum depth of 6 in. or ½ in. for each foot of backfill over the top of the pipe, whichever is greater. However, when the height of fill governs the cushion depth, it should not be necessary to overexcavate for a depth of more than three-quarters of the inside diameter of the pipe.

Unstable Material

The opposite situation from the unyielding rock foundation is that of yielding, unstable material. These conditions may be encountered in marshes or bogs where there is an overabundance of ground water. Such areas may often support the pipe alone, but when the overburden is replaced, there is a chance of unpredictable, uneven settlement.

In such cases it has been found satisfactory to drive 4×4-in. or 6×6-in. poles into the bog and to fasten wooden cradles. The pipe is then laid on the cradle and the backfill is placed. The cradle remains intact. Eventually, the wood may rot out, but by the time it does, the foundation and the backfill will have become stabilized and no settlement will occur.

There are variations in this method. One is to drive sheeting that fulfills the dual purpose of retaining the trench wall and acting as an anchorage for the cradle. Another is to actually build a plank flooring and place a sand cushion on it for supporting the pipe.

Tunnel Excavation

In some instances, the use of tunnel construction may be more practical and economical than an open trench, particularly where the excavation is very deep. For example, in closely built-up areas, trench construction may interfere with traffic or endanger building foundations.

Tunneling may be done by open-face mining with or without breasting or by use of tunnel shields. Tunnel construction introduces some hazards not normally encountered in an open ditch. Such problems as ventilation, lighting, falling rock, and cramped working space complicate this type of construction. In very soft clay or running sand, it will probably be necessary to use a shield with compressed air. The lining then will have to be strong enough to support the surrounding earth and serve as a backstop for jacking the shield forward.

Where the ground is more stable, open-face mining can be used (Fig. 17-9). The use of compressed air will depend on the amount of ground water and the type of soil. Lining may be of steel or wood, but wood presents a fire hazard in the presence of compressed air. Precast concrete segments have been used as a primary lining on some tunnel jobs.

In sewer construction the tunnel excavation is usually completed before the pipe is installed. Then the pipe is carried into place either on a special cart or by sliding over skids. The pipe should be blocked to line and grade. Lean concrete is placed closely behind the tunnel excavation, or the concrete lining may be delayed until all excavation is complete. Large boring machines that tunnel into the soil ahead of the pipe are used to excavate both in tunnel and jacking operations.

Jacking Concrete Pipe

When it is necessary to install a sewer beneath a highway, railroad, or city street without interrupting traffic, the pipe may be jacked into posi-

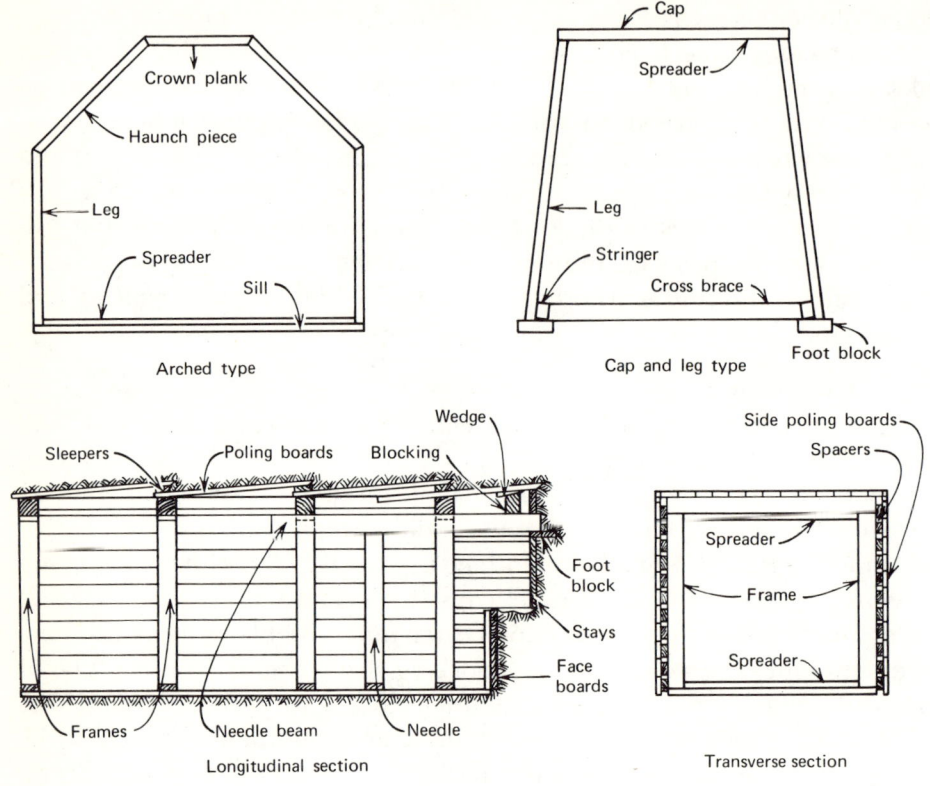

Figure 17-9 Typical tunnel bracing.

tion. This method is also used for deep installations where conventional open-cut and backfill methods may not be economical. In jacking, the pipe sections are pushed beneath the roadway by power-driven hydraulic jacks (Fig. 17-10). Because of their strength and smooth exterior, reinforced concrete pipe is well suited for this use. Pipe from 30 to 156 in. in diameter have been installed in this manner. Reinforced concrete boxes as large as 10 × 10 ft also have been successfully jacked into position.

In preparation for jacking, a working pit is excavated on one side of the work. This pit should be large enough to provide space for one or two sections of pipe, frames to support them, jacks, and the backstop.

Guide timbers for the support of the pipe to be jacked are very carefully installed so that the pipe will be at the correct line and grade. Rail-

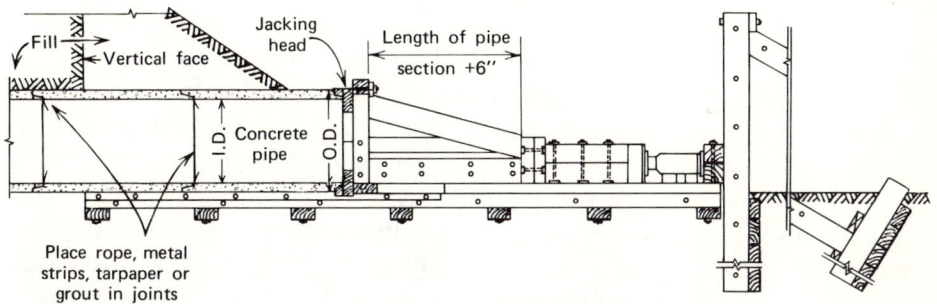

Figure 17-10 Typical layout for jacking concrete pipe.

road track set in concrete is sometimes preferred for large pipe. A steel cutting ring may be installed on the leading edge of the first pipe. Fittings for pumping friction-reducing lubricant to the outside of the pipe sometimes are installed on the first pipe and at intervals on subsequent pipe. Some contractors have found it useful to coat the outside of each pipe section with lubricant before jacking. Bentonite and asphalt emulsions are most commonly used.

A jacking head, consisting of bearing blocks (usually oak), is used to transfer the pressure uniformly from the jacks to the pipe. The number of jacks required depends upon their capacity, the pipe size, the nature of the soil, and the length of the pipeline to be jacked. The jacks must have a support at the rear of the jacking pit strong enough to withstand the huge thrust developed. Ingenious schemes have been devised to provide suitable backstops or thrust blocks. Timbers, concrete bulkheads, and in some cases the exposed end of the completed sewer have been used for this purpose.

As the pipe progresses, workmen excavate the material from the tunnel face just ahead of the pipe. Mechanical spades are used in hard ground. Usually the excavation is limited at the top and sides to about 1 in. larger than the pipe. As the operation progresses, a cart or sled may be used to remove the excavated material from the head to the jacking pit where it can be wasted. Mechanically operated belt conveyors also have been used for this purpose.

In recent years, boring machines have been used in jacking concrete pipe. A cutter head operating ahead of the first pipe section opens the hole into which the pipeline is jacked. An auger operating on the same shaft as the cutter head pulls the excavated material out through the pipe to the jacking pit. This procedure has generally been used for pipe less than 36 in. in diameter. However, 1200 ft of 120-in. reinforced concrete pipe were installed in California by this method.

Jacking is usually done upgrade so that seepage will drain to the working pit, which can be kept dry with well points or a slump pump. As the work progresses, the line and grade of the lead pipe should be checked frequently so that corrections can be made before large errors occur. Line and grade are often maintained with laser sighting systems to a tolerance of ± 1 in.

It is important to keep the pipe continually moving insofar as possible. When a long stop is made, there is danger that the soil will set up around the pipe, thus freezing it and making it difficult, if not impossible, to get the pipe started again. This is particularly true for pipe that is jacked under railroads where the vibration of passing trains may compact the soil around the pipe. It is often advisable, therefore, when jacking is once started, to continue day and night until completed. If, however, the work is stopped for any reason, arrangements should be made to occasionally move the pipe slightly to prevent freezing. If for any

Figure 17-11 Jacking 42-in. diameter reinforced concrete pipe under railroad. Note the drive shaft extending into pipe to turn auger. (U.S. Bureau of Reclamation photograph.)

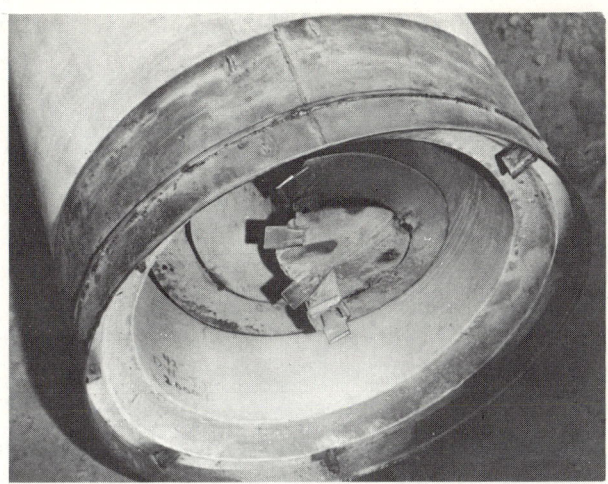

Figure 17-12 An auger is used in jacking 42-in. concrete pipe equipped with a steel cutting edge. (U.S. Bureau of Reclamation photograph.)

reason it becomes impossible to move the pipe during jacking, the operation may be completed from the other end of the project.

INSPECTION AND TESTING

When construction is completed, the inspection and testing must be satisfactorily passed before the work will be finally accepted. Manholes are inspected by direct visual observation, and each section of sewer line between opened manholes is also inspected visually. When the pipe is too small for direct observation, mirrors can be used to reflect sunlight or artificial light through the pipe between observers at the manholes. Any noticeable leaks or deviations from line and grade shown by this inspection must be repaired before quantity leakage tests are made.

Ideally, a pipeline is supposed to keep all the effluent in and everything else out. Specifications on leakage are expressed in many ways. Some refer only to quantity of leakage, length of line, and time. Most specifications incorporate pipe diameter into the specification. Some incorporate the hydrostatic head on the line also. The method of test may be either infiltration or exfiltration—water leaking in or water leaking out. When the pipe is substantially below ground-water level, an infiltration test is probably the most logical. Above the ground-water table, the exfiltration test determines the tightness of the line most efficiently. When flooding is used to consolidate backfill, it may be possible to check the infiltration concurrently.

Infiltration Test

This test measures the amount of ground water that comes into the sewer line. In order for the test to be effective, ground water must completely surround the pipe. This is a condition that is very difficult to achieve, especially during normal construction operations. Even when the trench is being compacted by means of water flooding, there is no assurance that water is actually surrounding all of the pipe being tested. Consequently, it may be an unreliable test.

After ground water completely surrounds the pipe, all flow is excluded from the section of pipe being tested by placing a plug in the influent side of the upstream manhole. Then the lower end of the section being tested is plugged with a plug that has a valved pipe through the center of it. When the flow through the valved pipe becomes constant, the discharge through the pipe can be checked by catching it in a container of known volume.

Infiltration might also enter the line being tested through the upper manhole; so, where excessive infiltration is noted during a test, it is well to check the manholes separately for infiltration. This can be done visually or by plugging both the inlet and outlet of the manhole and observing how much water infiltrates the manhole during a given period of time.

No nationwide specification has been adopted for measuring the leakage of a concrete pipeline. However, several national organizations have tried to clarify the requirements.

The American Public Works Association (APWA) recommends that infiltration not exceed 1000 gal per 1000 ft of sewer for 24 hours. Allowable limits for different pipe diameters have been suggested. For example, the limit is 100 gal per inch of diameter per 1000 ft in 24 hours for pipe from 12 to 48 in. in diameter.

The American Society for Testing and Materials' specification on rubber gasket joints contains a note that states that leakage should not exceed 0.6 gal per hour per inch of diameter per 100 ft of line. This is further qualified by setting a limit of 25 ft of hydrostatic head measured from the pipe center line.

Recently revised Corps of Engineers' specifications state that the maximum head should not exceed 10 psi. This is equal to 23.1 ft of head and is comparable to the 25 ft specified by the ASTM.

Actually, all of the requirements are similar when all qualifying considerations are taken into account. Leakage limits are determined by—

1. Flow rate (F) in gallons per hour.
2. Pipe diameter (D) in inches.
3. Pipe test length (L) in hundreds of feet.

Leakage limit $= \dfrac{F}{D \times L}$

Flow rate, F, is indicated here in gallons per hour (gph). APWA rate is in terms of gallons per 24 hours. As long as the principle is understood, the units are clear. Pipe diameter, D, is in terms of inches of internal diameter; and the length of pipe tested, L, is in terms of hundreds of feet. This varies from APWA, which defines the limit for 1000 ft.

Based on a limit of 0.4 gph per in. dia. per 100 ft, which is the specification for the State of Washington, 300 ft of 8-in. concrete sewer line would be allowed a leakage of 9.6 gal per hour—or about a half pint in 24 seconds. This is a very small amount of leakage and, if enforced, will certainly induce proper construction practices.

Exfiltration Test

An exfiltration test may also be made with a minimum of effort. To make the test, the upper end of the section to be tested is closed with a plug containing an open air vent. The lower portion of the line to be tested is then closed with a plug that has a pipe through it, which is in turn connected to a tee, a valved water connection, and a riser (preferably a transparent sight glass).

When the plugs are in place, water is introduced into the line through the valved connection. When all the air is exhausted, the air valve at the upper end of the section is closed. The water is then allowed to stand in the line till absorption of the pipe is complete (up to 24 hours). When absorption is complete, additional water is added to bring the head on the pipe to the specified level in the riser, and the water source is disconnected. Exfiltration from the line is then measured by the amount of water that needs to be added to maintain the water level in the riser at the specified height over a specified period of time.

When grades on a sewer line are such that an excessive head will be induced at the lower end of the line during exfiltration testing, the line should be tested in segments. The recommended maximum head that should be induced during exfiltration testing is 20 ft. Segmental testing may be achieved by installing special test tees and risers at points between manholes, so as to keep the maximum head on the pipe below 20 ft. A proper plug may be fashioned by using a rubber ball that can be inflated after it is positioned.

Whenever excessive exfiltration is indicated in a test that involves a manhole, the manhole should be tested before the sewer line is condemned. Poor house connections are also a source of leakage. Some areas require these to be tested before a sewer connection is made.

Joints

When a sewer line is of sufficient size to allow workmen to enter the completed line, the joints may be tested one at a time. Special rigs are available to seal the space around the inside of the joint. Then water is pumped into the sealed space to determine leakage, if any, of the joint. Joint-by-joint testing in large-size sewer lines eliminates water wastage in these large diameters and also is less time consuming than infiltration or exfiltration testing.

Air Test

The air test consists in subjecting the concrete pipe to internal air pressure and noting how quickly the air pressure is dissipated.

SECTION SIX
SOIL-CEMENT

CHAPTER 18
SOIL-CEMENT

SOIL-CEMENT AND ITS USES

Soil-cement is a mixture of pulverized soil and measured amounts of portland cement and water compacted to a high density. As the cement hydrates, the mixture becomes hard and durable. A bituminous wearing course is placed on the finished soil-cement base course to complete the pavement structure.

Soil-cement is used primarily as a base course for roads, streets, and airports. Soil-cement mixtures are also used for road widening, shoulders, parking and storage areas, linings for reservoirs, ditches and canals, and as an erosion-resistant face for earth dams.

There are three general types of cement and soil mixtures: (1) compacted soil-cement, (2) cement-modified soil, and (3) plastic soil-cement.

Compacted soil-cement contains enough cement to harden the soil and enough moisture for both adequate compaction and hydration of the cement. This is by far the most commonly used type of cement-soil mixture.

Cement-modified soil is an unhardened or semihardened mixture of soil and cement. It uses relatively small quantities of portland cement and water. The cement is added to granular or silty clay soils in order to change their physical and chemical characteristics. Cement reduces the plasticity and water-holding capacity of the soil, and this increases its bearing value. The degree of improvement depends on the amount of cement used and the type of the soil. In general, less cement is required than needed to produce a hard soil-cement. Cement-modified soils may be used for subbases and treated subgrades and as trench backfill material. Other terms such as "cement-treated soil," "cement-stabilized soil," and "stabilized aggregate" are sometimes used.

Plastic soil-cement is a mixture of soil and cement that at the time of placing contains enough water to produce a consistency similar to that of a plastering mortar. By comparison, compacted soil-cement is placed with only enough moisture to permit adequate compaction and cement hydration. Plastic soil-cement is used to pave steep, irregular or confined areas where it is difficult, if not impossible, to use road-building compacting equipment.

Because compacted soil-cement is by far the most common of the three types described above, the term *soil-cement* will refer to compacted soil-cement.

Soil-cement has the capacity to bridge local weak subgrades. It does not soften when exposed to extreme wetting and drying or freezing and thawing. Since soil-cement is a cemented, hardened material, it has a far greater load-carrying capacity than other low-first-cost paving materials.

Only three basic materials are needed in soil-cement, and low first cost is achieved mainly by using inexpensive local soil. The soil that makes up the bulk of soil-cement is either in place or obtained nearby. The water is usually hauled only short distances.

The word *soil*, as used in soil-cement, means almost any combination of gravel, sand, silt, and clay and includes such materials as cinders, caliche and chat, as well as many waste materials. Any earth material that is free of organic materials (except shale, bedrock, and top soil) can be considered for soil-cement.

The percentage of portland cement and water to be added and the density to which the mixture must be compacted are determined by laboratory test. Any type of portland cement that complies with requirements of the latest ASTM, AASHO, or federal specifications may be used. Type I (normal) portland cement is the most commonly used.

The water used in soil-cement should be relatively clean and free from harmful amounts of alkalines, acids, or organic matter. Any water fit to drink is usually satisfactory. Sea water has been used satisfactorily when fresh water was unobtainable.

Soils

Practically all soils and soil combinations can be hardened with cement. They do not need to be well graded because stability is obtained primarily through the hydration of cement and not through cohesion or internal friction of the soil. The usability of soils for soil-cement can be approximated before they are tested in the laboratory on the basis of their gradation and their position in the soil profile.

On the basis of gradation, soils for soil-cement construction can be divided into three broad groups.

1. Sandy and gravelly soils with about 10 to 35% silt and clay combined have the most favorable characteristics and generally require the least amount of cement for adequate hardening. Glacier- and water-deposited sands and gravels, crusher-run limestone, caliche, limerock, and almost all granular material work well if they contain 55% or more material less than ¼ in. in size. (Exceptionally well-graded soils may contain up to 65% gravel larger than ¼ in. and still have sufficient fine material for adequate binding.) These soils are readily pulverized and easily mixed, and can be used under a wide range of weather conditions.

2. Sandy soils deficient in fines, such as some beach sands, glacial, and wind blown sands, make good soil-cement although the amount of cement needed for adequate hardening is usually higher than for the soils in group 1 above. Because of poor gradation and the absence of fines, construction equipment may have difficulty in obtaining traction. These soils are likely to require extra care during final compacting and finishing to get a dense smooth surface.

Figure 18–1 Nearby "borrow" materials are sometimes used instead of in-place soils.

3. Silty and clayey soils will make satisfactory soil-cement, but those with a high clay content are harder to pulverize. Generally, the more clayey the soil, the higher the necessary cement content. Construction with these soils is greatly dependent on the weather.

The *soil profile* (Fig. 18-2) is a descending cross section of the earth's surface that shows the different soil horizons or layers. Each soil horizon is generally of a different texture, structure, and color. Color indicates the soil's chemical makeup. In some instances the gradation of the soil is secondary to chemical makeup insofar as the soil's reaction to portland cement is concerned. For instance, a red soil indicates the presence of iron that generally reacts exceptionally well with cement. Conversely, a black farmland soil may react rather poorly with cement because of the presence of organic material. Soil for soil-cement is usually taken from the "B" horizon.

Some sandy materials react poorly with cement, are slow hardening, and require exceptionally high quantities of cement. These are called poorly reacting sands. Two alternative corrective measures may be considered: (1) replacing or blanketing the poorly reacting sand with a normally reacting sand or (2) adding to the sandy soil a small percentage of calcium chloride, a friable silty or clayey soil, or a calcareous material such as limestone screening. Sodium chloride or sea water has also been effective in counteracting this problem.

Soils formed from similar parent material and under the same climatic conditions, topography, drainage, and vegetation are similar, and have the same soil profile, wherever they are found. These soils have been classified into soil series by the United States Department of Agriculture (USDA). Studies have shown that any soil taken from the same soil series and horizon will require the same amount of cement wherever it is found. Most areas have been surveyed and mapped by the USDA. These data are a valuable aid in soil survey work. Before construction begins, the soils that occur on the project should be identified, and representative samples of each should be taken to the laboratory to determine the optimum percentage of cement for the job.

Table 18-A gives the normal range of cement requirements for soils for the various AASHO soil groups. Preliminary estimates of cement requirements may be used for rough cost estimating and then confirmed or revised as laboratory test results become available.

The proper cement content is a requisite for hard, durable soil-cement and, in the discussion that follows, it is assumed that the optimum cement factors have been determined for the soil to be used.

	A_0	Organic debris.
Organic debris lodged on the soil; usually absent on soils developed by grasses.	A_1	A dark-colored horizon containing a relatively high content of organic matter but mixed with mineral matter. Thick in chernozem and very thin in podzol.
Zone of eluviation.	A_2	A light-colored horizon, representing the region of maximum leaching (or reduction) where podzolized* or solodized.** The bleicherde of the podzol. Absent in chernozem,† brown,† sierozem,† and some other soils.
	A_3	Transitional to B but more like A than B. Sometimes absent.
	B_1	Transitional to B but more like B than A. Sometimes absent.
Zone of illuviation. (Exclusive of carbonates or sulfates as in chernozem, brown, and sierozem soils. In such soils this horizon is to be considered as essentially transitional between A and C.)	B_2	A (usually) deeper-colored horizon representing the region of maximum illuviation where podzolized or solodized. The orstein of the podzol and the claypan of the solodized solonetz. In chernozem, brown, and sierozem soils, this region has a definite structural character, frequently prismatic, but does not have much, if any, illuvial materials; it represents a transition between A and C. Frequently absent in the intrazonal soils of the humid regions.
	B_3	Transitional to C.
The parent material.	C	Parent material.
Any stratum underneath the parent material, such as hard rock or a layer of clay or sand, that is not parent material but may have significance to the overlying soil.	D	Underlying stratum.

Figure 18–2 A hypothetical soil profile having all the soil horizons.

* Process of water leaching downward through A and B horizons.
** Process of accumulating surface minerals through leaching upward, produced by evaporation in areas of low rainfall causing moisture movements to be toward the surface.
† Members of great soil groups.

Table 18–A
NORMAL RANGE OF CEMENT
REQUIREMENTS FOR B- AND C-HORIZON SOILS[a]

AASHO Soil Group	Percent by Volume	Percent by Weight
A–1–a	5–7	3–5
A–1–b	7–9	5–8
A–2–4 A–2–5 A–2–6 A–2–7	7–10	5–9
A–3	8–12	7–11
A–4	8–12	7–12
A–5	8–12	8–13
A–6	10–14	9–15
A–7	10–14	10–16

[a] A-horizon soils (topsoils) may contain organic or other material detrimental to cement reaction and may require higher cement factors. For dark gray to gray A-horizon soils, increase the above cement contents 4 percentage points; for black A-horizon soils, 6 percentage points.

Table 18–B
AVERAGE CEMENT REQUIREMENTS OF MISCELLANEOUS MATERIALS

Material	Percent by Volume	Percent by Weight
Caliche	8	7
Chat	8	7
Chert	9	8
Cinders	8	8
Limestone screenings	7	5
Marl	11	11
Red dog	9	8
Scoria containing plus No. 4 material	12	11
Scoria (minus No. 4 material only)	8	7
Shale or disintegrated shale	11	10
Shell soils	8	7
Slag (air-cooled)	9	7
Slag (water-cooled)	10	12

The amount of water added depends on the moisture present in the soil and on the best moisture content of the soil-cement mixture. The most desirable moisture (determined by AASHTO test T134, "Standard Method of Test for Moisture Density Relations of Soil-Cement Mixtures") for soil-cement is a condition of being neither mushy nor dry but containing sufficient moisture to make a firm cast when squeezed in the hand. Water cannot be squeezed out of the material, but dampness will appear on the hand. With a little experience the correct amount of moisture can be approximated in the field by the hand-squeeze test. From a practical standpoint, the highest moisture should be maintained that permits packing and finishing without surface checking, shoving, rutting, or displacement during the compacting and finishing operation.

Sandy soil-cement mixtures require approximately 5 gal of water per square yard for 6-in. compacted thickness. Silty and clayey soil-cement mixtures may require about 7 gal per square yard. Losses by evaporation may amount to 2 or more gal per square yard. Evaporation losses must, of course, be replaced by spraying to assure adequate compaction and complete hydration.

CONSTRUCTION

In soil-cement construction the object is to mix pulverized soil and cement thoroughly in correct proportions with sufficient moisture to allow maximum compaction. Construction methods are simple and follow these steps.

1. Initial preparation
    a. Shape to crown and grade.
    b. If necessary, scarify, pulverize, and prewet the soil
    c. Reshape to crown and grade
2. Processing (mixed-in-place)
    a. Spread portland cement
    b. Mix and apply water
    c. Compact
    d. Finish
    e. Cure

During grading operations, all soft subgrade areas, springs, and frost heave areas should be located and corrected. The grade should be free of stumps and other debris. Most soil-cement is built from soils that require little or no pulverizing. Processing operations (mixing and compacting) after the cement is added must be completed the same working day. Soil, cement, and water can be mixed with any of several types of equipment.

1. Traveling mixing machnie
    a. Windrow type

    b. Flat type (single pass)
    c. Multiple pass rotary
2. Stationary mixing plants. (Soil, cement, and water are mixed at a central plant and hauled to the job site; this is often used when borrow soil is needed.)
    a. Batch type
    b. Continuous flow type

Old Pavement

The materials usually found in old gravel or stone roadways make excellent soil-cement. They are generally friable, mix easily, and require only a minimum amount of cement. Frequently, old bituminous mats can be pulverized and mixed with the existing base course material for processing with cement. Successful soil-cement has been built from old roadways containing as much as 70% old mat material.

Borrow Soil

It is sometimes less expensive to use borrow soil rather than use the soil in place. The existing soils may require a relatively high cement factor or considerable effort to pulverize them. Deposits of friable or granular soils

Figure 18-3 Old, failing, granular-base roads and streets can be salvaged and strengthened by adding relatively small quantities of portland cement. Here a motor grader rips up old pavement to prepare it for soil-cement processing.

that require less cement and very little pulverizing often can be found nearby and can be used to blanket the existing soil or can be combined with it to form soil-cement. Selective use of motor graders is often made to place the most favorable soils in the top of the grade. Comparative cost estimates will dictate the most economical method to use.

PROPERTIES OF SOIL

For construction purposes, soil can be considered to be any earth material encountered in the field except embedded rock and shale. Soil, then, varies from plastic clays to glacial gravel.

Most soils were originally solid rock. Climate and other erosive factors have, over long periods of time, broken the rock into progressively smaller particles. This process can be demonstrated in a laboratory by taking two or three pieces of stone and pulverizing them. First, the sand-size particles can be made, then silt-size, and finally the clay-size particles. However, as rock is naturally reduced into finer particles, chemical changes also take place, therefore the clay that is produced in the laboratory will be different from the clay produced over many years of time.

Any discussion of soil-cement must include some information about soils and the systems used to identify the various kinds of soils. While there are several systems in use today (United States Department of Agriculture, American Association of State Highway Officials, Federal Aviation Agency, and Unified Soil Classification), they all use the same physical characteristics of soil. These characteristics are:

1. Particle size (gradation).
2. Texture or feel.
3. Color.
4. Structure (characteristic shape of broken pieces).
5. Location in profile.
6. Water content.
7. Consistency. (How the soil behaves when moisture content is changed).
8. Density, porosity, void ratio, and degree of saturation.
9. Shrinkage and expansion.

Particle Size

Soil engineers have given names to the groups of particle sizes found in soil.

1. Gravel—from 80 mm down to 2 mm.

2. Sand—2 mm to 0.08 mm.
3. Silt—0.08 mm to 0.005 mm.
4. Clay—smaller than 0.005 mm.

Knowing how much of each of these four are in a sample is one of the major tools used in analyzing and classifying soils. This is found by a process called sieve analysis in which the soil sample is passed through a series of sieves to remove particles larger than 0.08 mm. The amount of material retained on each of the sieves is recorded. The fine fraction passing the No. 200 sieve is then mixed with water and allowed to settle for a given period of time. The larger silt particles will settle first; the clay will settle above the silt, and the finer material (sometimes called colloids) may remain suspended in the water indefinitely.

The distribution (percentage) of each of these grain sizes is called the gradation of the soil. Well-graded soils have some of each size with no sizes missing. Poorly graded soils have a predominance of one size group, and nongrade soils are all one size. Beach sand is nongraded. A mixture of clay and gravel without sand-sized particles would be considered poorly graded.

The fraction of material retained on each sieve is called a soil separate.

Soil Texture

The amount of each soil separate (smaller than $\frac{1}{4}$ in.) contained in a soil mixture will determine its texture or feel. The texture tells as much as possible about a soil in a few words. When the texture is known, estimates can be made on many properties of the soil, such as bearing value, water-holding capacity, liability to frost heave, and adaptability to soil-cement construction.

The primary field tests for texture of a soil are (1) a hand-formed cast of both a dry and a moist sample and (2) pressing a ball of moist soil between the thumb and forefinger.

Soil textures fall into three groups—sand, silt, and clay—with gradations in between. In the first group, *sand,* the individual grains can be seen and felt quite readily. Squeezed in the hand when dry, this soil will fall apart. Squeezed when moist, it will form a cast that will hold its shape when the pressure is released, but will crumble when touched.

Sandy loam consists largely of sand, but with enough silt and clay to give it a small amount of stability. Individual sand grains can be seen and felt readily. Squeezed in the hand when dry, this soil will fall apart when pressure is released. Squeezed when moist, it forms a cast that will not

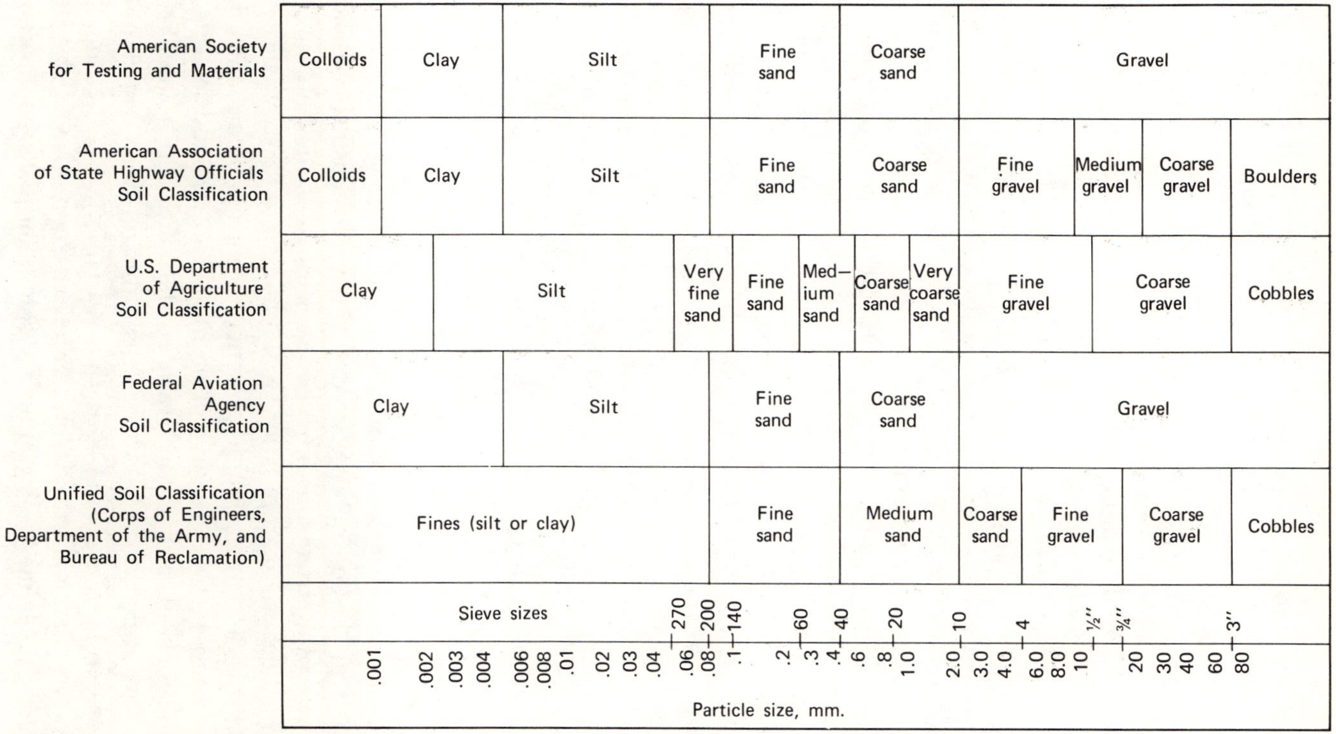

Figure 18-4 Soil-separate size limits of ASTM, AASHO, USDA, FAA, and Corps of Engineers, Department of the Army and USBR.

SOIL-CEMENT

only hold its shape when pressure is released, but will withstand careful handling without breaking. The stability of the moist cast differentiates this soil from sand.

Loam consists of a uniform mixture of sand, silt, and clay. It is easily crumbled when dry: has a slightly gritty yet fairly smooth feel; is slightly plastic; and when squeezed in the hand while dry, it will form a cast that will withstand careful handling. A cast formed of moist loam can be handled freely without breaking.

Silt loam consists of a moderate amount of fine grades of sand, a small amount of clay, and a large quantity of silt particles. Lumps in a dry undisturbed state appear quite cloddy but can be pulverized readily and feel soft and floury when pulverized. When wet, silt loam runs together in puddles. Either dry or moist casts can be handled freely without breaking. When a ball of moist silt loam is pressed between thumb and forefinger, it will not press out into a smooth ribbon but will have a broken appearance.

Clay loam is a fine-textured soil that breaks into clods or lumps that are hard when dry. When a ball of moist soil is pressed between the thumb and finger, it will form a thin ribbon that will break readily, barely sustaining its own weight. The moist soil is plastic and will form a cast that will withstand considerable handling.

Clay is a fine-textured soil that breaks into very hard clods or lumps when dry. It is plastic and usually sticky when wet. When a ball of moist clay is pressed between thumb and finger, it will form a long unbroken ribbon. More accurate determination of texture requires laboratory equipment.

Soil Color

The color of soil varies with its moisture content. It is standard to determine the color of the soil in a moist condition and record the amount of moisture. Color of mottled soils must be determined at their natural moisture contents because manipulation will blend and destroy individual colors. Color indicates the possible presence of certain compounds. Black to dark-brown colors are indicative of organic matter. Reddish soils indicate the presence of unhydrated iron oxides and are generally well drained. Yellow and yellowish brown soils indicate the presence of iron, perhaps hydrated iron, and are poorly drained (otherwise, the iron would be of a different chemical form with a different color—probably redder). Grayish blue, gray, and yellow mottled colors indicate poor drainage. White indicates the presence of considerable silica or lime or, in some cases, aluminum compounds.

Soil Structure

A moist or dry soil mass in its natural state tends to break into pieces of rather definite shape, called structure. Hence, soils are called prismatic, block, granular, crumb, or floury depending on their appearance. Structure is indicative of the drainage characteristics in the area where the soil was formed and is one of the tools used to locate different soils. Soil structure should not be confused with the structural strength characteristics of the soil.

Soil Profile

A vertical cross section of soil layers constitutes the soil profile, which is composed of three major layers designated A, B, and C horizons. The A and B horizons are layers that have been modified by weathering, while the C horizon is unaltered by soil-forming processes and is sometimes called the parent material.

The *A horizon*—the original top layer of soil having the same color and texture throughout its depth—is usually 10 to 12 in. thick but may range from 2 in. to 2 ft. (See Fig. 18-2.)

The *B horizon* is the soil layer just below the A horizon that has about the same color and texture throughout its depth but is usually darker than the A horizon. The B horizon is usually 10 to 12 in. deep, but it may range from 4 in. to 8 ft. In regions of humid or semihumid climate, the B horizon is a zone of accumulation in the sense that colloidal material carried in suspension from overlying horizons has lodged in it. It is often referred to as the subsoil.

The *C horizon* is the soil layer just below the B horizon and has a uniform color and texture throughout its depth. It is quite different from the B horizon. It is of indefinite thickness and may extend below any elevation of interest to the highway engineer. At the beginning of the soil profile development, the C horizon constituted the entire depth. However, weather and other soil-forming processes have changed some of it into the A and B horizons just described. The C horizon may be clay, silt, sand, gravel, stone, or any combination of these. The C horizon is also referred to as parent material or soil material.

The *D horizon* is a layer of solid rock, gravel, sand, or clay underlying the C horizon. It is of indeterminate depth and is in its original condition of formation.

Soil Water

The amount of water or moisture in the soil is always given as a percentage of the oven-dry weight of the soil. Percentage of moisture is determined by weighing the soil with the moisture in it, oven-drying the soil at 230°F to constant weight, and then reweighing. The difference between the moist and dry weights is the moisture lost. This figure divided by the oven-dry weight of the soil and multiplied by 100 gives the percentage of moisture; for example:

$$\text{Percentage Moisture} = \frac{\text{moist wt} - \text{dry wt}}{\text{dry wt}} \times 100.$$

This method of reporting moisture content applies not only to the types of water occurring in soil but also to all other moisture-content tests such as liquid limit, shrinkage limit, plastic limit, optimum moisture, and so on. The following example shows the determination of moisture content using the above formula.

236 g—weight of moist sample (as it comes from the ground)

176 g—oven-dried weight

60 g—weight of moisture in sample (236 − 176)

34%—moisture content found by dividing the weight of moisture by the dry weight and multiplying by 100 to find the percentage of moisture (60/176 × 100)

The type and amount of water found in a soil will have a great influence on how well a soil will perform as a construction material.

The soil engineer recognizes three different types of soil moisture: (1) gravitational water, that is, water free to move downward from the force of gravity, (2) capillary water, that is, water held in a soil by capillaries or pores—this is free water that can be removed from the soil only after the water table is lowered or when evaporation takes place at a rate faster than the rate of capillary flow, and (3) hygroscopic water, that is, water retained by the soil after gravitational and capillary moisture are removed. Hygroscopic water is a film of moisture held by each grain of soil and has both a physical and chemical affinity for the soil grain. It is also spoken of as the air-dry moisture content. This film is in equilibrium with the moisture content of the air and increases or decreases as the relative humidity goes up or down. The hygroscopic moisture content of a soil also varies with grain size. As the grain size decreases, the surface area, and therefore the hygroscopic moisture content, increases.

Soil-Water Consistency

Tests have been devised to determine the moisture content of a soil when it changes from one major physical condition to another. These tests show the water-holding capacity of soil under various controlled conditions and are conducted on material that passes the No. 40 sieve (a sieve having 1600 openings per square inch). They have been used as key factors in classifying soils for structural purposes. The Bureau of Public Roads has given the tests worldwide recognition in its A-1 to A-7 soil group classification, now referred to as the AASHTO Soil Classification. The two major physical conditions used for this are the liquid limit and the plastic limit.

```
← MOISTURE CONTENT OF SOIL, DRY TO SATURATED →
     SEMISOLID        PLASTIC        LIQUID
                  PL            LL
                  ←———— PI ————→
```

Liquid Limit. The liquid limit (LL) is the moisture content at which a soil passes from a plastic to a liquid state. The test used to determine the liquid limit is ASTM D 423 or AASHTO T89. The liquid limit testing apparatus consists of a small brass dish, hinged so that it will drop 1 cm when a crank is turned. The material passing the No. 40 sieve is used in the LL tests.

About 250 g of this minus 40 material are mixed with water and pressed into the brass dish. The cake is divided into halves with a special grooving tool. The crank is turned, dropping the dish once for each revolution, until the two halves flow together over a distance of ½ in., and the number of blows is recorded. A portion of the soil in the dish (usually about ¼) is weighed immediately and then oven-dried to determine the moisture content. More water is added to the sample and the test is repeated. A third test is run with still more water to give a final set of figures.

These three sets of figures are plotted on graph paper and the moisture content at the 25 blow line is the liquid limit.

High liquid limits indicate soils of high clay content and low load-carrying capacity. Sandy soils have low liquid limits on the order of 20. Silts and clays have liquid limits as high as 80 or 100. Most clays in the United States have liquid limits between 40 and 60.

Plastic Limit. The plastic limit (PL) of soils is the moisture content at which a soil changes from a semisolid to a plastic state and is governed by clay content. This change takes place when the soil contains just enough water so that it can be rolled into ⅛-in.-diameter thread without breaking. Silt and sandy soils that cannot be rolled into threads at any moisture content are termed nonplastic. Soils having plastic limits contain silt and clay, and the moisture content of such soils has a direct bearing on their load-carrying capacity. Soils will generally reach or exceed their plastic limit moisture content at some season of the year, even in semiarid climates; therefore, the bearing capacity of the subgrade and the pavement design requirements should be based on this low load-carrying capacity (high moisture content) condition.

Plasticity Index. The plasticity index (PI) is defined as the numerical difference between the liquid limit and the plastic limit: $PI = LL - PL$. The plasticity index gives the range in moisture contents at which a soil is in a plastic condition. A low plasticity index of five indicates that a small change in moisture content will change the soil from a semisolid to a liquid condition. This soil is very sensitive to moisture unless the silt and clay content combined is less than 20%. A high plasticity index of 20 shows that considerable water can be added before the soil becomes liquid. When the liquid limit or plastic limit cannot be determined or when the plastic limit is equal to or higher than the liquid limit, the plasticity index is reported as nonplastic (NP).

Other tests have also been devised to determine the moisture content of soils in some particular condition. These are the Field Moisture Equivalent (FME) test (AASHTO T93), where the minimum moisture content is determined by a smooth surface that will absorb no more water in 30 seconds when the water is added in individual drops; and the Centrifuge Moisture Equivalent (CME) test (AASHTO T94), where a saturated sample is centrifuged for 1 hour under a force of 1000 Gs and reweighed. Neither of these tests, however, is in common use.

Density, Porosity, Void Ratio, and Degree of Saturation

A soil mass is a porous material containing solid particles interspersed with pores or voids. These voids may be filled with air, water, or both. There are several terms used to define the relative amount of soil, air, and water in a solid mass; they are density, porosity, void ratio, and degree of saturation.

The density is the weight of a unit volume of the soil. It may be either a wet density, including both soil and water, or a dry density, with soil only, and is usually measured in pounds per cubic foot.

Porosity is the ratio of the volume of voids to the total volume of the mass regardless of the amount of air or water contained in the voids. Porosity is usually expressed as a percentage.

Void ratio is the ratio of the volume of the voids to the volume of the soil particles. The porosity and void ratio of a soil depend upon the degree of compaction or consolidation. Therefore, the porosity and void ratio will vary for a particular soil under different conditions, and they can be used to judge relative stability and load-carrying capacity, which increase as porosity and void ratio decrease.

Degree of saturation is the ratio of the volume of water to the volume of voids. It is usually expressed as a percentage.

Shrinkage and Expansion in Soil

Soils containing clay may exhibit serious volume changes (shrinkage upon drying and expansion upon wetting) as the moisture content varies. The shrinkage/expansion characteristics are of considerable importance in selecting soil to be used under pavements. Several terms are used to describe and measure the amount of shrinkage and expansion that accompany wetting and drying. Details of these are given in ASTM D 427 and in AASHTO T92 and T116.

Shrinkage Limit. As moisture leaves the soil, the soil shrinks (decreases in volume in direct proportion to the loss in moisture) until a condition of equilibrium is reached where shrinkage stops.

Linear Shrinkage. Linear shrinkage is the decrease in one dimension expressed as a percentage of the original dimension of the soil mass when the moisture content is reduced from a given value to the shrinkage limit. This information is used to give some indication of the amount of cracking that will take place in such soils and subgrades as a result of drying.

Expansion. There are several test procedures for evaluating the expansive qualities of soils. These include a measure of the pressure a soil will exert when it expands by soaking up water. Usually, this is done by compacting a specimen to a predetermined density at proper moisture content and making a supply of water available. The amount of expansion or pressure is recorded for a given period of soaking.

ASTM D2488 is a descriptive manual of soils that provides a great deal of information based on a visual examination of soils. The procedures are too detailed to include here, but for any students planning to work with soils or soil-cement the manual would be a worthwhile purchase.

ENGINEERING PROPERTIES OF SOILS

The basic characteristics of soils—internal friction, cohesion, compressibility, elasticity, capillarity, and permeability—combine to produce the mechanical and hydraulic properties that determine the suitability of soils for construction uses. In most applications the strength of the soil, that is, its load-carrying capacity and resistance to movement or consolidation, is most important. Depending on the proposed use, other properties, such as volume change characteristics or drainage, may be considered in evaluating suitability. Engineering properties are influenced most by the soil types, gradation, and composition. Thus it is possible to know, at least in a general way, whether a soil will be strong or weak, free draining or impermeable if we know its gradation, texture, or classification grouping. These properties are greatly affected by the degree of compaction and the moisture content. Therefore, a discussion of engineering properties and corresponding test methods will be preceded by information on soil compaction.

Soil Compaction

The term compaction refers to the practice of artificially increasing the unit weight (density) of a soil mass by rolling, tamping, vibrating, or pounding. There is no other single treatment that produces so marked a change in physical properties at so low a cost as does controlled compaction. The density of a soil is measured in terms of its unit weight and is usually expressed as pounds of wet or dry soil per cubic foot, designated as wet density or dry density. The three most important factors influencing density obtained by compaction are (1) the moisture content of the soil, (2) the nature of the soil, that is, its gradation and physical properties, and (3) the type and amount of compactive effort.

Moisture/Density Relationships. A basic principle in soil analysis is that for a given force of compaction and a given moisture content, a soil will have a corresponding density. When a soil's structural properties are being determined, its moisture content and density must be defined and controlled to permit accurate evaluation of the soil in that particular condition. The next step is to determine the most unfavorable moisture and density condition that will exist in the field and conduct tests simulating these conditions. These facts about the meaning of moisture/density relationships are learned by comparing the moisture/density tests for soils of all types. Tests on a soil and corresponding soil-cement mixtures show that the same principles apply to both soils and soil-cement. The moisture/density relationship is so important that the moisture content and density of a soil must be known when a particular load or bearing test is being conducted. Otherwise, it is difficult, if not impossible, to interpret the test results.

Moisture/Density Test. Tests to determine the density to which a soil can be compacted with various moisture contents have come into wide use. The greatest density obtained is termed *maximum density*, and the corresponding moisture content is termed *optimum moisture*. A high maximum density will range from 125 to 145 lb/cu ft (oven-dry weight), and a low maximum density will range downward from about 100 to about 85 lb/cu ft (oven-dry weight). A low optimum moisture (8%) coincides with a high maximum density. The 8% is the weight of water in a sample compared to the oven-dry weight of the sample.

The maximum density of a soil gives approximate information about its gradation; the optimum moisture gives approximate information on the silt and clay content. The shape of the moisture curve may vary from a sharply peaked curve to a flat parabolic curve to one sloping irregularly downward as the moisture content increases. The curve gives additional information about the influence of moisture on the load-supporting characteristics of the soil. For example, a flat curve indicates that a soil will have the same load-supporting capacity over a wide range of moisture contents.

Field Density Determination. The performance of pavement structures depends, to a great extent, upon proper uniform compaction of the subgrade and pavement components. Road-building agencies usually control compaction by specifying minimum requirements based on soil density, compactive effort, or a combination of the two. In most cases the maximum density and optimum moisture content as determined by AASHTO T99 or T180 form the basis for these specifications. The minimum acceptable density is usually specified as a percentage of maximum density. For example, 95% of maximum density at a moisture content of 90 to 110% of optimum moisture content as determined by AASHTO T99 may be specified.

The basic procedure used to determine the density of a soil consists of removing a sample of the compacted soil and determining its weight, its moisture content, and the volume of the cavity previously occupied by the soil. The soil sample is removed from an area approximately 5 in. in diameter and extending the full depth of the layer being tested. The cavity should be approximately cylindrical in shape. The moist soil is weighed and its moisture content determined by the formula:

$$\% H_2O = \frac{\text{wet wt} - \text{dry wt}}{\text{dry wt}} \times 100$$

The volume of the cavity is determined by accurately measuring the amount of material of a known unit weight required to fill the hole. The most common method uses the sand-density cone. However, the water balloon method and the oil method have also been used for this purpose. These methods are described in AASHTO T147 and T181.

In-place densities can also be determined by means of undisturbed samples. There are several methods involving nuclear devices that are used. These nuclear devices are gaining in popularity.

Field Determination of Soil-Bearing Value
Soil-bearing values are determined in the field for soils under buildings, bridges, dams, and pavements. Various direct loading procedures are used. In essence, these field tests are simply a measurement of the amount of force required to push a plate (usually 30 in. or more in diameter) into the soil. The information obtained will give a unit load and a time-deformation curve, which in turn will give some idea of how much weight can be put on the soil. Bearing value field tests of subgrade soils for planned or existing pavements involve certain standard procedures that are outlined in ASTM D 1195 and D 1196. It is important when making these tests to know the moisture content of the soil. For instance, a clay soil having a moisture content below the shrinkage limit may be almost as hard as kiln-dried brick and will have a very high supporting power. Yet when it has a moisture content near the liquid limit, it is almost a fluid and will have little or no supporting power. Extensive tests of this nature have been conducted by the Bureau of Public Roads, the Corps of Engineers, many state highway departments, and others.

Estimating the Bearing Value of Soils. Much study and experience are required to arrive at a final figure for the bearing value of a subgrade soil for use in pavement design at a particular location. However, a general idea of a soil's bearing value can be obtained from published data and from a general correlation of soil classification with bearing values. Thus, after laboratory tests are available to permit a close estimate to be made as to where a soil will fall in a specific classification, it is possible to estimate the bearing value providing considerations of drainage, rainfall, and other factors that influence subgrade performance are taken into account.

Elasticity and Compressibility

Elasticity and compressibility are the properties that cause soil either to rebound or remain compressed after compaction. Most soils are compressible and do not present special design problems. Elastic soils—those that rebound after loading—are not considered desirable for pavement subgrades. When a measurement of elasticity is required, it is determined by special tests that simulate moisture and loading conditions expected in the field.

The consolidation test is used to estimate the settlement that may take place in soils under large structures, such as buildings, bridge piers, and very high earth embankments. The test consists of a cylinder filled with soil that is placed between two porous stones. The soil is consolidated by a piston placed on the upper porous stone. Any moisture forced from the specimen may escape through the stones because they are porous. The piston is mounted on a lever arm with weights. Gages are used to measure the amount of consolidation.

Permeability and Capillarity

Permeability is that property of a soil allowing it to transmit water. It depends on the size and number of continuous pores. Determined by test on a representative sample of soil, permeability is expressed as the coefficient of permeability. It equals the apparent velocity of water flow under a hydraulic gradient of 1, which exists when the pressure or the height of the water head on the specimen equals the depth of the specimen. The coefficient of permeability (k) is used to determine the quantity of water that will seep through a particular cross section of soil, in a given time, under a given amount of pressure.

Capillarity is the action by which a liquid rises in a channel in any upward direction from a supply of free water. The number and the size of the channels in the soil will determine its capillarity. This soil property is measured as the distance (ranging from 0 to 30 ft) that moisture will rise above the plane of the water table. Moisture in clay soils may be raised by capillary action for vertical distances as great as 30 ft. The capillary rise in gravel or coarse sand varies from 0 to a maximum of a few inches. Complete soil saturation seldom occurs at the upper limits of the rise of capillary moisture. Capillarity of a soil and the elevation of the water table under the pavement determine whether the subgrade will become saturated. The bearing value of the soil is dependent on this saturation capacity. Subgrade saturation by capillary action also determines whether frost heaves and similar occurrences need to be considered in design requirements for subgrade and pavement.

CHAPTER 19
USING SOIL-CEMENT

ROAD CONSTRUCTION

Soil-cement road construction involves two general operations: preparation and processing. Both can be divided into definite construction steps. Regardless of the equipment and methods used, it is essential to have an adequately compacted, thorough mixture of pulverized soil that contains the proper amount of cement and moisture in order to get strong, durable soil-cement.

Preparation
Before construction starts, the crown and grade of the roadway should be checked and any fine grading needed should be done. The grade at start of construction will determine the final grade to a major extent. If borrow soil is to be used, the subgrade should be compacted and shaped to proper crown and grade before the borrow is placed.

Guide stakes should be set to control the width of treatment and to guide the operators during construction.

Arrangements should be made to receive, handle, and spread cement and water efficiently. The number of cement and water trucks that will be required depends on length of haul, condition of haul roads, and anticipated rate of production. The extent of the different soils on the roadway

Figure 19-1 Since there is little displacement of soil during processing, the roadway should be at proper crown and grade at the start of soil-cementing.

and their corresponding cement requirements will be established by the project engineer prior to construction.

Scarifying, Pulverizing, and Prewetting. Soils that are difficult to pulverize when dry can often be broken down readily if water is added and allowed to soak in, whereas sticky soils can be pulverized more easily when they are dried out a little. Proper equipment will help to reduce pulverization work to a minimum. Pulverizers, rotary mixers, disk harrows, and scarifying plows are commonly used.

Soil should be pulverized or broken up so that there are no soil lumps larger than 1 in. and no more than 20% of the total soil sample larger than ¼ in.

Prewetting by adding moisture before cement is applied often helps in scarifying and pulverizing the soil and often saves time during actual processing operations.

Handling and Spreading Portland Cement

Portland cement is spread by any of several means to the specifications of the designer. This can be done with mechanical spreaders or it can be done with bag cement by hand.

Spreaders that are pulled by a dump truck filled with portland cement are often used to distribute a uniform layer of cement over the area to be

Figure 19-2 On most jobs, bulk cement is placed on the roadway with a mechanical spreader. The spreader may be adjusted to regulate the amount of cement being spread.

worked. This layer is then thoroughly mixed with the soil and water and compacted.

The same sort of arrangement is used with windrows, except that the cement is placed in measured amounts in the top of the windrow.

With hand spreading, some simple but exact method for properly spotting the bags is necessary. Bags can be spotted correctly by flags or markers fastened to chains at proper intervals to mark transverse and longitudinal rows. The spacing of the bags should be practically square and such that the proper percentage of cement will be added. When the bags are opened, cement is dumped so that it forms a fairly uniform transverse windrow across the area being processed. A drag should make at least two round trips over the area to spread the cement uniformly. A spike-tooth harrow, a nail drag, a length of chain-link fencing, or the tailgate of a rotary mixer can be used for leveling the spread of cement.

Mixing
Soils that contain excessive amounts of moisture will not mix readily with cement. Sandy soils can be mixed with a moisture content at optimum or slightly above, while clayey soils should have a moisture content

Figure 19-3 On small projects, bags of cement may be used. The bags are positioned evenly on the roadway and opened. A spike-tooth harrow or similar equipment spreads cement evenly.

slightly below optimum for efficient mixing. If the soil is excessively wet, it should be aerated and dried before cement is applied.

After the proper amount of cement has been distributed on the area to be treated, it is thoroughly mixed with the soil in any of several ways.

Flat-Type Mixers (Multiple-Shaft). Flat-type mixers have a series of rotors that (1) pulverize the soil, (2) dry-mix the soil and cement, and (3) mix water into the dry mix, all in one pass. Since flat-type traveling mixing machines have a high-speed pulverizing rotor, preliminary pulverization usually is unnecessary. The only preparation required is shaping the soil to approximate crown and grade. Occasionally, an old roadbed may be extremely hard and dense; however, in such cases prewetting and scarifying will facilitate processing.

Processing is done in lanes 350 to 500 ft long with a width equal to that of the machine. Cement is spread over the soil ahead of the mixing machine. Either bulk cement or bag cement can be used. Cement spreading should be completed in the first working lane and be under way in the second lane before mixing operations are started. This assures a full-width cement spread without a gap between lanes; also, cement-spreading equipment is out of the way of mixing equipment.

Figure 9-4 This single-pass traveling machine is shown mixing soil-cement shoulders. Dry-mixing, addition of water, and wet-mixing are accomplished in one pass.

The machine mixes the soil and cement in place. The first rotors of the mixing machine pulverize the soil, and dry-mix soil and cement. Water is measured through a meter and injected into the mix through a spray bar in the mixing chamber. The remaining rotors mix the soil, cement, and water, and the mixed material is left flat on the roadway ready for compaction. Water is supplied to the mixer by tank trucks. A small exchange water tank on the machine permits continuous operation while water trucks are being changed.

After initial compaction, the roadway surface is shaped and final compaction is completed full width in one operation. Shaping consists of lightly shaving off any small ridges and filling depressions left by the finishing equipment. If compaction planes occur, they can be removed with a nail drag, harrow, or cultivator.

Rotary Mixers (Single-Shaft). Soil-cement construction with multiple-pass mixers is slightly different from the preceding example. Because they have only one set of blades, multiple-pass mixers will pulverize the soil in one pass, dry-mix the soil and cement in the second pass, then mix in the water on the third pass. The basic principles and objectives are the same, however.

Shaping the roadway, scarification, and pulverization are the first steps of preparation. Since most rotary mixers were not designed to scarify, usually the soils must be loosened with a scarifier. Prewetting the soil during scarification and pulverization is common practice.

Applying water at this stage of construction saves time during actual processing operations because most of the required water has already been added to the soil when cement is spread. In very granular soils, prewetting prevents cement from sifting to the bottom of the mix by causing it to adhere more readily to the sand and gravel particles. Mixing of soil and cement is easier if the moisture content of the raw soil is two or three percentage points below optimum. Sandy soils can be mixed even if the

Figure 19–5 Soil-cement may be mixed with any of several machines. Shown here is a multiple-pass single-shaft mixer.

moisture content is one or two percentage points above optimum. Moisture should be applied uniformly during prewetting. By mixing it into the soil immediately, evaporation losses are reduced. Because of the hazard of night rains, some contractors prefer to do the prewetting in the early morning. After scarifying, pulverizing, and prewetting, the loose, moist soil is shaped to grade and crown.

Cement is spread through a mechanical cement spreader or in bags. Cement should not be applied when there are puddles of water nor should it be applied to soils that are extremely wet.

Occasionally, the prewet soil becomes compacted by cement-spreading equipment. In such cases mixing may be facilitated by loosening the soil again after cement is spread, usually with a scarifier or a cultivator. Care should be taken so the scarifier teeth are extended, enabling cement to flow through the teeth and not be carried forward or displaced by the scarifier frame.

Then cement and soil are mixed. The object at this stage is to distribute the cement evenly throughout the soil mass. Only enough mixing is required to prevent cement balls from forming when water is applied. Complete and thorough mixing is not necessary until the water is added.

Next, sufficient water is applied in increments to bring the mixture to optimum. Each increment is mixed with the soil and cement. Water is usually applied directly to the mixture by water distributors or by a pump through a spray bar mounted in front of the rotor on the mixer and supplied by a water truck. After the last increment of water has been applied, mixing is continued until the soil, cement, and water are thoroughly mixed throughout the full depth and width of treatment. The material is then ready for compacting and finishing.

Windrow mixers are used almost exclusively with borrow soil. The soil is spread on the area to be covered with soil-cement in a neat row about 6 ft wide and 4 ft high. The cement is placed in the top of the windrow by special spreaders. This windrow is then picked up, mixed, and replaced in a continuous operation.

Check on Cement Spread. A check on the accuracy of the cement spread is advisable to insure that the proper quantity is actually being applied. When bulk cement is being used, this is done in two ways:

1. Place a canvas, usually 1 sq yd in area, on the roadway ahead of the cement spreader. After the spreader has passed, pick up the canvas carefully and weigh the cement collected on it.

Figure 19–6 Soil-cement is mixed from a windrow with this one-pass pugmill-type mixer. The mixed windrow is then leveled and compacted.

2. Check the area over which a known weight of a truckload of cement is spread.

Given: Required cement content, by volume, 8%. Depth of compacted soil-cement, 6 in. Width of spread, 8 ft. Weight of cement on truck, 7520 lb

Required: Linear distance that one truckload of cement should travel to spread the required amount of cement (8% in this example).

Procedure: Find 8% in the left-hand column of the above table. Proceed horizontally to the 6-in. compacted depth column. Find the weight of cement per sq yd (33.8 lb) and divide this by 9. This will give you the weight per sq ft (3.76 lb). Multiply this weight by the width of the spread (8 ft) to get the weight per linear ft required—30.0 lb (3.76×8).

Answer: The required distance of travel for this truckload to obtain the specified cement spread equals the total weight of cement on the truck divided by the pounds per linear foot required: 7520 = 250.7 ft.

Bulk cement that is spread mechanically in the top of a windrow of soil may be checked by pushing two triangular metal plates into the top of the windrow exactly 1 ft apart. All cement between the plates is carefully scraped out and weighed. This method gives directly the quantity of cement spread per linear foot.

Table 19–A
CEMENT SPREAD REQUIREMENTS PER SQUARE YARD

Percent Cement by Volume	Compacted depth (in.)							
	5		6		7		8	
	Pound	Bag	Pound	Bag	Pound	Bag	Pound	Bag
4	14.1	0.15	16.9	0.18	19.75	0.211	22.55	0.24
5	17.6	0.188	21.2	0.225	24.8	0.263	28.2	0.30
6	21.4	0.225	25.4	0.27	29.7	0.315	33.9	0.36
7	24.7	0.263	29.6	0.315	34.6	0.368	39.5	0.42
8	28.2	0.30	33.8	0.36	39.5	0.421	45.1	0.48
9	31.8	0.338	38.1	0.405	44.6	0.474	50.8	0.54
10	35.2	0.375	42.3	0.45	49.5	0.527	56.4	0.60
11	38.8	0.413	46.5	0.495	54.4	0.579	62.0	0.66
12	42.3	0.45	50.8	0.54	59.4	0.632	67.7	0.72
13	46.8	0.488	55.0	0.585	64.4	0.684	73.3	0.78
14	49.4	0.525	59.2	0.63	69.3	0.737	78.9	0.84
15	53.0	0.563	63.5	0.675	74.3	0.79	84.6	0.90
16	56.4	0.60	67.7	0.72	79.2	0.842	90.2	0.96

Generally, checking the cement on the square-yard basis is used to adjust the spreader, while a final check is made by figuring the quantity of cement spread per unit area from the area covered by a truckload of cement. Bag cement is checked by counting the number of bags actually placed per 100-ft section. It is imperative to keep a continuous check on cement-spreading operations.

Correct Amount of Moisture. Proper moisture content is one of the three basic control factors of soil-cement construction. The approximate optimum moisture content as determined from the laboratory tests is used in starting processing operations. A moisture-density test made in the field on a representative sample taken at the conclusion of moist-mixing determines the optimum moisture and maximum density for field control. This procedure takes into consideration two factors.

1. Any changes in the optimum moisture and maximum density resulting from minor changes in soil.
2. Any changes in the optimum moisture and maximum density resulting from lengthy mixing.

The optimum moisture and maximum density, as determined in the laboratory, may be used to govern field control if the mixing cycle is short (less than 30 min.), if compaction starts immediately, and if the soil is identical to that used in the tests.

In order to estimate water requirements, representative moisture samples are obtained from the raw soil prior to mixing, or from the dry soil-cement mix before water is applied.

The samples are weighed, and then dried and reweighed. The moisture content is computed as follows:

$$\text{Percent moisture} = \frac{\text{wet weight} - \text{dry weight}}{\text{dry weight}} \times 100$$

The approximate percentage of water required is equal to the difference between the optimum moisture content and the moisture content of the dry soil-cement mix, as determined above. Approximately 2% additional moisture must be added to compensate for the dry cement added to the soil if the moisture test was made on the raw soil prior to the addition of cement.

With a little experience, the moisture content of a soil-cement mixture can be estimated and judged closely by eye and feel. For instance, a mixture near or at optimum moisture content is just moist enough to dampen the hands when packed in a tight cast. Mixtures above optimum will leave excess water on the hands, whereas mixtures below optimum will crumble and cannot be cast. If the mixture is near optimum, it is possible to break the cast into two pieces with very little crumbling.

The hand-squeeze test is not a replacement for the field-laboratory, moisture-content test, but it does reduce the number of these tests required during construction. The moisture determination test validates what has been determined by visual inspection and the hand-squeeze test.

At the start of compaction the moisture content of the soil-cement mixture must be at optimum moisture or slightly above. A final check of moisture is made at this time. Proper moisture is important because it assists in compaction and is necessary for hydration of the cement. It is more practical to have a slight excess of moisture than a deficiency when compaction begins. During compaction and finishing, the surface may become dry, as evidenced by a graying of the surface. Should evaporation losses be noticeable, very small applications of water are made to bring the moisture content to slightly above optimum. A water distributor is used to make these fog applications of water. Proper surface moisture is evidenced by a smooth, moist, tightly knit surface free of checks, cracks, or ridges.

Uniformity of Mix: Depth and Width of Treatment. A thorough mixture of pulverized soil, cement, and water must be obtained to make quality soil-cement. The uniformity of mix is checked by digging trenches or a series of holes at regular intervals for the full depth of treatment across the area being processed and then inspecting the color of the exposed material. When the mixture is of uniform color from top to bottom, a satisfactory mix has been attained. Material that has a streaked appearance is not thoroughly mixed.

Depth of mixing is usually checked at the same time. Usually 8 to 9 in. of loose mix will produce about 6 in. of compacted thickness. This varies slightly with the type of soil being processed. Routine depth checks should be made during mixing operations to assure that the specified thickness is attained.

Line stakes set 1 ft outside the desired roadway edge are used to control width of processing.

Compaction

After the mixture has been moistened to optimum and thoroughly mixed, it should be compacted to maximum density and finished immediately. Moisture lost by evaporation during compaction, as indicated by the

Figure 19-7 A sheepsfoot roller is often used for initial compaction of most soil-cement bases. Other types of rollers may be used, depending on the type of soil.

graying of the surface, should be replaced with light applications of water.

There are numerous types of compaction equipment. Those most commonly used in soil-cement construction are sheepsfoot, pneumatic-tired, or steel-wheel rollers. Many types of soil compaction equipment have recently been developed; for example, grid-type rollers, segmented steel-wheel rollers, plate vibratory compactors, vibratory steel-wheel rollers, vibratory pneumatic-tired rollers, heavy pneumatic-tired rollers, and tampers. Plate vibrators, and grid and segmented rollers have been used to satisfactorily compact soil-cement made of nonplastic granular soils.

Sheepsfoot rollers are generally used to compact all but the most granular soils. To obtain adequate compaction it is best to operate the rollers ballasted to give high unit pressures. The general rule is to use the heaviest roller that will not overstress the soil mixture and that will compact to the desired density in the least number of passes. Friable silty and clayey sandy soils will compact best with rollers having unit pressures of 75 to 125 psi. Clayey sands, lean clays, and silts that have low plasticity will pack best with 100- to 200-psi rollers. Medium to heavy clays and gravelly soils require higher unit pressures—150 to 300 psi. An 8-in. compacted thickness is about the maximum that can be compacted satisfactorily in one lift with most rollers.

Pneumatic-tired rollers are used to compact very sandy soils containing little or no binder material—such as dune or blow sand. Sand and gravel having very little or no plasticity can also be compacted satisfactorily with pneumatic-tired rollers. These rollers are usually heavily loaded and pulled with a crawler tractor equipped with street plates. A second and lighter pneumatic-tired roller pulled by a wheeled tractor is used for final rolling on most soils. Cohesionless sand may also be compacted with a large track-type tractor with street plates. Compaction is obtained by the weight and vibration of the tractor. A 6-in. compacted depth is about the maximum that can be compacted in one lift with these methods.

Twelve-ton, three-wheel steel rollers are commonly used in some areas to compact granular soils containing little or no binder material. Gravelly soils that contain less than 20% material passing the No. 200 sieve and have a low plasticity index (PI) are best suited for compaction with these rollers. Compacted depths of 6 or 7 in. are about the maximum thickness that can be compacted satisfactorily with this equipment.

When sheepsfoot rollers are used for initial compaction, the mixed material must be in a loose condition so that the compaction feet will penetrate to the bottom and gradually "walk out." In some instances, penetration may not be obtained, and a cultivator or rotary mixer may be

Figure 19–8 Pneumatic-tire rollers are used for final finishing and initial compaction of coarse sand-gravel and sandy soil-cement mixtures containing little or no binder or plasticity.

used to loosen the mix during sheepsfooting, thus allowing the feet to penetrate. Such a procedure will also increase densities.

Occasionally, during compaction and finishing, a localized area may yield under the compaction equipment. This may be due to one or more of the following causes: (1) the soil-cement mix has much more than optimum moisture, (2) the subsoil may be wet and unstable, or (3) the roller may be too heavy for the soil. If the soil-cement mix is too damp, it should be aerated with a cultivator, rotary mixer, or motor grader. After it has been dried to near optimum moisture it is then compacted. If the subsoil is unstable, it should be removed and replaced or perhaps stabilized with portland cement.

Some silty soils may become overpacked. This is very unusual, but the roller may be too heavy and have too high a unit pressure for the soil being compacted. Overpacked soil will show surface cracks much like "wrinkled skin."

For best results compaction should start immediately after the soil, cement, and water have been mixed. Densities are obtained more readily, there is less water evaporation, and production is increased. Most specifications require that soil-cement be compacted to within 5 lb (some specify 95%) of the maximum dry density as determined on a representative field sample taken from the moist mix.

Degree of Compaction and Final Depth Check. The density of a section built the first day should be determined at several locations after final rolling is completed. Comparison of these densities with the results of the field moisture-density test will indicate any adjustments in compaction procedures that may be required. Generally, specifications require that the density obtained shall not be less than 5 lb below the maximum density as determined by the field moisture-density test. After compaction procedures have been corrected, only routine daily density checks are required.

A density test is made by drilling or digging a 5-in.-diameter hole the full depth of processing. All material removed is carefully salvaged. The wet weight, moisture content, and oven-dry weight of this material are determined. The volume of the excavated hole can be determined by various methods that are basically the same. The hole is filled with a material of known density, and the volume is calculated from the weight of the material used to fill the hole. The density is determined by dividing the dry weight of soil-cement removed from the hole by the volume of the hole.

While a density determination is being made, the depth of processing is measured. Usually the bottom of treatment is quite apparent because of the difference in color between the subgrade and the wet soil-cement mixture. However, sometimes it is difficult to distinguish the bottom of treatment by color. In this event, water is poured into the density hole and allowed to stand. The subgrade will be softened by the soaking, while the full depth of treatment will remain firm. Then the bottom of treatment can be determined by probing with a pointed instrument.

The most common methods used for the volume determination of the density hole are:

1. Sand-cone method.
2. Balloon method.
3. Oil method.

If the test is performed with care, these and other methods can be used satisfactorily to determine the degree of compaction obtained. Various types of apparatus are available for all these methods. Oven-dry densities are used for comparison with the degree of compaction obtained. However, a rough check on the degree of compaction can be made quickly by comparing wet densities. The sand-cone method is one of the most common and is a standard method of test. It has an AASHTO designation of T147.

The nuclear method of density determination is becoming increasingly popular because (1) the necessary equipment is compact and portable, (2) it can be operated reliably by a technician, and (3) it is a nondestructive method that requires no digging. These advantages may well cause nuclear testing to become the standard for density measurement in the future.

Finishing

There are several acceptable methods of finishing soil-cement. The exact procedure depends on the equipment, job conditions, and soil characteristics. Regardless of the method used, the fundamental requirements of adequate compaction at optimum moisture and of removal of all surface compaction planes must be made to produce a high-quality surface. Surface compaction planes are smooth areas left near the surface by wheels of heavy equipment. A thin surface layer of compacted soil-cement may not adhere properly to these areas and in time may loosen and spall. For good bond the area must be rough and damp. A scratcher, such as a weeder, nail drag, or spike-tooth harrow is used to remove potential surface compaction planes. The surface should be smooth, dense, and free of ridges or cracks. Following are outlines of two of the methods in use today. Either will produce proper compaction and a satisfactory surface finish if the fundamentals are adhered to.

Methods of Finishing.

1. Here is the finishing procedure for most soil-cement mixtures previously compacted with a sheepsfoot roller:
   a. Remove compaction planes with weeder, nail drag, or spike-tooth harrow while shaping with motor grader. (Apply water lightly as needed. Broom dragging sometimes used to level ridges.)
   b. Roll with pneumatic-tired roller.
   c. Shave with motor grader. (Broom drag to level ridges.)
   d. Roll with pneumatic-tired roller. (Apply water lightly as needed. Tandem steel-wheel roller is sometimes used prior to final rolling with pneumatic-tired roller.)
2. Here is the finishing procedure for very sandy mixtures containing little or no fines (e.g., cohesionless dune or blow sand containing 0 to 10% minus No. 200 sieve material) and compacted with heavy pneumatic-tired roller pulled by track-type tractor with street plates:
   a. Remove compaction planes with weeder, nail drag, or spike-tooth harrow while shaping with motor grader. (Apply water lightly as needed.)

    b. Roll with pneumatic-tired roller and drag with broom.
    c. Shave with motor grader.
    d. Broom drag.
    e. Roll with pneumatic-tired roller (Apply water lighty as needed.)

Both are acceptable methods of finishing soil-cement. The objective is to produce a smooth, dense, moist surface that is free of cracks, ridges, and compaction planes.

When shaping is done during finishing, all smooth surfaces such as tire imprints and blade marks should be lightly scratched with a weeder, nail drag, or spike-tooth harrow. A thin layer of soil-cement may not adhere to an underlying smooth area unless this has been dampened and roughened. The surface should be kept moist (optimum moisture) during finishing operations. Steel-wheel rollers have a particular advantage when rock is present in the surface. A broom drag is sometimes used advantageously to pull binder material in and around pieces of gravel that have been punched down by the steel-wheel roller. In lieu of using a steel roller, surfaces are commonly shaved with the motor grader. Only a very thin depth ($\frac{1}{8}$ in. $\pm$) is shaved, and all material clipped is bladed to the edge of the road and wasted. The final operation usually consists of a light application of water and rolling with a pneumatic-tired roller to seal the surface.

Curing

Curing of soil-cement is based on the same principles as any other mixture of portland cement, water, and aggregates. The strength of the material is dependent on the hydration of the cement. The process of hydration depends on the availability of moisture. This means that if water is lost through freezing or drying, hydration will stop and the strength expected will not be realized. Assuming that the proper amount of water was added initially, and that water lost by evaporation was replaced with a light fog-spray, all that is needed for proper curing of soil-cement is a coating that will act as a moisture barrier. One of the most economical and convenient cure coats is a thin layer of bituminous material.

Construction Joint

At the end of each day's construction, a vertical transverse construction joint is formed by cutting back into the completed soil-cement. This usually is done the last thing each day. The material in front of the joint is prepared for processing in the next day's work. After mixing has been completed at the joint, it is cleaned of all dry and unmixed material and

Figure 19-9 A bituminous cover holds moisture in a finished soil-cement base to allow proper curing. If local traffic is to use the base immediately, the cover should be sanded to prevent pickup.

retrimmed if necessary. Mixed moist material is bladed into the area. The material next to the joint is then compacted thoroughly. During this stage of construction the joint is left slightly high. During final rolling it is trimmed to grade with the motor grader and rerolled. If bituminous material is used as a curing agent, it is applied right up to the joint and sanded to prevent pickup.

Multiple-Layer Construction
Where the specified thickness of compacted soil-cement exceeds the depth (usually 6 in.) that can be thoroughly mixed, it must be constructed in multiple layers. No compacted layer should be less than 4 in. thick. Final finishing of the lower layer does not have to be exact, nor do surface compaction planes need to be removed since they are too far from the final surface to be harmful. The completed lower layer can be cured with moist soil that is then used in building the top layer. The top layer may be constructed immediately, on the followiing day, or sometime later.

STREET CONSTRUCTION

Construction of soil-cement streets is fundamentally the same as for roads. A 6-in. thickness of soil-cement is generally used. Where subgrade conditions are good and traffic is light, such as on some subdivision streets, a

5-in. thickness is adequate. The type of bituminous surface used depends on several factors. For most residential streets a double bituminous surface treatment is satisfactory. Asphalt is sprayed on the roadway using a special truck. Before this bituminous material dries, a fine (pea-sized) gravel is spread on top. There are several of these compounds, and future street usage will dictate which one is sprayed on. Thickness will vary with the method and speed of application.

The type and thickness of surfacing depends upon climate, traffic volume, availability of materials, cost, and local practices. In general the thickness of the wearing course required on a soil-cement base will be less than that required on a granular-type base. A common type of wearing course for lightly traveled roads is a double surface treatment about $3/4$ in. thick. As traffic volumes increase, thicker and higher-quality surfacings are needed. If snowplows will be operated on the street (or road), a minimum protective coating would be $1\frac{1}{2}$ in.

The same basic control factors apply to soil-cement street construction as to road construction. Proper cement content, moisture content, and density are essential. Laboratory testing of representative soil samples to establish an adequate cement content is a prerequisite to construction; simple field-control tests made during construction assure quality and eliminate guesswork.

Details of Street Construction

Concrete curbs and gutters are often included in a street paving project. They should be built before paving is started.

Soil materials should be shaped to approximate crown and grade before soil-cement processing is started. Crown and grade should conform to any existing or planned structures such as curb and gutters, manholes, driveways, and street intersections. This procedure simplifies and speeds up final finishing.

Most engineers and contractors prefer to remove manhole covers and frames before processing and to place a temporary cover over the manhole just below the depth to be processed. This permits processing over the manhole without any difficulty or delays. After final finishing and before soil-cement has hardened, the temporary cover is removed, the manhole frame and cover is replaced, and soil-cement is hand-tamped tightly around the structure. (Manholes should be accurately stacked for later location.)

Sometimes it is expedient to process around manholes. When this is done, their locations should be marked clearly with flags or barricades to avoid damage by construction equipment. Other utility structures should be marked carefully ahead of processing so that equipment can be lifted over or guided around them.

Processing should be organized so that the full street width is completed each day of construction. To hasten construction, processing is done usually in successive construction lanes, which are usually some multiple of the mixing machine's width, until the full street width is mixed. This procedure avoids construction of longitudinal joints.

Special attention to adjacent curbs and gutters and utility structures is necessary to insure that mixing is thorough. All soil and cement should be cleaned away from the gutter section for the full depth of processing to insure satisfactory mixing. The point of the motor-grader blade, a plow, or special cleaning devices are used for this purpose.

Preliminary and final shaping of the section should be done with care to obtain proper grade adjacent to curb and gutter, manholes, driveways, and street intersections. During shaping, a special guide is sometimes attached to the end of the motor-grader blade so that when it is resting on the gutter, the proper space for the subsequent bituminous surface will be obtained between the bottom of the gutter and the top of the soil-cement paving. If no curb and gutter exist, it is important to build the edge of the pavement on line and at uniform grade to facilitate the placing of curb and gutter or utility structures at some later time.

The moist soil-cement must be compacted thoroughly adjacent to the curb and gutter and utility structures. The rear wheels of a motor grader frequently are used to obtain additional compaction along the gutter line.

Generally, construction is planned so that the turnaround area for equipment is at the end of the block within the street intersection. Construction joints are located at the end of the block being paved but should not be within the intersection unless paving is to end there. Sometimes the soil-cement paving is extended a few feet beyond this point and cut back when the cross street is paved. The entire intersection is then paved as a part of the construction of the intersecting street. Miscellaneous areas to be paved, such as corners of intersections and approaches to alleys or unpaved streets, may require separate mixing and finishing but are usually handled without undue delay at the time the street is built.

The bituminous wearing course may be placed as soon as the curing material has dried. Often, all streets scheduled for construction are processed first. Then all surfacing can be placed in one continuous operation.

SOIL-CEMENT SLOPE PROTECTION

Engineers concerned with the design and construction of earthfill dams have spent considerable time and effort designing protection for embankments from the persistent attack of winds and waves. Many materials have been used, with considerable range in both cost and durability.

Riprap (a layer of broken stone) was the most successful method of slope protection through the first half of this century. Few failures were experienced as long as the embankment was properly designed and constructed; as long as rock of suitable size, specific gravity, and durability was used; and as long as a gravel filter bedding was used that prevented pumping of the embankment soil through the interstices of the riprap. However, it was found that riprap did require periodic maintenance.

The rapid expansion of water resource development projects in the last two decades showed that high-quality riprap was not available in many areas where reservoirs were desperately needed. Long hauls of the material to these sites involved costs in some instances up to six or eight times usual costs. Obviously, a method of slope protection that was economical over a wider range of site conditions, yet equally durable, needed to be developed. Through research, field tests, and competitive bids, this need has been successfully met with soil-cement.

In 1951, the Bonny Reservoir test section was designed and built by the U.S. Bureau of Reclamation. On a special embankment section 350 ft long and 22 ft high, soil-cement was placed in 7-ft-wide overlapping horizontal layers stair-stepped up the face of the 2:1 slope to give a 2.7-ft minimum thickness, measured perpendicular to the slope (Fig. 19-10). On-site soils were placed on the embankment slope, mixed in place with portland cement, compacted with a sheepsfoot roller and pneumatic tire equipment, and cured with an earth cover.

Two separate soil-cement test sections were constructed. One used fine silty sand and required 12% cement; the other used medium silty sand and required 10% cement by volume of compacted soil-cement.

Although the Bonny Reservoir test site was selected for conditions of extreme exposure to freezing and thawing, wetting and drying, and wave action, damage to the soil-cement sections has been negligible. Ten years after construction, cores drilled from the two facings showed average compressive strengths of 2000 psi and 2160 psi, respectively, for the fine silty sand and medium silty sand soils used.

The Bonny test proved that durable, low-cost slope protection for a dam can be built with natural soils and portland cement, using construc-

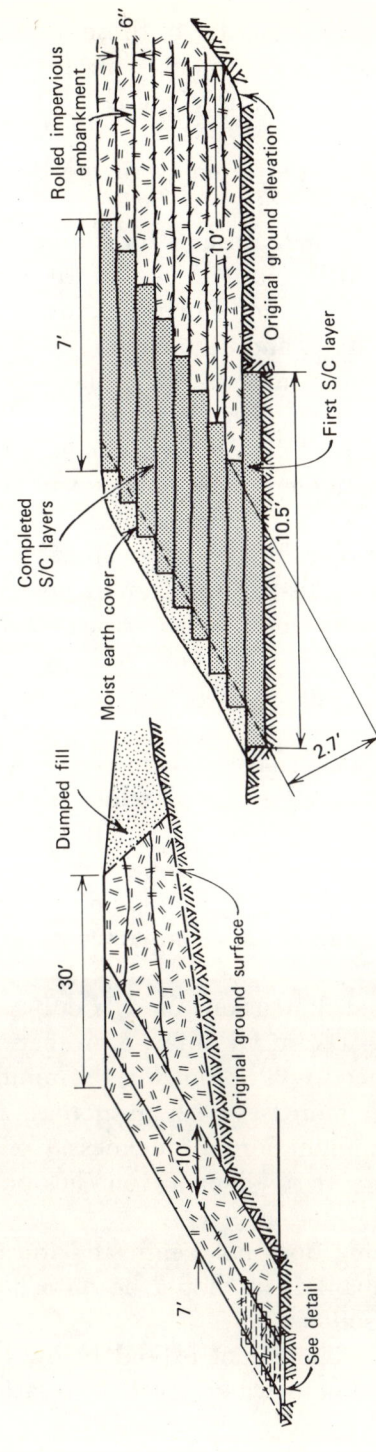

Figure 19-10 Soil-cement slope protection cross section showing layered design.

USING SOIL-CEMENT 437

tion procedures and equipment similar to those used in soil-cement road construction. The prototype soil-cement dam facing at Bonny Reservoir now has an appearance strikingly similar to that of a natural rock formation.

As a result of the excellent performance of the Bonny test section, the Bureau of Reclamation took bids on Merritt Dam near Valentine, Nebraska, in 1961, with alternate designs for soil-cement and riprap. The bid for soil-cement was little more than half the bid for riprap. Soil-cement slope protection has become a major factor in earth-dam planning and design in the United States and Canada.

No unusual considerations are necessary in the design of the dam itself to make use of the soil-cement protection method. Proper embankment design procedures should be followed to meet individual project conditions and to prevent embankment subsidence or any other type of earth-fill-dam distress.

A practical soil-cement dam-facing design specifies that soil-cement be placed and compacted in stair-step horizontal layers to give maximum placing and compacting efficiency. With typical embankment slopes of from 2:1 to 4:1, horizontal layer widths of 7 to 10 ft will provide minimum protective thicknesses of about 2 to 2½ ft, measured normal (perpendicular) to the slope.

Beginning at the lowest layer of soil-cement, each succeeding layer is stepped back a distance equal to the product of the compacted layer thickness and the embankment slope. For example, if the slope is 2:1 and the compacted thickness is 6 in., the step-back is $2 \times \frac{1}{2}$, or 1 ft. The usual compacted layer thickness is 6 in. and soil-cement layers of these dimensions can be effectively placed and compacted with standard highway equipment.

Depending on the exact dimensions used, a cubic yard of soil-cement yields approximately 1 to 1½ sq yd of slope protection. About 8 to 12% of the soil-cement so placed will lie outside the minimum normal thickness line. This percentage increases with an increase in the vertical layer thickness, for a given minimum normal thickness. Conversely, the thinner the layer, the less volume of soil-cement that lies outside the minimum normal thickness line.

If the soil-cement facing does not begin at natural ground level, the lower portion of the embankment should be on a flatter slope than the portion protected by the soil-cement.

It is essential that the soil-cement extend below the minimum water level and above the maximum water level. The latter elevation should have a freeboard allowance of at least 1.2 times the anticipated maximum wave height, or about 5 ft, whichever is greater. The edges of the completed soil-cement layers should not be trimmed as the stair-step effect is beneficial in retarding wave runup.

The construction of a soil-cement patrol road along the top of the dam will provide additional protection against erosion of the crest.

Soils

The principal criterion for determining soil type is gradation. Coarse, sandy, or gravelly soils containing about 10 to 25% material passing the No. 200 sieve are ideal. If the amount of material smaller than the No. 200 sieve exceeds 35%, efforts to find a coarser material may be justified to reduce the amount of cement required. Soils containing 50% or more material passing the No. 200 sieve are not recommended for use in their natural state; but they can be utilized in combination with a more granular material hauled in from another source.

Soils that are bright in overall coloration (white, yellow, and red-brown) generally are excellent soil-cement aggregates; those that are dark (gray, dark brown, and black) are more apt to contain excessive fines and organic matter. Potentially suitable soils near the project site can usually be selected by visual observation and samples taken for design tests in the laboratory.

Cement

Type I, or normal portland cement, is most commonly used, as the special properties of other types of portland cement are not usually required for soil-cement construction. Where sulfate soils or waters are encountered, however, use of Type II or Type V portland cement may be desirable.

Water

Water from most sources, whether raw or treated, and even sea water, is suitable for mixing with soil and sement. The only requirement is that it be free from excessive amounts of alkalis, acids, or organic matter that might inhibit its reaction with the soil and cement. Laboratory analyses should be made of any proposed water source of dubious quality.

Laboratory Design Tests

Standard laboratory tests (wet-dry, freeze-thaw) are necessary to verify the acceptability of the soil and to determine proper cement content, optimum moisture, and maximum density of the soil-cement. If on-site soils contain excessive fines, mixtures of these soils with nearby granular materials can be tested for required cement contents to determine the most economical combination. All design tests should be made in a qualified soils laboratory.

Soil-cement mixtures with a compressive strength of about 450 psi or more at 7 days will generally pass the wet-dry and freeze-thaw compressive strength tests. Using cement contents of about 10%, 7-day compressive strengths of 500 to 1000 psi are common for a wide range of soils. In some cases, coarse, well-graded, sandy soils used in conjunction with even lower cement contents have developed 7-day compressive strengths in excess of 1500 psi. The strength gain between 7 and 28 days is usually less than 50%, but the 28-day strength may be doubled in 10 years.

Other tests, such as the short-cut procedures for sandy soils, are used for special situations. The short-cut procedure takes only two weeks, instead of six weeks for the standard procedures described. The short-cut method will determine a safe cement content but not necessarily the minimum amount to be used. The more accurate results given by the standard procedure are preferable except for emergency, preliminary, or small-project tests.

Construction
Construction of soil-cement slope protection is the same as any other soil-cement construction except for two points: (1) the "roadways" are built one on top of the other in a stair-step fashion and (2) the amount of cement is increased 2 to 4% for greater resistance to freezing and the pounding of waves.

Moisture control is important in minimizing shrinkage cracking. Shrinkage cracking is normal evidence of satisfactory hardening in all soil-cement construction. The amount of cracking varies directly with original moisture content and rate of moisture loss. These factors are of primary importance in field control. Shrinkage cracking will be held to a minimum by placing the mixture at optimum moisture content and preventing rapid loss of the moisture by proper curing. At the Bonny Reservoir test section, the pattern of narrow transverse shrinkage cracks, spaced about 10 to 20 ft apart, developed during the first year and did not change significantly thereafter.

Weather Conditions. Soil-cement should not be placed in freezing weather and should be protected from freezing for at least 7 days after placement.

Figure 19-11 Soil-cement slope protection on Bonny Dam, Colorado, test section after 14 years of service. Compacted soil-cement was used in place of riprap.

Rain is not usually a problem (the mixing water required is generally equivalent to more than an inch of rain). The only critical time for rainfall is the brief period (in mixed-in-place projects) between the spreading and mixing of the cement. This time can be held to a minimum by keeping the mixer close to the spreader.

On central-plant projects, trucks hauling the mixed soil-cement to the placing area should have protective covers. Mixed and placed soil-cement can be protected from sudden heavy showers by rapid rolling with all available equipment to seal the surface. Once the soil-cement has been compacted and finished, it will not be damaged by rainfall.

Mixing. Soil-cement slope protection has been successfully constructed with both central-plant and mixed-in-place procedures. Central-plant provides more accurate control, uniform mixing, and a smoother-running sequence of operations.

For small projects the mixed-in-place method may be more economical. When mixed-in-place procedures are used, close and constant attention must be given to verify accurate depth control of the traveling mixer. The mixer blades should continuously skim the completed surface below the layer being processed.

Suitable central-mix plants with capacities of 200 to 600 tons per hour are produced by several manufacturers. The plant should be of the twin-shaft pub-mill type with either continuous flow or batch operation. Specifications for the plant should provide that all materials will be proportioned within 3% by weight of the design mix quantities.

Accurate calibration of the central plant before the start of construction is essential. The soil feed rate will vary according to the size, gradation, and moisture content of the stockpile. Surge hoppers and compressed air agitation are important in obtaining a uniform cement feed rate. Uniform soil feed is simplified by keeping a reasonably constant head of soil above the hopper, sometimes by using two bulldozers or other machines simultaneously on the stockpile. Shuttling or reciprocating plates in the raw soil hopper are also helpful in maintaining a steady flow of material.

The quantities of soil, cement, and water produced at various control settings should be collected separately and weighed. Any necessary adjustments to the plant and stockpile operations may be made then in order to bring all feed rates to the required accuracy before starting production. The plant calibration should be rechecked periodically or whenever a change is noted in the soil-cement mixture or the stockpile. Close observation of the stockpile moisture content is necessary for continuous proportioning accuracy.

The mixing time (usually about 30 seconds) necessary to achieve an ultimate uniform combination of the soil, cement, and water will vary with the soil gradation and the mixing plant being used.

Figure 19–12 Twin-shaft, continuous-flow stationary plant mixing soil, cement, and water.

Transporting. The mixture should be hauled from the plant to the placing area in trucks with smooth, clean, tight beds. Protective covers should be available as a precaution against sudden showers or other weather extremes.

On sizable projects, a smooth traffic flow for the hauling trucks is essential for efficiency and uniformity of construction. This can be achieved by the use of a multilevel road system with intermediate earth ramps to protect the edges of the dam facing. The hauling time, from completion of moist-mixing to placement for compaction, should be limited to 30 min.

Placing. It is important that the surface on which the fresh soil-cement is to be placed be firm, clean, and surface-moist in order to secure layers that are well bonded.

The soil-cement should be uniformly placed with a mechanical spreader. The loose thickness required for a 6-in. compacted layer is

Figure 19–13 Dump truck deposits soil-cement mixture into a mechanical spreader pushed by a crawler tractor.

usually 8 or 9 in. The spread layer should be compacted as soon as possible, and in no case should it be left undisturbed for more than 30 min.

Compaction. Most soils can be effectively compacted with a sheepsfoot roller. Pneumatic-tire and steel-wheel rolling equipment can be used for final compaction. Some of the more granular soils may be satisfactorily compacted with pneumatic-tire or steel-wheel rolling only.

When sheepsfoot rollers are used, the length of the feet should be less than the loose layer thickness. This will avoid damage to the layer below and still gain full compaction in the fresh layer. Compaction of any portion of a layer to the required density should be completed within 1½ hours after the material has been spread.

Finishing. It is not necessary to trim and roll the completed layer to exact line and grade, although excessive step-backs reducing the normal thickness and feathering out of the compacted vertical thickness to less than 4 in. should be avoided. Any surface "fluff" should be rolled down, but irregularities left by pneumatic-tire treads on a well-compacted surface need not be removed. The entire surface of a completed layer or section should be kept moist by fog-spraying or should be covered by additional soil-cement or curing material for at least 7 days.

Bonding. In order to secure good bond between the layers of the slope protection, it is essential that the lower bonding surface be clean and moist and the upper bonding surface uniformly mixed and solidly compacted.

Care should be taken to see that no loose or liquid material other than water is allowed to get on surfaces that are to be bonded. Although the best bond is achieved when the lower surface is still fresh, it is not necessary to keep an exposed bonding surface continuously moist for more than 7 days. When construction of the next layer is necessarily delayed beyond this time, an adequate bond can be obtained by cleaning and moistening the hardened soil-cement surface just before the new layer is placed.

Curing. All permanently exposed surfaces should be cured for a minimum period of 7 days. The curing material, which should be applied within 24 hours, may be a covering of moist earth or any other material that will prevent loss of moisture.

Heavy curing papers and insulating blankets on bonding surfaces are desirable from the standpoint of speed of application and removal. If earth or liquid curing materials are used, a surface-moist condition of the soil-cement layer at the time of application is particularly essential to prevent penetration of the curing materials into surface voids. Where earth, straw, or burlap curing covers are used, they should be kept continuously moist for the 7-day period. If a bituminous curing membrane is used, it should be applied at the rate of about 0.25 gal per square yard and maintained intact for at least 7 days.

Proprietary curing compounds and products should be applied following the manufacturer's recommendations. Careful consideration should be given to the fact that they may prevent the next layer from bonding to the cure-coat.

Inspection and Field Control. Inspection and field control are even more critical in slope protection than in roadways because the portion of soil-cement doing the "work" is exposed to extreme wave action and heavy icing. The three essentials of proportioning, compacting, and curing must be kept to close tolerances. Careful control and frequent inspection can keep the proportions of the mix to 1% of those specified. If variations exceed 3% by weight of the design mix quantities, operations should be stopped and necessary adjustments made.

Careful attention should be given to the basics of soil-cement construction.

1. Proper amount of cement.
2. Proper amount of water.
3. Adequate mixing.
4. Adequate compaction.
5. Good bonding.
6. Proper curing.

These will produce sound and durable protection for a slope that must withstand severe wave action and the destructive forces of repeated freezing and thawing.

BACKFILLING WITH SOIL-CEMENT

It is often necessary to make cuts in street paving to excavate for utilities. Settlement of backfills may be prevented at these locations by using soil-cement or cement-modified soil. The excavated material is mixed with small quantities of cement, moistened to optimum and used for backfilling the excavation. Premixed soil-cement from a central-mixing plant may also be used. Hand tampers, pneumatic air tampers or various mechanical tampers may be used to compact the material tightly in thin layers. After the backfill is in place, the pavement may be rebuilt immediately without danger of settlement.

Soil-cement or cement-modified soil may be used also as a backfill material for walls and bridge abutments to keep settling to a minimum.

AIRPORT PAVING

A 6-in.-thick soil-cement base with a suitable bituminous surfacing has been used to pave small and medium-sized airports. An analysis of subgrade conditions should be made to determine whether a greater pavement thickness is needed, particularly over clay subgrades and for heavier wheel loads. The Corps of Engineers, Civil Aeronautics Administration, and the U.S. Navy, for example, have published design manuals to serve as guides in design of airport pavements.

Details of Airport Construction

The same principles of construction and the same fundamentals of control that apply to road construction apply as well to airport construction. A few details, however, require special comment. These are (1) the necessity of shaping the material to proper crown and grade prior to processing; (2) a workable plan for processing; and (3) proper construction of joints.

Shaping Material to Crown and Grade. Since there is little longitudinal and transverse displacement of soil during processing, accurate grading before construction starts will save time and will make finishing easier. This is particularly true of large areas such as airport runways. A system of grade stakes is used for grade control.

Plan of Processing. The most practical plan of processing is one that reduces longitudinal and transverse construction joints to a minimum and permits finishing to be done longitudinally, thus providing a smooth surface. There are many processing plans but the one most generally used is to divide the area into processing sections of convenient length and width. Turnaround areas 40 to 50 ft in width are left between successive sections and are later processed transversely. It is important that the end of each longitudinal section be built to grade. At the end of each day's construction, the material next to the end and within the turnaround area is loosened to prevent its hardening.

Joint Construction. There are three types of joints that should be mentioned: (1) a longitudinal joint constructed adjacent to only partially hardened soil-cement, that is, adjoining the preceding day's processing; (2) a longitudinal joint adjacent to hardened soil-cement, that is, adjoining soil-cement processed three or four days previously; and (3) a transverse construction joint made when turnaround areas are processed, adjoining hardened soil-cement. The material next to all joints must be thoroughly pulverized, mixed with cement, moistened, and tightly compacted. Compaction close to the joint can be accomplished with motor-grader wheels and by operating the compaction rollers as close to the joint as possible. The location of all joints should be properly marked by stakes and string lines or by pegs so that streaks of unprocessed material will not occur along joints.

A longitudinal joint adjacent to partially hardened soil-cement can be constructed with most mixing equipment by merely cutting back into the previously constructed area a few inches. The amount of overlap is determined by digging back into the completed work until solid material is reached. Peg stakes or string lines are set as a guide for cement spreading and mixing.

If longitudinal and transverse joints are to be built after the soil-cement has hardened to the point that mixing equipment cannot cut it, the joint must be made by cutting to a string line with the point of the grader blade. A disk mounted on the end of a blade makes a good edge-cutter. With a windrow-type mixing machine this method of joint construction is used for all joints.

Regardless of the method of construction, the joint must be cut back to solid soil-cement; and the material next to the joint must be properly pulverized, mixed with cement, moistened, compacted, and finished level with the adjoining section. Care must be taken to maintain proper grade.

STORAGE AND PARKING AREAS

A 6-in. thickness of soil-cement is usually built for storage and parking areas, although a 5-in. thickness is satisfactory if subgrade conditions are favorable.

The same soil-cement processing procedures and controls discussed in previous chapters apply to construction of storage and parking areas. Accurate grading before processing and good grade control during processing will provide for rapid runoff of water and will prevent puddles. Large storage and parking areas are processed by following the same general plan outlined for airport runways.

Most parking areas are surfaced with bituminous material, while storage areas for coal and similar materials are generally not surfaced. If the soil-cement is to be left unsurfaced, the cement factor should be increased to provide additional abrasion resistance. This increase should be two percentage points for granular soils containing less than 50% silt and clay and four percentage points for other soils.

WIDENING AND SHOULDERS

Equipment and procedures used for constructing soil-cement widening and shoulders are basically the same as those for road construction except that equipment widths must conform to the more limited space. Thorough mixing and compaction along the juncture of the old roadway edge are essential.

Widening

Pavement widening is generally built thicker than ordinary pavement because it is subjected to greater stresses. Soil-cement widening in service today varies in thickness from 6 to 12 in., depending on traffic and subgrade conditions. Thicknesses greater than 8 in. should be built in two layers.

Construction. In cases where edges of the old roadway are found to be weak, the edge material may be ripped up and pulverized; then both edges and widening are processed in a single operation. Widening of 4 ft or more can be processed in place. For narrow widening the material may be bladed or trenched out and placed on the surface of the roadway for processing. The material is then mixed with a rotary mixer and motor grader or with some other type of traveling mixing machine. The processed material is bladed back into the trench, compacted, and finished.

When the in-place soil is not suitable, borrow soil may be used. The area to be widened is trenched, and the excavated material is used for widening the shoulder. Borrow material, cement, and water may be mixed in the trench or on the surface of the old roadway. Sometimes they are mixed in a central-mixing plant, hauled to the site, and dumped through a spreader box into the trenched area. If a spreader box is not available, the mix may be dumped in a windrow near the edge of the pavement and bladed into the excavated area with a motor grader. After the mixture is in place, it is compacted and finished. Single-drum sheepsfoot rollers, trench rollers, job-made rollers, and even the dual rear wheels of heavily loaded trucks have been used for compaction in narrow areas. The completed widening is cured and surfaced with a suitable bituminous mixture. Frequently, the entire old roadway and the widening are surfaced as one operation.

Shoulders

In contrast to widening, which is designed and built to carry continuous traffic, shoulders are improved to provide better pavement performance and a safer emergency stopping or parking area. Improved shoulders increase the safety and traffic capacity of a road.

Soil-cement shoulders are stable under all weather conditions, provide rapid runoff of surface water, and resist erosion and growth of vegetation.

Construction methods for shoulders are the same as those used for widening. Soil-cement shoulders are usually built 4 to 6 in. thick. Bituminous surfacings usually are of a lower quality than the pavement surface and have different texture and color so that shoulders may be readily distinguished from the roadway. This encourages moving vehicles to remain in the traffic lanes.

SALVAGING FLEXIBLE PAVEMENTS

One of the most important uses of soil-cement is for salvaging or reconstructing granular base courses that have failed. Incorporation of portland cement with the base-course materials, often including the old bituminous surfacing, provides an economical means of salvaging and strengthening worn-out granular-base pavements. Cement binds the granular particles together to form a paving material capable of withstanding moisture infiltration and frost action and of bridging over localized soft spots in the subgrade. Portland cement has been used with gravel, crushed stone, caliche, limerock, clay-gravel, sand-clay, and similar granular materials to produce soil-cement. These materials require a minimum of cement for adequate hardening.

If the old bituminous mat is "dead" (oxidized), has lost most of its flexibility, and can be readily pulverized, it can be considered satisfactory for inclusion in the soil-cement mixture. If, on the other hand, the mat is "alive" and the bituminous material retains most of its original viscosity, it should not be used in soil-cement.

Relatively thin surface treatments present no particular construction problem and can usually be readily scarified, broken up, and pulverized. Thick, heavy mats may sometimes be difficult to handle and, in some instances, it may be uneconomical and impractical to break them up for use in the soil-cement mixture. The old mat must be pulverized sufficiently to meet the soil-cement gradation requirement of 55% passing a No. 4 sieve. Exceptionally well graded materials may contain up to 65% gravel and have sufficient fine material for adequate binding. Up to 75% has been used successfully. The largest pieces of mat in the mixture after pulverization should not exceed 3 in.

Construction equipment used to scarify, break up, and pulverize old surfacings includes rippers, motor-grader scarifiers, rotary mixers, pulverizers, traveling mixing machines, disk harrows, and various types of rollers. The hardness, thickness, and type of surface will dictate the choice of equipment for a specific job.

SUBBASES FOR CONCRETE PAVING

Several factors are responsible for the increasing use of cement-treated subbases for concrete pavements. One major factor is the rapidly dwindling supply of good granular material at a time when ever-increasing quantities are needed. Another factor is the demand for higher daily production rates in highway construction which require a stable working base unaffected by changes in weather. Finally, there is the factor of constant striving for improved performance, longer life, and lower maintenance costs for concrete highways.

Most of the early experimentation done in soil stabilizing consisted of treating clay subgrade soils with relatively small amounts of cement for the purpose of controlling volume changes.

These cement-stabilized soils generally met neither the criteria (freeze-thaw, wet-dry tests) for soil-cement because of the low cement contents, nor the criteria for cement-treated subbases because of the nongranular soils used.

Experimental work in Indiana and Ohio has shown that soil-cement made with clay soil is unsatisfactory as a subbase for concrete pavement. Therefore, only granular soils are recommended for cement-treated subbases.

In areas where satisfactory granular subbase materials are expensive or nonexistent, soil-cement subbases offer an economical and satisfactory alternative. Frequently, cement treatment of in-place soils or nearby substandard granular materials results in a highly stable subbase of equal or lower cost than does importing subbase materials such as crushed stone.

Cement-treated subbases are stronger than untreated bases, which permits use of a more economical concrete pavement cross section than would otherwise be required. The saving in pavement costs partly compensates for the cost of cement treatment. In addition, the uniformity of support and the strength of the pavement foundation are greatly increased—valuable assets to any pavement.

Treating soil with cement reduces its plasticity and forms a more stable mass that is resistant to further consolidation and moisture erosion (pumping). Consolidation of subbases under the action of traffic is greatly reduced because the material is densified during construction and stabilized in that state by the cement.

Hundreds of millions of square yards of cement-treated subbases for concrete pavement have been built in the U.S. and Canada since 1945. In most cases, granular soils have been used to make a 4-in.-thick cement-treated subbase. Practically all of the concrete pavement on California expressways is built on cement-treated subbase.

Swelling and Shrinkage of High-Volume-Change Subgrade Soils
Rough-riding pavements with high joints and with loss or gain in crown are characteristics of this condition. Many investigations have led to the conclusion that the cure for this condition is to control the moisture content of the expansive subgrade soil before paving. Swelling is at a minimum when these heavy clay subgrade soils are compacted to 90% and 95% of standard density at a moisture content slightly in excess of optimum—near the plastic limit. In dry climates a blanketing course of granular material will reduce the moisture loss from the clay by evaporation and capillarity and also will serve as a subbase to prevent pumping.

Where gravel is expensive, cement can be used to reduce the volume changes in the upper portion of expansive subgrades. Generally a 6-in. depth of treatment is adequate except for soils with excessive volume-change characteristics, in which case greater depths are desirable. Only enough cement is added to change the physical characteristics of the subgrade soil. This is cement-modified soil rather than soil-cement. The cement content required to control volume change is determined by laboratory tests.

Damage from Frost Action

When soil-cement is used to prevent damage from differential frost heave, it should be constructed to the depth and width normally required for granular materials. Construction need not be for the full depth of frost penetration because the amount of heave is not equal to the depth of frozen material. Except for the most severe conditions, a combined thickness of pavement and soil-cement equal to half the depth of frost penetration should give satisfactory results.

SECTION SEVEN
FIRE RESISTANCE OF CONCRETE

CHAPTER 20
FIRE RESISTANCE OF CONCRETE

The ability of concrete to resist the destructive effects of fire has been well known for many years. For example, in 1904, 2000 buildings were destroyed in the great Baltimore fire. Only one building was left standing; it was a four-story structure made of concrete. During the San Francisco earthquake and fire in 1906, the Bekins Van and Storage Building —the only sizable building made of reinforced concrete—was the only building relatively unharmed. A fire in a warehouse in Dresden, Germany, in 1911, was hot enough to melt the glass lighting fixtures, but the reinforced concrete building was structurally undamaged. In 1916, the entire business section of Paris, Texas, was wiped out. Three reinforced concrete buildings were the only ones to come through virtually unharmed (1440 other buildings were destroyed).

In Stockholm, Sweden, a warehouse that was under construction burned for several hours when insulating material that had been stockpiled in the basement caught fire. In some areas the concrete spalled on the fire side of the floor enough to expose all the reinforcing steel. Tests showed that after the replacement of the steel and shotcreting, the building was usable.

A reinforced concrete warehouse near Chicago, built with a thin-shell barrel roof, was subjected to an intense fire for 8 hours. High-stacked flammable cartons burned at a temperature of about 2000°F for a considerable length of time. Several beams were cracked completely through because of the heat and yet remained in place. Several beams and columns were replaced, but the roof was tested and found to be able to carry more than its rated load, despite the punishment of the fire.

By contrast, the large multimillion dollar McCormick Place Exhibition Hall in Chicago, constructed of heavy steel framing, was almost a total loss from a fire in January, 1967. The National Housewares Show was prepared to open the morning of the fire, and the exhibition hall was filled with flammable plastic, fabric, and paper. The fire spread rapidly and burned with enough heat to cause the roof to collapse in about half an hour. The total loss of the McCormick Place fire—including the building, material lost by exhibitors, and the loss to the city of Chicago in business because the exhibition place was not usable—exceeded $300 million.

TESTING FIRE RESISTANCE

Just what happens in a burning building is largely unknown due to the obvious difficulty in measuring heat, stresses, and deformations during a fire. Because of this, a system for measuring temperatures, stresses, and deformations has been devised that simulates actual fires. As a yardstick for performances in a "real" fire, a time-temperature curve (Fig. 20-1) for what is known as a "standard fire" is used. ASTM E 119 specifies that in a standard fire the temperature will rise in the first 5 min to 1000°F; in 10 min it will be 1300°F; in 30 min 1550°F; in 1 hours 1700°F; and in 4 hours 2000°F. This is the specified rise in furnace temperature when a structural member is being tested. In order to accurately control this temperature rise, large furnaces have been built with rather elaborate controls. The standard fire is the temperature rise that would be the result of burning 10 lb. of ordinary combustibles (wood, paper, and cloth, etc.) for each square foot of floor area per hour of fire.

Slabs are tested in a furnace that is built so that the slab to be tested forms the top section of the fire area of the furnace. ASTM E 119 specifies that a floor specimen must have at least 180 sq ft exposed to fire with neither dimension less than 12 ft. During the test the slab must carry the maximum load for which it is designed. The underside is exposed to the standard fire until one of the following three end points is reached. (1) If the specimen collapses during the test, it has reached the *struc-*

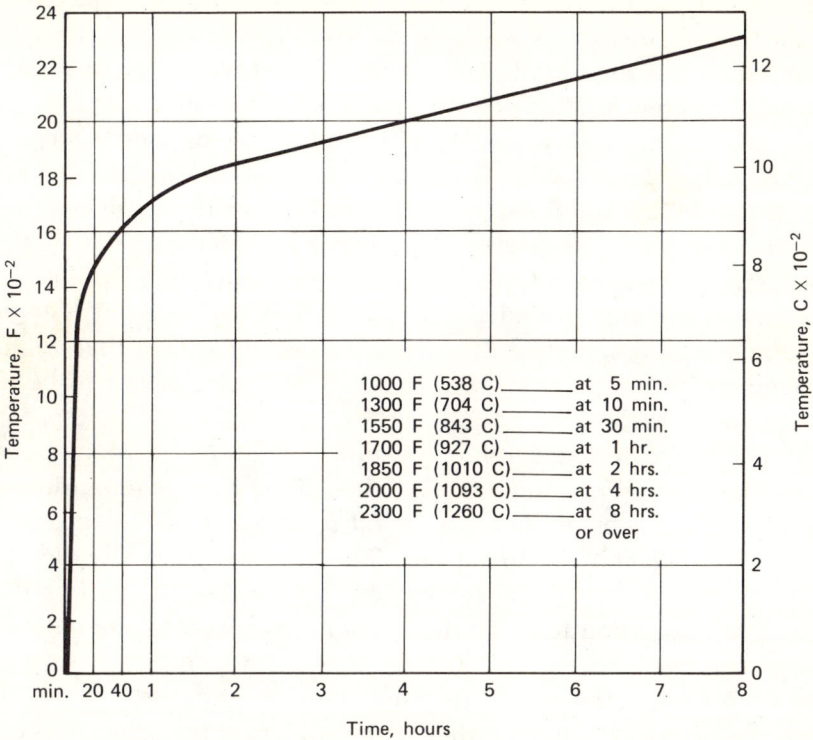

Figure 20-1 Time-temperature relationship for heating fire-testing furnace in the United States and Canada.

tural end point. (2) If an opening occurs that will allow hot gasses to get through, the *flame passage end point* has been reached. (3) If the temperature of the surface away from the fire goes up an average of 250°F (or if the temperature at any point rises 325°F) above its initial temperature, the test slab is said to have reached the *heat transmission end point.* The limit on temperature rise of the unexposed surface is meant to be a safeguard against the ignition of flammable materials on the side of the floor, roof, or wall away from the fire. However, wood or cotton do not show signs of ignition until temperatures exceed a temperature of 400 to 450°F (a 325 to 375°F rise). The *heat* transmission end point is the one usually reached first with reinforced concrete slabs.

The simply supported slab (Fig. 20-2) is free to move without restraint. Simply supported slabs are a rarity in buildings today. In almost every case the slab is restrained from expanding by the adjacent slabs, walls, beams, or columns. Tests have shown that when a slab is restrained with anything less than total (100%) restraint, it will carry its rated load long after it has reached the heat transmission end point.

Slabs are tested in furnaces large enough to accommodate at least a 12×15-ft slab, which is the minimum size required by ASTM E 119. Some, of course, are much larger. The slab furnace below shows some of the typical features. The overhead framework supports the 16 hydraulic rams (8 are shown) that supply the pressure to simulate a live load.

The heavy framing around the slab (the slab is shown with 8 holes to locate the 8 missing live-load rams) is a restraining system that will allow any amount of restraint from zero to 100%. The slab about to be tested is resting on the firebox.

Beams are also tested for fire resistance. PCA has a special "beam" furnace that is capable of maintaining the ASTM time-temperature curve while supporting a beam as much as 60 ft long and 4 ft deep, with an applied load of 2800 lb per linear foot. The live load is usually distributed over the length of the heated section. In this furnace the firebox is 40 ft long and has provision for a 10-ft section at each end to simulate a continuous beam.

For beams, only the structural end point is used. Beams are almost always reinforced with steel, either with ordinary reinforcing bars or prestressing steel. In the fire testing of beams, steel temperature is one of the critical factors. This is because cold-drawn steel will retain about 75% of its strength at 600°F and only about 35% of its strength when it reaches 900°F. The thermal insulating properties of concrete can keep the steel from reaching its critical temperature in a fire and, thereby, enhance the fire endurance.

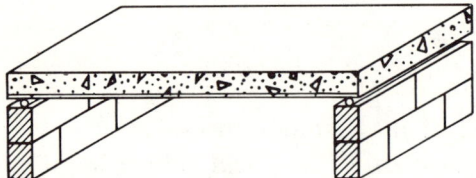

Figure 20-2   A simple supported concrete slab.

THERMAL FACTORS

Factors affecting the transmission of heat through a concrete structure are type of aggregate used; amount of free moisture present; and the volume of concrete per square foot of exposed area (thickness).

Because the heat transmission end point is the end point that usually governs, it is worthwhile to look into the factors that affect heat transmission in concrete. Figure 20-3 shows several factors affecting fire endurance ratings for slabs. The kind of concrete used (insulating, lightweight, or normal-weight) will make a great deal of difference. Slab thickness will also affect fire rating. To have a 4-hr rating (the test slab is subjected to a standard fire for 4 hours without reaching any of the three end points), you will need about 3½ in. of insulating concrete, about 5½ in. of lightweight concrete, or about 6¾ in. of normal-weight concrete. These are the thicknesses necessary to prevent the temperature of the unexposed surface from rising an average of 250°F (the heat transmission end point).

Type of Aggregates

Different aggregates will give different fire ratings. For example, a 5-in. slab of concrete made with calcareous aggregate will not reach the heat transmission end point for about 2½ hours. A slab of the same size, shape, and mass and made of siliceous aggregate, will probably reach the heat

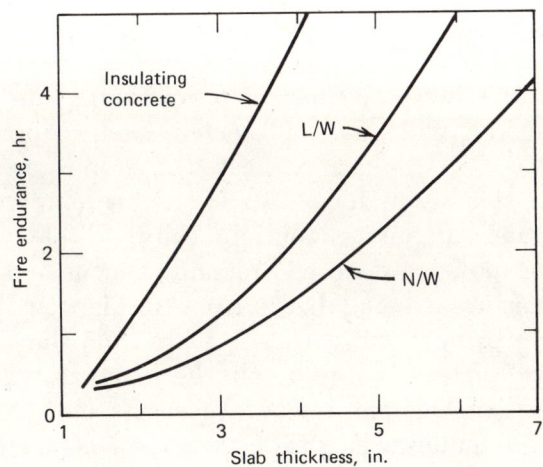

Figure 20-3

transmission end point in 2¼ hours—not a serious difference, but one that does exist.

Explaining why these aggregates give different results when tested in the same way requires some use of chemistry and physics. As the temperature rises in a concrete slab made of siliceous aggregates, the silicon dioxide (SiO_2), a major component of these aggregates, is chemically stable enough to withstand high temperatures and will simply change its physical structure. For example, it might change from sand to a material very much like glass at temperatures somewhere above 1000°F, but still have the composition SiO_2. This material passes the heat along and gives the shortest fire rating because the slab fails in the heat transmission end point.

Calcareous aggregates (limestone, for example) contain large amounts of calcium carbonate ($CaCO_3$). When heated, the $CaCO_3$ breaks down into CaO (quicklime) and CO_2 (carbon dioxide) and, in so doing, absorbs a great deal of heat. (The reaction is very different of course, but the result is the same as when water is heated enough to change to steam.) As the $CaCO_3$ gives up CO_2 to become CaO, it takes heat away from the mass of the concrete and thereby increases the fire endurance by increasing the time required to reach the heat transmission end point.

The lightweight aggregates give better fire endurance because of their physical properties—the lighter they are, the more tiny air spaces they contain. More air spaces mean better insulation properties, which in turn mean longer fire ratings for concrete made of these materials.

Moisture Effect

The moisture content of concrete at the time of the test will have a marked effect on the unexposed surface temperature. The more moisture present, the longer it will take for the temperature of the unexposed surface to rise 250°F. As the slab is heated, the unexposed surface temperature will rise steadily to about 212°F. At this temperature, moisture is driven from the slab fast enough to cause visible wisps of vapor (steam). The unexposed surface will remain at or near this temperature as long as the moisture is being driven from the slab. As it is changed to steam, water absorbs heat and carries it away from the slab. Obviously, the more moisture present in a test slab, the longer it will take to reach the heat transmission end point. Because excess moisture can give an artificially high fire endurance, ASTM specifies a maximum relative humidity of 75% in the center of the slab at the time of the test. Excess moisture has also caused explosive spalling and artificially low fire ratings in some of the early tests.

STRUCTURAL FACTORS

Compressive resistance is the most often used indicator of concrete strength. Strengths in Figure 20-4 are shown as a percentage of original compressive strength for each type of concrete after having been heated to the temperatures shown.

Several interesting facts are displayed in Figure 20-4.

1. The lightweight concrete had essentially the same strength up to about 500°F and kept more of its strength at almost every temperature.
2. The carbonate aggregate lost more strength than either of the others up to 800°F., but between 800°F and about 1000°F, it surpassed them both and held its lead over siliceous aggregate to about 1500°F.
3. The siliceous aggregates did comparatively well to about 800°F (all three were almost the same at 800°F) and then dropped off sharply. This sudden drop in compressive strength is caused by several factors but mainly by the inversion of quartz. This is a change in the crystalline structure that causes both a weakening of the quartz and an expansize force capable of exerting great pressure. The expansion of the quartz-bearing (siliceous) aggregates coupled with the shrinkage of the hardened cement paste is probably the major cause of the loss of compressive strength above 800°F.
4. All three of these concretes had lost about 20% of their strength at 800°F.

The chart illustrated in Figure 20-4 summarizes the effects of various temperatures. Three aggregates are shown: (1) sanded lightweight (an expanded shale with some sand, 30% by weight); (2) carbonate aggregate (a dolomitic sand and gravel); and (3) siliceous sand and gravel.

Concrete Stress Level

Very high stress levels (not to be confused with *restraint*, mentioned earlier) cause lower fire endurance ratings. Table 20-A shows the effect of applied stress on a particular series of concrete columns. This is caused by the fact that both concrete and steel lose strength as temperature increases and are not able to carry as great a load. The reduced fire rating under increased load is more pronounced in columns and beams than in slabs because slabs usually fail in the heat transmission end point.

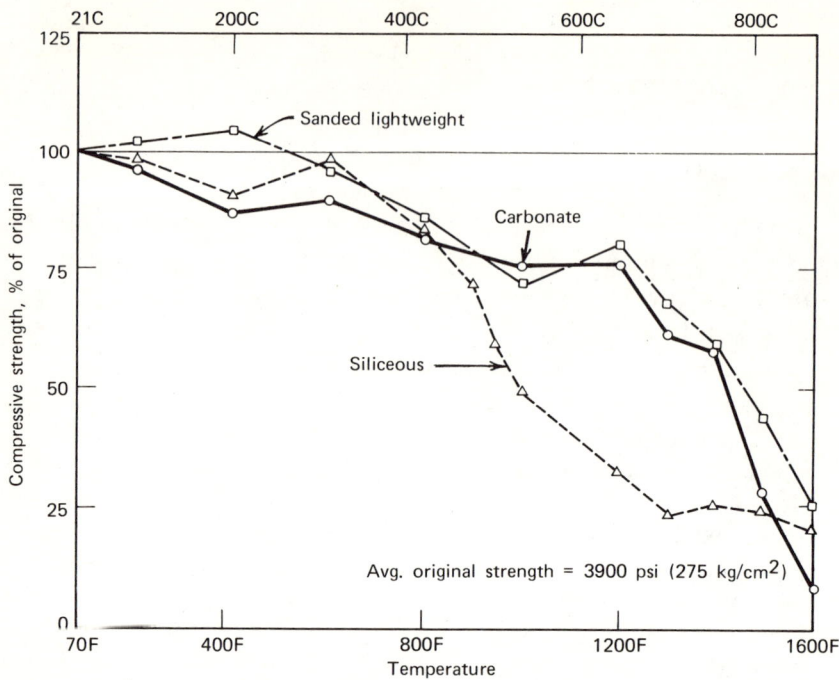

Figure 20-4 Effect of aggregate type on compressive strength of concretes tested unstressed during heating.

At high temperatures the materials in concrete may undergo transformations such as calcination of carbonate aggregates, inversion of quartz, or dehydration of the cement paste. Any of these transformations will affect the strength of concrete. To find out what does happen, the Portland Cement Association conducted a series of tests on 3×6-in. cylinders. Three types of aggregates (carbonate, siliceous, and expanded shale) were used.

Table 20-A
EFFECT OF APPLIED LOAD ON FIRE ENDURANCE

Applied Load Percent of Design Load	Fire Endurance (minutes)
150	68
100	124
75	198
50	248
30	358

The cylinders were heated in an electric furnace to a specified temperature somewhere between 70 and 1600°F and then compressed to failure (crushed). Some were heated without load, then crushed; some were loaded to a percentage of unheated compressive strength and then crushed; while a third group was heated, allowed to cool slowly, and then tested 7 days later.

Results of the test showed the following.

1. Carbonate and expanded shale retained more than 75% of their original strength when heated to 1200°F under the most unfavorable conditions (heated without load and tested hot). The siliceous aggregate concrete retained about 75% of its strength at a temperature of 800°F.
2. Specimens loaded to $\frac{1}{4}$ to $\frac{1}{2}$ of their ultimate strength were somewhat stronger then those that were not loaded. (The amount of loading did not seem to make much difference.)
3. Cylinders that were heated and cooled were not as strong as those tested hot.
4. The original strength had little effect on the percentage of strength retained.

Figure 20-5 shows the percentage of strength retained for the three conditions of testing the carbonate aggregate concrete.

Concrete Mix

When concrete is heated to temperatures of about 750°F, dehydration of the calcium hydroxide begins. The paste tries to shrink as the absorption, capillary, and hydration water are driven out. On the other hand, the aggregate tends to expand while the paste is shrinking. These contrary actions markedly affect the strength of concrete at this temperature level. A rich mix will show greater internal stress than a lean mix at temperatures over 750°F.

Concrete Spalling

The shrinkage of the paste and, more importantly, the expansion of the aggregates, together, set up large strains that cause fracture planes that result in pieces of concrete being sloughed off the surface of the heated member in a process called spalling. The mechanism is not completely understood, but there is evidence that spalling is caused primarily by the mechanical strain of excessive restraint coupled with the great reduction in paste strength from dehydration.

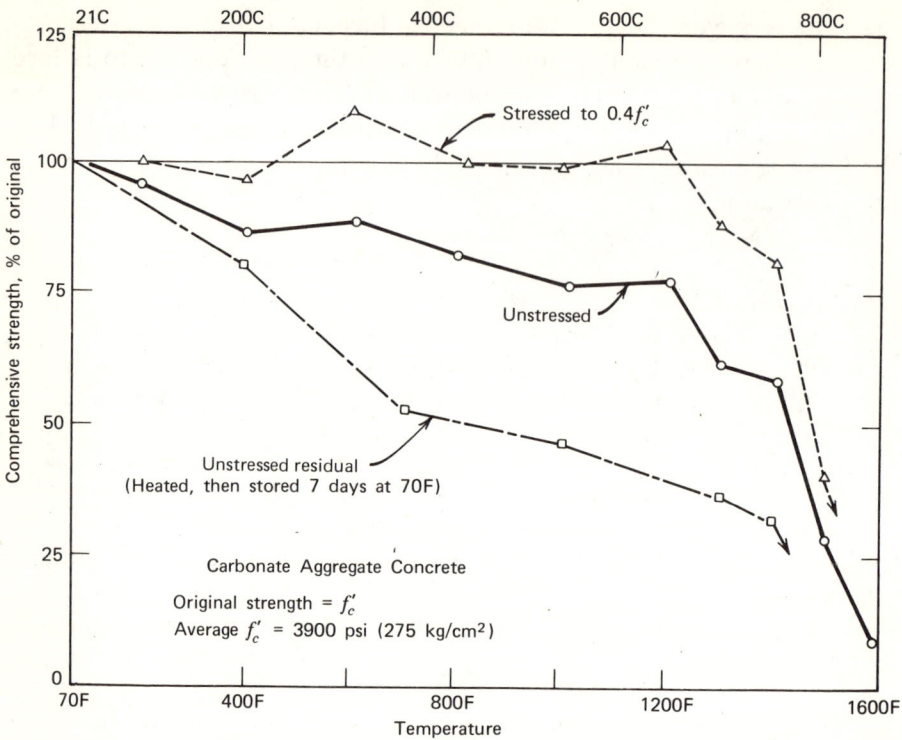

Figure 20–5 Effect of temperature on compressive strength of carbonate aggregate concrete.

STEEL REINFORCEMENT

Nonreinforced concrete is seldom used today as a structural material. Steel and concrete work together so well—each imparting its best qualities—that the combination is hard to beat for strength, low cost, durability, and fire resistance. When full-sized structural members are fire tested in a laboratory, they are usually constructed in the same way that they will be on the job. The four ingredients of concrete (fine and coarse aggregates, cement and water) are specified as to kind and amount; the steel size, type, amount and location, as well as the type of concrete are detailed by the engineer and are carefully followed in the construction of the test member.

Steel is bedded in concrete to perform one of three jobs: (1) usually, to resist pulling forces (tension steel); (2) sometimes, because of space limitations, to resist squeezing forces (compression steel); and (3) almost always, by means of steel added near the surface, to resist the small random movements caused by differences in temperature in various parts of the member (temperature steel).

The closer to the tension side of a structural member the steel is placed, the more effective it will be. The most effective location from a standpoint of structural strength *only* would be on the tension side with a very thin cover of concrete (about the diameter of the bar).

A thin cover is not used because there are other important considerations that require a heavier cover, in most cases between ¾ and 2 in. One such consideration is for protection from exposure to corrosive or eroding elements (for example, weathering of the outside of buildings), another is for development of bond, and a third is fire protection. Fire protection is the one that will be discussed in detail in this chapter.

Concrete Cover for Fire Protection

Laboratory studies show how concrete loses some if its compressive strength, elasticity, and shear strength, for instance, at temperatures over 800°F. Hot-rolled steel (the kind most often used for reinforcing) also loses yield strength over about 300°F. Yield strength is the amount of pull that can be put on steel and still have it return to its original shape—like stretching a rubber band and letting it snap back. If steel is stressed (pulled) with more force than its yield strength, it will stretch and stay that way. Yield strength is expressed in pounds per square inch (psi). Most reinforcing steel has a specified yield strength of either 40,000 psi or 60,000 psi.

Assuming that a 12×15-ft, 6-in.-thick concrete slab has been properly designed, constructed, and cured, we could expect the following to occur in a fire test while the slab is carrying its design load.

As the furnace is heated, following the time-temperature curve of a standard fire, the slab would also begin to warm up. In 30 min the furnace would be at 1550°F and the hot side of the slab at about 1300°F. About ½ in. from the surface the temperature is about 800°F. The steel (1½ in. from the fire) is only about 300°F, has not lost any strength, and is safely carrying the test load.

After 1 hour of fire test:

1. The furnace temperature is 1700°F.
2. The hot surface of the slab is 1500°F.
3. One-half in. away from the fire, the temperature is 1000°F.
4. The steel temperature is about 500°F, still not hot enough to be seriously damaging.

After 2 hours of fire test:

1. The furnace temperature is 1850°F.
2. The hot surface of the slab is 1650°F.
3. One-half in. away from the fire, the temperature is 1300°F.
4. The steel is now up to 800°F and losing strength. About 70% of the yield strength is retained and, because of safety factors designed into the slab, there is still more than enough strength remaining to carry the load.

After 3 hours of fire test:

1. The furnace temperature is 1925°F.
2. The hot surface of the slab is over 1700°F.
3. One-half in. away from the fire, the temperature is 1500°F.
4. One and one-half in. into the concrete, it is about 1050°F. The steel has about 50% of its yield strength. The slab is carrying its design load even though the unexposed surface temperature will soon be high enough to cause the slab to fail the fire test in the heat transmission end point, and thereby be given a 3-hr. fire rating.

After 4 hours of fire test: the slab has reached the heat transmission end point where most tests end. Few slab tests are continued to the point of structural collapse. However, hypothetically:

1. The furnace temperature is 2000°F.
2. The hot surface of the slab is over 1800°F.
3. One-half in. away from the fire, the temperature is about 1600°F.
4. One and one-half in. into the concrete the temperature is about 1250°F.

Steel at the 1250°F temperature retains only about 25% of its yield strength. If the stress (pull) on the steel is less than 25% of the yield strength, the slab would still withstand the load and fire. Since we started with intermediate grade steel (40,000 psi) and it retains 25% of its strength, there is only about 10,000 psi available to withstand the slab bending. If the slab were restrained, it is likely that the stress would be redistributed and be below the existing strength level. If the slab were unrestrained or if the test were continued, the slab would soon collapse.

An unprotected all-steel floor would have softened and collapsed like warm taffy after 20 min. It would reach the heat transmission end point in less than 5 min. However, the reinforced concrete used in modern construction is capable of this kind of dramatic durability, and seldom are real fires this severe.

The fuel consumed in a 4-hr-plus fire is greater than what would be available in most buildings to feed a real fire. In 4 hours, 40 lb of ordinary combustibles would have been burned for each square foot of area. Only a warehouse (or perhaps a library) would have a concentration like this. Even a crowded office or apartment building simply would not have enough fuel to create such a fire.

Roofs and Beams

Slabs are tested in much the same way whether they are to be used for floors or roofs. Beams are tested in a furnace with the heat reaching both sides as well as the bottom. The heat transmission end point is of no importance (and is never reached). Only the structural end point is used. In some cases a deflection at the center of the span of about 15 in. is considered to be total collapse.

SUMMARY

The following factors affect the fire resistance of concrete.
1. *Temperature.* The higher the temperature, the more strength lost.
2. *Type of aggregate.* Lightweight aggregates give a better fire endurance rating than normal-weight ones. For normal-weight concretes, those made with carbonate aggregates do better than those made with siliceous material.
3. *Moisture.* Excess moisture in mature concrete will increase the fire rating by delaying the temperature rise of the unexposed surface.
4. *Restraint.* In general, restraint of thermal expansion will increase the load-carrying capacity of concrete structures during free tests.
5. *Mix proportions.* The water-cement ratio seems to have little significance in the percentage of strength lost by heating. It does, however, greatly influence initial strength.
6. *Yield strength of reinforcing steel.* When the reinforcing steel is heated above about 600°F, its yield strength is reduced. At 1200°F, more than half its strength is gone. The time-temperature relationship necessary to reach the critical temperature* of the steel in concrete is governed by the average cover of concrete and the mass of the structural member. An increase in cover will reduce the rate of temperature rise of the steel; an increase in the mass of the structure will also slow the rate of temperature increase.

* Temperature at which the yield strength drops down to equal the existing stress.

GLOSSARY

axle steel reinforcing bars. Reinforcing bars rolled from carbon-steel axles for railroad cars.

bar chart or graph. A graph used by builders for scheduling construction work. Each element of the job is represented by a bar or line with one end representing the starting date and the other the completion date. In this way, the progress of the job is visually represented.

bar fabricator. A bar company equipped for the storage, shearing, bending, bundling, tagging, loading and delivery of reinforcing bars, and for the preparation of placing drawings and bar lists.

bar list. A bill of bar materials where all quantities, sizes, lengths, and bending dimensions are shown.

bar number. A number (approximately, the bar diameter in eighths of inches) used to designate the bar size. For example, A #5 bar is approximately $5/8$ in. in diameter.

bar placing subcontractor. A contractor or subcontractor who handles and places reinforcement and bar supports, often colloquially referred to as a "bar placer" or "placer."

bar spacing. Spacing of reinforcing bars measured from center-to-center of the bars.

bar supports. Devices, usually of formed wire, to support, hold, and space reinforcing bars.

base course (road construction). A compacted soil-cement mixture, usually 4 to 6 in. thick, on which a bituminous wearing course is placed to make a roadway or parking area.

bearing value. The load a soil will sustain without detrimental deformation.

belled. Having a butt or bottom end shaped like a bell, often applied to concrete piers or caissons.

bending moment. The tendency of a structural member to rotate about an axis; for example, the midsection of a beam has a tendency to bend downward.

bent bar. A reinforcing bar bent to a prescribed shape such as a truss bar, hook bar, stirrup, or column tie.

billet steel reinforcing bars. Reinforcing bars rolled from steel billets, rather than from rail or axle steel.

block. 1. A concrete masonry unit. 2. (soil-cement) A soil texture resembling blocks or cubes.

butt-welded splice. A reinforcing bar splice made by welding the butted ends.

caissons. Piers extending through soil to earth having a high bearing capacity, or to solid rock. Also, cast-in-place, drilled-hole poles of relatively large diameter.

camber. A slight (usually) upward curvature of a truss, beam, or form to improve appearance or to compensate for anticipated deflection.

cantilever. 1. A free part of any horizontal member projecting beyond a support. 2. An extended bracket for supporting a balcony, cornice, or the like.

cantilever beam. A beam that extends beyond the supports in an overhanging position with the extended end unsupported. Similarly, such a slab is called a **cantilever slab**.

capillary water. Water capable of moving in any direction in the soil by capillary action.

clay. Soil, usually considered to be 0.002 mm or less in particle size.

click test. A rough estimate of the hydration of a soil-cement by knocking two pieces together and noting the clarity of the sound.

cohesion. The mutual attraction of particles of soil caused by molecular attraction and moisture films. The cohesion of a particular soil will vary with the moisture content.

cold-drawn steel. Steel that is rolled to final shape at relatively low temperatures.

column ties. Bars bent into square, rectangular, U-shaped, circular, or other shapes for the purpose of holding column vertical bars laterally in place.

compression bars. Steel used to resist compression forces.

concrete cover. The clear distance from the face of the concrete to the reinforcing steel, also referred to as **fireproofing, clearance,** or **concrete protection**.

continuous high chairs. Welded wire supports consisting of a top longitudinal supporting wire with evenly spaced legs welded to it and used to support bars near the top slabs. See also **bar supports, individual high chairs**.

creep. Time-dependent deformation due to load.

crumb. A soil texture resembling bread crumbs.

deformed bar. A reinforcing bar manufactured with deformations (bumps, lugs, or ridges) to provide a licking anchorage with the surrounding concrete.

floury. A soil texture in which the materials are quite fine and resemble flour.

footing. That portion of the foundation of a structure that spreads and transmits load directly to the piles, to the soil, or to the supporting grillage.

foundation. Substructure through which the loads are carried to the earth or rock.

friable. Capable of being easily broken into small particles.

glacial gravel. "Clean" gravel left by a glacier as it retreats. It is a nongraded material since it has no fines.

gradation or **particle-size distribution.** The distribution of particles of granular material among various sizes.

grade. 1. The prepared surface in a highway system on which the base or subbase is placed. 2. The design engineer's term to indicate leveling and inclining of the surface prior to placing a highway system on it.

grade mark. A marking rolled onto the bar to identify the grade of steel. See **bar number.**

grade of steel. The means by which a design engineer specifies the strength properties of the steel he requires in each part of a structure, generally using ASTM designations to distinguish them.

granular. Soils containing 35% or less material passing the No. 200 sieve.

gravel. Hard, stony material varying from about ¼ in. up to 3 in.

gravitational water. Water able to move downward through the soil due to the pull of gravity.

hanger. A device used for suspending one object from another such as the hardware attached to a building frame to support forms.

heavy bending. Bars sizes #4 through #11, which are bent at not more than six points, radius bent to one radius and bending not otherwise defined. See **light bending.**

hickey. A hand tool with a side opening jaw used in developing leverage for making bends on bars or pipes in place.

high chair. See **individual high chairs, continuous high chairs.**

hook. A semicircular (180°) or a 90° turn at the free end of a bar to provide anchorage in concrete. For stirrups and column ties only, turns of either 90° or 135° are used.

hygroscopic water. A film of moisture surrounding each soil particle, varying with the relative humidity.

individual high chairs. Welded wire bar supports used under a support bar to provide support for top bars in slabs, joints, or beams; also used to support upper mats of bars without support bars. See **continuous high chairs.**

inserts. Devices buried in concrete to receive a bolt or screw to support shelf angles and machinery, for example.

internal friction. Resistance to sliding within the soil mass.

inversion of quartz. The realignment of SiO_2 crystalline structure (at 1050°F) from alpha to beta, accompanied by large and rapid expansion.

joist. A horizontal structural member, such as that which supports deck form sheathing; it usually rests on stringers or ledgers.

joist chairs. Bent or welded wire supports that hold and space the two bars in the bottom of a joist.

joist schedule. A table on the placing drawing that gives the quantity and mark of the joists; the quantity, size, length and bending details of bars; and usually the quantity of joist chairs in each joist.

light bending. All #3 bars and all stirrups and column ties; and all #4 to #11 that are bent at more than six points, bent in more than one plane, radius bent with more than one radius in any one bar, or a combination of radius and other bending. See **heavy bending.**

liquid limit. The point of moisture content at which a soil material changes from a solid to a liquid state.

maximum density. The highest unit weight to which a material can be compacted.

mesh. See **welded wire fabric.**

modulus. A coefficient or constant number that serves as a measure of one or more of the properties of a material.

GLOSSARY 471

optimum moisture. The moisture content at which soil can be compacted to maximum density, expressed as a percentage.

oven-dry weight. The weight of the soil sample after oven drying for 12 hours or to a constant weight. The point at which all of the hygroscopic moisture has been driven from the soil material.

pick test. A rough estimate of the hydration of cement in a soil-cement sample by careful picking with a relatively sharp instrument such as a dull ice pick.

plasticity index (PI). The numerical difference between the liquid limit and the plastic limit, a rough indication of clay content.

plastic limit. Point of moisture content at which soil material loses its plasticity (cannot be rolled into $\frac{1}{8}$-in. threads).

posttensioning. A basic method of stressing concrete after it has been cured.

prestressing. A way of adding stress to a material to offset the stresses experienced when the member is loaded.

pretensioning. Tension applied with reinforcing steel prior to the placing of concrete.

radial bending. Reinforcing bars bent to a radius larger than that specified for standard hooks; a bar curved to fit into circular walls, as the horizontal bar in a silo.

rail steel reinforcing bars. Deformed reinforcing bars rolled from selected, used railroad rails.

rebar. Abbreviated term for reinforcing bar, usually steel.

sand. Granular material varying from 0.08 mm for very fine sand up to almost $\frac{1}{4}$ in. for coarse sand.

scalp. To remove a small portion of the surface of a material.

scarify. A process of breaking up compacted or hardened earth material prior to pulverizing by drawing a tooth steel bar through the soil at a depth of 4 to 6 in.

schedule. A table on placing drawings (or elsewhere) giving size, shape, and arrangement of similar items; for example, beam schedule, column schedule, joist schedule, and slab schedule.

shear. To tend to cut off as by two equal offset forces.

shearhead. An assembled steel unit in the top of the columns of flat slab or flat plate construction to transmit loads from slab to column.

signed to resist shearing forces, usu-
shear reinforcement. Reinforcement deally consisting of stirrups or truss bars that are bent and located as required.

shrinkage limit (SL). The moisture content at which a soil sample will no longer reduce in volume even if more moisture is driven off.

silt. Earth material usually considered to be between 0.08 and 0.005 mm in particle size.

simple beam. A beam supported at each end (two points) without bending restraint at either end.

slab bolster. A bar support with corrugated longitudinal wire and supporting legs, used to support bottom slab bars.

slab spacer. A bar support with straight longitudinal top wire and supporting legs spaced to match the spacing of slab bats that they support.

soil-cement. A hard, durable material made by mixing a specified amount of cement with soil and water, compacting to specified density and allowing to cure.

spiral. A continuously coiled bar or wire.

spiral column. A column in which the vertical bars are enclosed within a spiral.

spreader bar. A steel beam suspended from a crane, having two or more hooks or shackles to which slings are attached to eliminate the possibility of bending bars in a bundle due to handling. Sometimes called **strong back**.

spread footings. Footings that support one or more columns or piers or a wall by bearing on earth or rock. Sometimes a simple mat footing is called a spread footing.

staggered splices. Splices in bars that are not made in the same line.

standee. A term used in some localities to designate a special bar bent to a U-shape with 90° bent legs extending in opposite directions at right angles to the U-bend. It is used as a high chair resting upon a lower mat of bars and supporting an upper mat.

stirrups. Reinforcing bars used in beams for shear reinforcement, typically bent into a U-shape and placed perpendicular to the longitudinal steel.

stringer. A horizontal structural member usually (in slab forming) supporting joists and resting on vertical supports.

strips. Bands of reinforcing bars in flat slab or flat plate construction. The **column strip** is a quarter-panel wide each side of the column centerline and runs either way of the building from column to column. The **middle strip** is a half a panel in width, filling in between column strips, and runs parallel to the column strips to fill the center part of a panel.

subbase. A layer in a pavement system between the pavement and the soil that is prepared to support it.

subgrade. The soil prepared and compacted to support a structure or a pavement system.

tamp. To compact freshly placed concrete or dirt by repeated blows.

temperature bars. Bars distributed throughout the concrete to resist tension stresses due to temperature changes and concrete shrinkage.

tensile strength. The maximum stress which a material is capable of resisting under axial tensile loading based on the cross-sectional area of the specimen before loading.

terrazzo concrete. Marble-aggregate concrete that is cast-in-place or precast and is ground smooth for decorative surfacing purposes on floors and walls.

thermite welding. A proprietary system employing a crucible containing metallic powder which, when ignited, melts and flows between two bar ends and fuses them to form a butt-welded splice.

tie. 1. A closed loop of reinforcing bars encircling the longitudinal steel in columns. 2. A tensile unit that holds concrete formwork secure against lateral pressure of unhardened concrete.

tie bars. 1. Bars at right angles and tied to main reinforcement to keep it in place. 2. Bars extending across a construction joint.

tie wire. Wire (generally, No. 16, 15, or 14 gauge) used to secure intersections of reinforcing bars for the purpose of holding them in place until concreting is completed.

trussed bars. Bars bent up to act as both top and bottom reinforcement (at the ends and middle of the span, respectively).

upperbeam bolster. A welded wire support for the upper layer of bottom bars in beams.

void content. The ratio of the volume of the voids to the volume of the solid particles.

wall spreader. An accessory, usually fabricated from reinforcing bar to a "Z" or "U" shape, used to separate and hold apart two faces or curtains or reinforcement in a wall.

weep hole. A drainage opening in a wall.

welded wire fabric. Wire mesh fabricated by welding the crossing joints, usually in rolls (sometimes flat sheets), and often used for temperature reinforcement in joist slabs, slabs on ground, and in highway pavements.

wet-dry test. A test devised to measure the durability of a soil-cement material to the stress imposed by repeated wetting and drying.

wrapping. Reinforcing bars or mesh surrounding a structural steel column or beam to reinforce concrete or plaster fireproofing.

yield strength. The load limit to which a steel bar can be stretched and still return to its original length.

INDEX

A horizon soil, 407
Absorption of lightweight aggregates, 14–15
Absorption tests for concrete pipe, boiling, 363
   ten-minute soaking, 363
Accelerators for concrete pipe, 354–355
Admixtures, for concrete pipe, 354–355
   of heavyweight concrete, 46–47
   for making mortar, 320–321
Aggregate types of insulating concrete, 25–32
Aggregates, for concrete pipe, 350–354
   exposed, architectural uses of, 69–78
   for heavyweight concrete, 41–45
      handling of, 45
   for lightweight concrete, 8–11
   for nonplastic mixes, proportions of, 53, 56
   relation to fire resistance, 459–460
   for stucco and plaster, 91–93
      void content of, 92–93
Agricultural pipe, 349
Air-entraining agents for concrete pipe, 355
Air-entraining cements for making mortar, 317
Air entrainment of cellular type insulating concrete, 34
Airport pavements, placing of bars in, 213–214
   with soil-cement, 446–447
Air tests of concrete pipe lines, 392
American Concrete Pipe Association, 347
American Society for Testing Materials specifications for concrete masonry, 294
Arbeton method for precast panel production, 76–77
Arches, placing of bars in, 207
Architectural concrete, 69–87
Atmospheric pressure curing, 283–286
Autoclaving, 286

B horizon soil, 407
Backfilling, with soil-cement, 445–446
   of trenches for concrete pipe, 377–378
Barite as aggregate for heavyweight concrete, 43
Barrel shell roofs, placing of bars in, 207
Bars, 138–143
   bending of radially, 142–143
   checking of, 159

concrete cover for, 159
fabrication of, 139–140
grades of, 138–139
identification of, 139–140
lateral spacing of, 157–158
lists of, 151–152
off-set column, 143
placing of, 153–159
   in other structures, 207–215
   tolerances for, 153, 157
placing drawings for, 143–145, 148, 151
for prestressing, 253–254
sizes of, 139
spliced, 161–168
   mechanical couplings for, 166–168
supports for, 152–153
tying of, 158, 168–171
Bases for stucco or plaster, 93–101
Batch weights, field conditions for, for nonplastic mixes, 65–66
Batching and mixing for concrete masonry manufacture, 281–282
Beam blocks, reinforcement with, 337
Beams, and compression reinforcement, 138
   continuous, steel in, 133–134
   fire tests for, 467
   plain, forces in, 236–241
   prestressed, eccentric, 243–245
      forces in, 242–249
   prestressing tendons of, 245–249
   simple longitudinal steel in, 132–133
Bearing values of soils, estimation of, 414–415
   field determination of, 414–415
Bedding, of concrete pipe, types of, 367–372
   effect of on field supporting strength of concrete pipe, 367–372
   in trench construction, 376
Bell end joints for concrete pipe, 378
Bell-and-spigot joints for concrete pipe, 379
Bending of bars, types of, 151
Bending moments in plain beams, 237–241
Bends, in bars, types of, 143
Bins, placing of bars in, 210–211
   rectangular, placing of bars in, 210–212

475

Bituminous joints for concrete pipe, construction of, 382
Black pigment materials, 116
Blocks, *see* Concrete masonry
Blue pigment materials, 115–116
Boiling absorption test for concrete pipe, 363
Bonding of soil-cement in slope protection construction, 444
Bonds to reinforcing steel of lightweight concrete, 19–20
Boron-containing aggregates for heavyweight concrete, 44–45
Borrow soil for soil-cement, 402–403
Brick, simulated, for finishes, 112
Brown pigment materials, 116
Buildings, layout of, 325–326
Bulk specific gravity of lightweight aggregates, 15
Bulk unit weight of lightweight aggregates, 14
Bullnose corner blocks, 298
Bushhammering, 77–78
By-product aggregates, 9

C horizon soil, 407
Cables for prestressing, 254
Camber in prestressing, 254–255
Cantilever footings, placing of bars in, 182
Cantilevered retaining walls, placing of bars in, 189–191
Capillarity of soils, 415–416
Cast-in-place concrete as a base for stucco or plaster, 99
Cast pipe, manufacturing of, 356
Casting beds for pretensioning, 255–257
Cavity walls, 308
Cellular type insulating concrete, 32–38
  history of, 32–33
  properties of, 34
  uses of, 33
Cement, for concrete pipe, 350
  for soil-cement slope protection, 439
  spreading of for soil-cement road construction, 423–425
Cement joints for concrete pipe, construction of, 381–382
Cement mortar for joints in concrete pipe, 380
Cement-modified soil, 395–396
Centrifugal pipe, manufacturing of, 358
Checking of bars, 159
Chemical aggregates, 11
Chemical method for mixing cellular type insulating concrete, 36
Chemical prestressing, 267
Chimney blocks, 298
Circumferential prestressing, 266
Clay masonry as a base for stucco or plaster, 98–99
Closed sheeting, 374
Closure blocks, placing of, 333
Coarse aggregates for concrete pipe, 354
Cold-weather masonry work, 341–342
Collars, concrete, for concrete pipe, 379
Colors, pigments for, 115–116

mixing of, 116–117
  tests for quality of, 117
  of soil, 406
Column dowels, 178–182
Column reinforcement units, assembled-in-place, 193
  preassembled, 191–193
Columns, and compression reinforcement, 137
  placing of bars in, 191–195
Combined footings, placing of bars in, 182
Compacted soil-cement, 395
Compacting factor method for stiff concrete, 61
Compacting ratio method for stiff concrete, 61–63
Compaction, in soil-cement road construction, 427–431
  degree of compaction and tests of, 430–431
  of soil-cement slope protection construction, 444
  of soils, 412–413
Compressibility of soils, 415
Compression reinforcement, of beams, 138
  of columns, 137
Compressive strength of hardened state of lightweight concrete, 18–19
Concrete, architectural, 69–87
  decorative, 67–126
  fire resistance of, 453–467
  for form liners, 85, 87
  heavyweight, 39–48
  insulating, 23–38
  loading formula for, 366
  painting of, 118–124
    application, 122
    coverage, 122
  precast, 217–234
  prestressed, 235–267
  reinforced, 129–171
    precasting, 127–267
    purpose and location of, 130–138
  special, 1–66
  white, 125–126
Concrete blocks, *see* Concrete masonry
Concrete collars for concrete pipe, 379
Concrete cover of bars, 159
Concrete cradle bedding, 367
Concrete flow test for stiff concrete, 58
Concrete masonry, 269–342
  applications of, 325–342
  applying mortar to, 331–332
  ASTM specifications for, 294
  as a base for stucco or plaster, 97–98
  batching and mixing in manufacture of, 281–282
  cold-weather work, 341–342
  control joints for, 335–337
  corner block laying, 329–331
  curing of, 283–286
  joints in, tooling of, 339–341
  machinery for producing, 287–290
  materials for, 271–276
  materials handling for plants producing, 275–276
  mixes for, 276–281
  molding of, 282–283

mortars for, 314–323
patterns created with, 302–307
placing of closure blocks, 333
precarbonation of, 290
properties of, 291–292
reinforcement of with beam blocks, 337
relation to wall dimensions, 298–302
shapes of, 294–298
sizes of, 294–298
storage of, 326–327
weight of, 294
Concrete mixes and fire resistance, 463
Concrete pipe, 343–392
  cement for, 350
  curing processes for, 358–358
  field-supporting strength, effect of bedding type, 367–372
  history of, 345–347
  jacking of, 385–389
  joints in, 378–382
  laying of, 376–377
    problems with, 383–389
  loading formulas for, 366
  manufacturing of, 345–363
  materials for, 350–355
  strength of, 366–372
  tests for, 359–363
  three-edge bearing strength, 366–367
  types of, 349–350
Concrete pipe construction, 365–392
Concrete pipe-making equipment, development of in the U.S., 347–349
Concrete spalling and fire resistance, 463
Concrete stains, 124–125
Concrete stress level and fire resistance, 461–463
Connections for precast units, 228
Consistency of nonplastic mixes, 51–52
Consistency tests for stiff concrete, 56–63
Construction, with concrete pipe, 365–392
  with soil-cement, 401–403
  of soil-cement slope protection, 440–445
Construction joints for soil-cement road construction, 432–433
Continuous beams, steel in, 133–134
Continuous wall footings, placing of bars in, 176–177
Control joints for concrete masonry, 335–337
Corner blocks, 295
Corners, inside, reinforcement of, 136
  laying blocks at, 329–331
Couplings, mechanical, for spliced bars, 166–168
Cracking in prestressed units, 255
Creep of hardened state of lightweight concrete, 20
Cubing, 291
Culvert pipe, 349
Curing, of cellular type insulating concrete, 38
  of concrete masonry, 283–286
  in pipe manufacturing, 358–359
  in pretensioning, 262–263
  in soil-cement road construction, 432

  in soil-cement slope protection construction, 444–445
Curtain walls, precast, 79–81

D horizon soil, 407
Decorative concretes, 67–126
Deflection strands in pretensioning, 262
Degree of saturation of soils, 411
Deicer scaling, resistance to of hardened state of lightweight concrete, 21
Density, of cellular type insulating concrete, 34–35
  of soils, 410–411
    field determination of, 413–414
Detensioning in pretensioning, 263
Detensioning deflected strands in pretensioning, 264
Domes, placing of bars in, 207
Door frames, construction of, 338
Double corner blocks, 298
Dowels for columns, 178–182
Drawings for placing reinforcing bars, 143–145, 148, 151
Dry mixes, *see* Nonplastic mixes
Dry tamp concrete, *see* Nonplastic mixes
Drying shrinkage of hardened state of lightweight concrete, 20

Eccentric prestressing of beams, 243–245
Elasticity, modulus of, 251
  of hardened state of lightweight concrete, 19
  of soils, 415
Engineering properties of soils, 412–416
Entrained air, amount of in nonplastic mixes, 66
Erection, of precast concrete units, 230
  of precast sections, 218–219
Exfiltration tests of concrete pipelines, 391–392
Expansion of soils, 411–412
Exposed aggregate, architectural uses of, 69–78
External strength test for concrete pipe, 359–363
Extrusion process for precast concrete, 229

Fabrication in precasting plants, 226–227
Face-down method for precast panel production, 72–73
Face-up method for precast panel production, 75–76
Farm drainage pipe, 349
Fiberglass-reinforced plastic form liners, 83–84
Field control in soil-cement slope protection construction, 445
Field-supporting strength of concrete pipe, effect of bedding type, 367–372
Fine aggregates, for concrete pipe, 351–353
  normal weight, use of in lightweight concrete, 21–22
Finishing, of precast concrete, 227
  in soil-cement road construction, 431–432
  of soil-cement slope protection construction, 444
Fire resistance, of concrete, 453–467

and steel reinforcement, 464–467
  structural factors of, 461–464
  testing of, 456–458
  thermal factors of, 459–461
Fire tests, for beams, 467
  of reinforced concrete, 465–467
  for roofs, 467
Flame passage and point, 457
Flat-type mixers (multiple-shaft) for soil-cement road construction, 420–421
Flexible pavements, salvaging of with soil-cement, 449–450
Flexural strength of hardened state of lightweight concrete, 19
Floors, concrete, painting of, 124
  placing of bars in, 195–207
Foamed aggregates, 11
Foamed method for mixing cellular type insulating concrete, 37
Folded plates, placing of bars in, 206
Footings, construction of, 327–329
  placing of bars in, 173–174
  reinforcement of, 135
Forces, in plain beams, 236–241
  bending moments, 237–241
  in prestressed beams, 242–249
Form liners, concrete for, 85, 87
  types of, 81–87
Forming of pretensioned units, 257–258
Forms for precasting, 225–226
Foundation mats, placing of bars in, 174–175
Frame roofs, attachment of, 339
Freeze-thaw resistance of hardened state of lightweight concrete, 21
Frost damage to subgrade soil for soil-cement paving, 452
Full cut header blocks, 295

Gaskets, compression-type rubber-ring, for joints for concrete pipe, 379–380
Glossary, 469–474
Gradation of lightweight aggregates, 15
Grading of soil-cement airport pavements, 446
Grain elevators, placing of bars in, 209–212
Green pigment materials, 116
Groined vaults, placing of bars in, 207
Ground-water table, and pipe laying problems, 383
  and trench construction, 383–384

Hardened state of lightweight concrete, 18–22
Hayde, Stephen J., 4, 9
Haydite, 4, 6, 9
Heat transmission end point, 457
Heavy aggregates for heavyweight concrete, 41, 43
Heavyweight concrete, 39–48
  admixtures of, 46–47
  mixing of, 46
  placing of, 46
  properties of, 47–48
  uses of, 39–40
High pressure steam curing, 286

Holding time in atmospheric pressure curing, 283
Hooks, 143
Hume, W. R., 348
Hydrostatic tests for concrete pipe, 363
Hydrous aggregates for heavyweight concrete, 43–44

Impermissible bedding, 368, 372
Infiltration tests of concrete pipelines, 390–391
Inside corners, reinforcement of, 136
Inspection, of concrete pipelines, 389–392
  of reinforcing steel, 215–216
  of soil-cement slope protection construction, 445
Insulating concrete, 23–38
  types of, 25–38
Integral color pigments, 115–126

Jacking of concrete pipe, 385–389
Jamb blocks, 295
Job errors in placing of steel, 216
Joint construction for soil-cement airport construction, 447
Jointing compounds, 380
  installation of, 382
Jointing materials for concrete pipe, 379–382
Joints, between precast panels, 227–228
  for concrete masonry, tooling of, 339–341
  for concrete pipe, 378–382
    construction of, 380–382
    materials for, 379–382
  of sewer lines, 392
Joist bars, placing of, 198–200

Laboratory tests for soil-cement slope protection, 439–440
Lapped splices, 162, 164
Lateral spacing of bars, 157–158
Laying of concrete pipe, 376–377
Layout of buildings for masonry construction, 325–326
Lift slab construction, 222–223
Lightweight aggregates, 8–11
  comparison with other types, 11–12
  handling of, 12
  properties of, 13–15
Lightweight concrete, characteristics of, 7
  hardened state of, 18–22
  history of, 4–7
  plastic state of, 15–16
  properties of, 12–13
  uses of, 3–4
  weight variations of, 7–8
Lime for making mortar, 320
Linear shrinkage of soils, 411
Lintels, construction of, 338–339
Liquid limit of soils, 409–410
Loading formulas for concrete pipe, 366
Loads on concrete pipe in a narrow trench, 372–373
Longitudinal steel in simple beams, 132–133

Machinery for concrete masonry

production, 287–290
Manufacturing of concrete pipe, 345–363
Marblecrete, 110–111
Maroon pigment materials, 116
Marston, Dean, 365
Masonry, as a base for stucco or plaster, 95–99
   good bonds for, 96–97
   types of, 97–99
  concrete; *see* Concrete masonry
Masonry cements for making mortar, 317
Masonry mortars, types of, 322–323
Mat slab foundations, placing of bars in, 182–184
Materials, for concrete masonry, 271–276
   for concrete pipe, 350–355
   for joints for concrete pipe, 379–382
   for mortar, 317–321
     proportioning of, 321
Materials handling for concrete masonry plants, 275–276
Maximum density of soils, 413
Mechanical couplings for spliced bars, 166–168
Mechanical method for mixing cellular type insulating concrete, 36–37
Metal reinforcement for bases for stucco or plaster, 94–95
Mix design, effect of on yield, 17–18
   for lightweight concrete, 22–24
   sample problem for nonplastic concrete, 63–66
Mix methods for cellular type insulating concrete, 36–38
   mix-foam method, 37
Mixes, for concrete masonry, 276–281
   for stucco and plaster, 90–93
   uniformity of in soil-cement road construction, 427
Mixing, of concrete, relation to fire resistance, 463
   of concrete paint, 120
   of heavyweight concrete, 46
   of mortar, 322
   in soil-cement road construction, 419–427
   in soil-cement slope protection construction, 441–442
Modified tongue-and-groove joints for concrete pipe, 379
Modulus of elasticity, 251
   of hardened state of lightweight concrete, 19
Moisture, control of in soil-cement road construction, 425–426
Moisture density relationships of soils, 412–413
Moisture density tests for soils, 413
Moisture effect, relation to fire resistance, 460–461
Molding of concrete masonry, 282–283
Mortar, applied to blocks, 331–332
   for concrete masonry, 314–323
     materials used for, 317–321
     properties of, 315–317
     types of, 322–323
   for joints in concrete pipe, 380
   mixing of, 322

   proportioning of materials, 321
   retempering of, 342
Multiple-layer construction for soil-cement road construction, 433
Multiple-strand release in pretensioning, 263

Natural aggregates, 8–9
Natural cements, for making mortar, 317–318
Naturbetong method for precast panel production, 76–77
No-fines concrete, 31–32
No-slump concrete, *see* Nonplastic mixes
Nonplastic mixes, 49–66
   aggregate proportions for, 53, 56
   consistency and workability of, 51–52
   fundamentals of, 49–51
   mix design for, sample problem, 63–66
   water-cement ratio for, 52–53

Off-set column bars, 143
Oil paints for concrete, 123–124
Old stucco or plaster as base for new, 99–100
On-site casting forms, 219–221
On-site precasting, 217–223
One-way slabs, placing of bars in, 200–201
Optimum moisture of soils, 413
Organic paints for concrete, 122–123

Packerhead process of pipe manufacturing, 356–357
Painting, of concrete, 118–124
   application, 122
   coverage, 122
   of concrete floors, 124
Parking areas using soil-cement, 447–448
Particle sizes in soil, 403–404
Partition walls, 308
Patterns using concrete masonry, 302–307
Pavement, use of in soil-cement, 402
Pavements, airport, soil-cement type, 446–447
   flexible, salvaging of with soil-cement, 449–450
   placing of bars in, 213–214
   placing of wire fabric in, 215
   subbases for using soil-cement, 450–452
Perlite, 7–8, 25–27, 29, 31
Permanent tension in prestressing, 251–252
Permeability of soils, 415
Permeability test for concrete pipe, 363
Pier blocks, 298
Pigments, integral color, 115–126
   materials for, 115–116
   mixing of, 116–117
   tests for, 117
Pile caps, placing of bars in, 177–178
Piles, precast concrete, 231–233
Pipe, concrete, *see* Concrete pipe
Pipe laying, construction problems in, 383–389
   unstable material effects, 384–385
Pipelines, concrete, inspection and testing of, 389–392
   construction of, 365–392
Placing, of bars, 153–159

in footings, walls and columns, 173–195
   in other structures, 207–215
   with straight bars only, 202–203
   with straight and truss bars, 203
of cellular type insulating concrete, 38
of closure blocks, 333
of heavyweight concrete, 46
of soil-cement in slope protection construction, 443–444
of steel, 173–216
   job errors in, 216
   tolerances in, 216
Plant recasting, 223–234
Plaster, 89
  application of, 101–110
    hand methods, 102–106
    machine methods, 106–110
  bases for, 93–101
  mixes for, 90–93
  old, as base for new, 99–100
Plaster of Paris form liners, 83
Plastic limits of soils, 410
Plastic soil-cement, 396
Plastic state of lightweight concrete, 15–16
Plasticity index of soils, 410
Polyethylene film form liners, 82
Porosity of soils, 410–411
Portland cement, handling and spreading of in soil-cement roads, 418–419
  for making mortar, 317
Portland-pozzolan cements for making mortar, 317
Posttensioning, 250–251, 264–266
Powers' remolding test for stiff concrete, 58–59
Pozzolans for concrete pipe, 355
Precarbonation of concrete masonry, 290
Precast concrete, 217–234
  connections for, 228
  erection of units, 230
  finishing of, 227
  joint construction, 227–228
  miscellaneous units, 233
Precast concrete piles, 231–233
Precast curtain walls, 79–81
Precast panels, production of, 72–75
Precasting, on-site, 217–223
    erection of sections, 218–219
    special forms, 225–226
  in plants, 223–234
  of reinforced concrete, 127–267
Precasting plants, fabrication techniques in, 226–227
Prefoamed method for mixing cellular type insulating concrete, 37–38
Preset time in atmospheric pressure curing, 283
Pressure pipe, 350
Prestressed beams, forces in, 242–249
Prestressed concrete, 235–267
Prestresses, computation of in pretensioning, 260–262
Prestressing, 249–267
  chemical, 267

circumferential, 266
cracking in, 255
materials for, 251–255
methods for, 249–251
and permanent tension, 251–252
types of steel for, 251–252
Prestressing tendons of beams, 245–249
Pretensioned units, handling and transportation of, 264
Pretensioning, 249–250, 255–263
Prewetting of soils in soil-cement road construction, 418
Processed aggregates, 9–11
Processing plans for soil-cement airport pavements, 446
Profiles of soil, 407
Pulverizing of soils in soil-cement road construction, 418

Radial bending of reinforcing bars, 142–143
Raft slab foundation, placing of bars in, 182–184
Red pigment materials, 116
Reentrant corners, reinforcement of, 136
Reinforced concrete, 129–171
  fire tests of, 465–467
  precasting of, 127–267
  purpose and location of, 130–138
Reinforcement using beam blocks, 337
Reinforcing bars, *see* Bars
Reinforcing steel, inspection of, 215–216
Resistance of cellular type insulating concrete, 35
Retempering of mortar, 342
Road construction with soil-cement, 417–433
Roads, preparation for soil-cement construction, 417–418
Rocks in pipe laying, problems with, 384
Roofs, fire tests for, 467
  frame, attachment of, 339
  placing of bars in, 206–207
Rotary mixers (single-shaft) for soil-cement road construction, 421–422
Rubber gasket joints for concrete pipe, construction of, 383
Rubber matting form liners, 81–82

Safety in trench construction, 373–376
Sand for making mortar, 318–320
Sand-bedding method for precast panel production, 73, 75
Sandwich panels, 228–229
Sanitary pipe, 349
Scarifying of soils in soil-cement road construction, 418
Schlick, Prof., 365
Schokbeton process, 229–230
Sewer lines, joints of, 392
Sgraffito, 113
Shearhead, 202
Sheet metal form liners, 82
Shoulders of pavements using soil-cement, 449
Shrinkage, of cellular type insulating

concrete, 35–36
  relation to reinforcement, 137
  of subgrade soils for soil-cement pavements, 451
Shrinkage limit of soils, 411
Sills, construction of, 338–339
Simple beams, longitudinal steel in, 132–133
Simulated brick finishes, 112
Simulated stone finishes, 112
Single-strands release in pretensioning, 263
Slab foundations, placing of bars in, 182–184
Slabs, on the ground, placing of bars in, 213
  one-way, placing of bars in, 200–201
  two-way flat, placing of bars in, 202–203
  two-way waffle type, placing of bars in, 206
Slag cements for making mortar, 318
Slip-form construction for circular tanks, placing of bars in, 209–210
Slope protection with soil-cement, 436–445
Soaking period in atmospheric pressure curing, 285
Soil-cement, 393–452
  for airport pavements, 446–447
  backfilling with, 445–446
  for flexible pavements, salvaging of, 449–450
  for parking areas, 447–448
  road construction with, 417–433
  for road widening and shoulders, 448–449
  slope protection with, 436–445
  soils for, 397–401
  for storage areas, 447–448
  street construction with, 433–435
  for subbases of concrete pavements, 450–452
  uses of, 395–452
Soil-cement construction, 401–403
Soil profiles, 398
Soil water, 407–408
Soils, bearing value of, estimation of, 414–415
  field determination of, 414–415
  engineering properties of, 412–416
  mixing of with cement in soil-cement road construction, 419–427
  preparation of for soil-cement road construction, 417–418
  properties of, 403–412
  for soil-cement, 397–401
  for soil-cement slope protection, 439
Spalling and fire resistance, 463
Spangler, Prof., 366
Special concretes, 1–66
Specifications, ASTM, for concrete masonry, 294
Spiral spacers for columns, 191–192, 193–194
Spirals, 142
Spliced bars, mechanical couplings for, 166–168
Splices, of bars, 161–168
  lapped, 162, 164
  welded, 164–166
Stacked casting in on-site precasting, 218
Stains for coloring concrete, 125–125

Stairs, placing of bars in, 207–209
Steel, in continuous beams, 133–134
  in simple beams, 132–133
Steel frames as a base for stucco or plaster, 101
Steel reinforcements, bonds to hardened state of lightweight concrete, 19–20
  and fire resistance, 464–467
  placing of, 173–216
Stiff concrete, consistency tests for, 56–63
Stirrups, 141
  spacing of, tolerances for, 150
Stone, simulated, for finishes, 112
Storage areas using soil-cement, 447–448
Storage of concrete masonry, 326–327
Storm sewer pipe, 349
Strands for prestressing, 254
Straub, F. J., 4, 6
Street construction with soil-cement, 433–435
Strength, of cellular type insulating concrete, 35
  of concrete pipe, 366–372
Strength test, external, for concrete pipe, 359–363
Stretcher blocks, 295
Structural end point, 456
Structural factors of fire resistance, 461–464
Structure of soil, 407
Stucco, 89
  application of, 101–110
    hand methods, 102–106
    machine methods, 106–110
  bases for, 93–101
  mixes for, 90–93
  old, as a base for new, 99–100
Subbases for concrete pavements using soil-cement, 450–452
Subgrade soils, frost damage to, in soil-cement pavements, 452
  swelling and shrinkage of in soil-cement paving, 451
Supports for bars, 152–153
Surface preparation for concrete paint, 120–121
Swelling of subgrade soils for soil-cement pavements, 451

Tamped pipe, manufacturing of, 357
Tan pigment materials, 116
Tanks, circular, slip-form construction for, 209–210
  placing of bars in, 209–210
Temperature effect in pretensioning, 259
Temperature tension, relation to reinforcement, 137
Ten-minute soaking absorption test for concrete pipe, 383
Tensile strength, 251
  of hardened state of lightweight concrete, 19
Tensioning method in pretensioning, 259
Tests, for concrete pipe, 359–363
  of concrete pipelines, 389–392
  of fire resistance of concrete, 453–467

for stiff concrete consistency, 56–63
Texture of soil, 404–406
Textured plastic form liners, 84–85
Thaulow drop table for stiff concrete, 58
Thermal factors of fire resistance, 459–461
Thermal insulation of hardened state of lightweight concrete, 21
Thermal properties of cellular type insulating concrete, 34
Three-edge bearing strength for concrete pipe, 366–367
Tie wires for reinforcing bars, 168–171
   types of, 169–170
   uses of, 170–171
Tied column units, 192–193
Ties, 142
Tilt-up construction, 221–222
Tolerances in placing of steel, 216
Tongue-and groove joints for concrete pipe, 379
   modified types, 379
Tooling of joints in concrete masonry, 339–341
Transporting of soil-cement for slope protection construction, 443
Trench boxes, 374
Trenches, backfilling of, 377–378
   bedding aspects of, 376
   construction of, safety aspects, 373–376
   loads on concrete pipe in, 372–373
Trial mixtures, adjustments of, for nonplastic mixes, 66
Truss bars, 141, 203
Tunnels, excavation of, 385
Two-way flat slabs, placing of bars in, 201–205
   sequence of placing, 202–203
Two-way waffle slabs, placing of bars in, 206
Tying of bars, 158, 168–171

Unit weight of plastic state of lightweight concrete, 15

Vebe test method for stiff concrete, 59–61

Vermiculite, 7–8, 25–26, 31
Vibrated pipe, manufacturing of, 356
Void content in stucco or plaster aggregates, 92–93
Void ratio of soils, 411

Wall dimensions for use of concrete masonry, 298–302
Wall patterns with concrete masonry, 302–307
Walls, masonry, special types of, 333–335
   placing of bars in, 184–191
   reinforcement of, 134
   types of, 308–313
Water, for concrete paint, 120
   for concrete pipe, 354
   for making mortar, 320
   in soil, 407–408
      consistency of, 409–410
   for soil-cement slope protection, 439
Water-cement ratio for nonplastic mixes, 52–53
Weather conditions and soil-cement slope protection construction, 441
Weight of concrete masonry, 294
Welded splices, 164–166
White concrete, 125–126
Widening of pavements using soil-cement, 448–449
Window frames, construction of, 338
Wire for prestressing, 251–252
Wire-reinforced concrete, placing of wire in, 215
Wood form liners, 81
Wood frames as a base for stucco or plaster, 100–101
Workability, of nonplastic mixes, 51–52
   of plastic state of lightweight concrete, 15–16

Yellow pigment materials, 116
Yield, effect of mix design on, 17–18
Yield strength, 251